Biosurfactants and Sustainability

Biosurfactants and Sustainability

From Biorefineries Production to Versatile Applications

Edited by

Paulo Ricardo Franco Marcelino
Lorena School of Engineering
University of São Paulo
Brazil

Silvio Silverio da Silva
Lorena School of Engineering
University of São Paulo
Brazil

Antonio Ortiz Lopez
Department of Biochemistry and Molecular Biology
University of Murcia
Spain

This edition first published 2023
© 2023 John Wiley & Sons Ltd

All rights reserved. No part of this publication may be reproduced, stored in a retrieval system, or transmitted, in any form or by any means, electronic, mechanical, photocopying, recording or otherwise, except as permitted by law. Advice on how to obtain permission to reuse material from this title is available at http://www.wiley.com/go/permissions.

The right of Paulo Ricardo Franco Marcelino, Silvio Silverio da Silva and Antonio Ortiz Lopez to be identified as the authors of the editorial material in this work has been asserted in accordance with law.

Registered Offices
John Wiley & Sons, Inc., 111 River Street, Hoboken, NJ 07030, USA
John Wiley & Sons Ltd, The Atrium, Southern Gate, Chichester, West Sussex, PO19 8SQ, UK

For details of our global editorial offices, customer services, and more information about Wiley products visit us at www.wiley.com.

Wiley also publishes its books in a variety of electronic formats and by print-on-demand. Some content that appears in standard print versions of this book may not be available in other formats.

Trademarks: Wiley and the Wiley logo are trademarks or registered trademarks of John Wiley & Sons, Inc. and/or its affiliates in the United States and other countries and may not be used without written permission. All other trademarks are the property of their respective owners. John Wiley & Sons, Inc. is not associated with any product or vendor mentioned in this book.

Limit of Liability/Disclaimer of Warranty
In view of ongoing research, equipment modifications, changes in governmental regulations, and the constant flow of information relating to the use of experimental reagents, equipment, and devices, the reader is urged to review and evaluate the information provided in the package insert or instructions for each chemical, piece of equipment, reagent, or device for, among other things, any changes in the instructions or indication of usage and for added warnings and precautions. While the publisher and authors have used their best efforts in preparing this work, they make no representations or warranties with respect to the accuracy or completeness of the contents of this work and specifically disclaim all warranties, including without limitation any implied warranties of merchantability or fitness for a particular purpose. No warranty may be created or extended by sales representatives, written sales materials or promotional statements for this work. The fact that an organization, website, or product is referred to in this work as a citation and/or potential source of further information does not mean that the publisher and authors endorse the information or services the organization, website, or product may provide or recommendations it may make. This work is sold with the understanding that the publisher is not engaged in rendering professional services. The advice and strategies contained herein may not be suitable for your situation. You should consult with a specialist where appropriate. Further, readers should be aware that websites listed in this work may have changed or disappeared between when this work was written and when it is read. Neither the publisher nor authors shall be liable for any loss of profit or any other commercial damages, including but not limited to special, incidental, consequential, or other damages.

Library of Congress Cataloging-in-Publication Data
Names: Marcelino, Paulo Ricardo Franco, editor. | Silva, Silvio Silvério da, editor. | Ortiz Lopez, Antonio, editor.
Title: Biosurfactants and sustainability : from biorefineries production to versatile applications / edited by Paulo Ricardo Franco Marcelino (University of São Paulo, Lorena School of Engineering, Lorena, São Paulo, Brazil), Silvio Silverio da Silva (University of São Paulo, Biotechnology Department, Lorena, São Paulo, Brazil), Antonio Ortiz Lopez (University of Murcia, Departamento de Bioquímica y, Murcia, Spain).
Description: Hoboken, NJ : John Wiley & Sons Ltd., 2023. | ECIP galley contains incomplete author affiliation for Antonio Ortiz Lopez. | Includes bibliographical references and index.
Identifiers: LCCN 2023000550 (print) | LCCN 2023000551 (ebook) | ISBN 9781119854364 (hardback) | ISBN 9781119854371 (adobe pdf) | ISBN 9781119854388 (epub) | ISBN 9781119854395 (ebook)
Subjects: LCSH: Biosurfactants.
Classification: LCC TP248.B57 B563 2023 (print) | LCC TP248.B57 (ebook) | DDC 668/.1--dc23/eng/20230213
LC record available at https://lccn.loc.gov/2023000550
LC ebook record available at https://lccn.loc.gov/2023000551

Cover Image: © süleyman Ibişov/EyeEm/Adobe Stock Photos
Cover Design: Wiley

Set in 9.5/12.5pt STIXTwoText by Integra Software Services Pvt. Ltd, Pondicherry, India

Contents

List of Contributors *xi*
Foreword *xv*

Introduction *1*
Paulo Ricardo Franco Marcelino, Carlos Augusto Ramos, Guilherme de Oliveira Silva, Ramiro Reyes Guzman, Silvio Silverio da Silva, and Antonio Ortiz Lopez
Biosurfactants: Concept, Biological Functions, Classification, General Properties and Applications *1*

1 Microorganisms Producing Biosurfactants in the Current Scenario *11*
Fernanda Palladino, Rita C.L.B. Rodrigues, Yasmim Senden dos Santos, and Carlos A. Rosa
1.1 Introduction *11*
1.2 Microbial Biosurfactants *12*
1.2.1 Structure and Classification of Biosurfactants *12*
1.2.2 Biosurfactants Producing Yeasts *14*
1.2.3 Biosurfactants Produced by Extremophile Microorganisms *17*
1.3 Industrial Applications of Biosurfactants *18*
References *20*

2 Selection of Biosurfactant-Producing Microorganisms *29*
Julio Bonilla Jaime, Luis Galarza Romero, and Jonathan Coronel León
2.1 Introduction *29*
2.2 Traditional Methods of Detection *30*
2.2.1 Direct Measure of Surface/interfacial Activity *31*
2.2.2 Indirect Measure of Surface/interfacial Activity *32*
2.2.3 Effects of Culture Media Based on Agro-industrial By-products on Properties of BS *34*
2.3 High-throughput Analysis Method for the Screening of Potential Biosurfactants Producers *35*
2.4 Screening of Microorganisms Biosurfactants and Lipases Producers *40*
2.5 Conclusion and Future Perspectives *45*
References *46*

3 Metabolic Engineering as a Tool for Biosurfactant Production by Microorganisms *61*
Roberta Barros Lovaglio, Vinícius Luiz da Silva, and Jonas Contiero
3.1 Metabolic Engineering and Biosurfactants *61*
3.2 Regulation and Heterologous Production of Biosurfactants *63*
3.3 Extension of Substrate Range for Biosurfactant Production *67*
3.4 Improvement of Overall Cellular Physiology *68*
3.5 Elimination or Reduction of By-product *69*
3.6 Future Perspectives *69*
3.7 Conclusions *70*
References *71*

4 Biosurfactant Production in the Context of Biorefineries *77*
Paulo Ricardo Franco Marcelino, Carlos Augusto Ramos, Maria Teresa Ramos, Renan Murbach Pereira, Rafael Rodrigues Philippini, Emily Emy Matsumura, and Silvio Silvério da Silva
4.1 Biorefineries in Contemporary Society *77*
4.2 Biomass and Biorefineries: Industrial By-products as Raw Materials for Biorefineries *78*
4.3 Biosurfactant Production in the Context of Lignocellulosic Biorefineries *80*
4.4 Biosurfactant Production in the Context of Oleaginous Biorefineries *85*
4.5 Biosurfactant Production in the Context of Starchy and Biodiesel Biorefineries *87*
4.6 Conclusion *88*
References *88*

5 Biosurfactant Production by Solid-state Fermentation in Biorefineries *95*
Daylin Rubio-Ribeaux, Rogger Alessandro Mata da Costa, Dayana Montero Rodríguez, Nathália Sá Alencar do Amaral Marques, Gilda Mariano Silva, and Silvio Silvério da Silva
5.1 Introduction *95*
5.2 Advantages of Biosurfactant Production by Solid-State Fermentation *96*
5.3 Suitable Biomasses for Biosurfactant Production in Biorefineries *96*
5.4 Microorganisms Used in Biosurfactant Production by Solid-state Fermentation *98*
5.5 Raw Materials Used in Solid-state Fermentation for Biosurfactant Production *99*
5.6 Pretreatment of Raw Materials for the Production of Biosurfactants in Solid-state Fermentation *101*
5.7 Physicochemical Factors of Solid-state Fermentation *103*
5.8 Strategies for Scaling-up of Solid-state Fermentation for Biosurfactant Production *105*
5.9 Conclusion *108*
References *108*

6	**An Overview of Developments and Challenges in the Production of Biosurfactant by Fermentation Processes** *117*	

F.G. Barbosa, M.J. Castro-Alonso, T.M. Rocha, S. Sánchez-Muñoz, G.L. de Arruda, M.C.A. Viana, C.A. Prado, P.R.F. Marcelino, J.C. Santos, and Silvio S. Da Silva

6.1	Introduction *117*	
6.2	Current Market and Potential Applications of Biosurfactants *118*	
6.3	Biosurfactant as a Sustainable Alternative: Factors Influencing its Production *118*	
6.3.1	Factors Involved in the Biosurfactant Production *119*	
6.4	Strategies and Main Challenges for Biosurfactant Production *122*	
6.4.1	Process Configurations as Strategies for Biosurfactant Production *123*	
6.4.2	Bioreactors Used in the Biosurfactants Production: Types, Advantages, and Disadvantages *125*	
6.4.3	Biosurfactant Separation Processes *128*	
6.5	Future Perspectives and Conclusion *132*	
	References *132*	

7	**Enzymatic Production of Biosurfactants** *143*	

Ana Karine F. de Carvalho, Heitor B.S. Bento, Felipe R. Carlos, Vitor B. Hidalgo, Cintia M. Romero, Bruno C. Gambarato, and Patrícia C.M. Da Rós

7.1	Introduction *143*	
7.2	What are the Biosurfactants Produced Enzymatically? Esterification Reactions of Sugars and Fatty Acids Catalyzed by Enzymes *144*	
7.2.1	Esterification Reactions of Sugars and Fatty Acids Catalyzed by Enzymes *144*	
7.3	Enzymes and Methods for Biosurfactant Production: Bioreactors and Ways of Conducting Enzymatic Processes *145*	
7.4	Advantages and Disadvantages of Enzymatic Biosurfactant Production *148*	
7.5	Potential Use of Enzymes for the Production of Biosurfactants *149*	
7.6	Production of Biosurfactants by the Enzymatic Route in Biorefineries: Demand for More Modern Production Processes *150*	
7.7	Conclusion *153*	
	References *153*	

8	**Co-production of Biosurfactants and Other Bioproducts in Biorefineries** *157*	

Martha Inés Vélez-Mercado, Carlos Antonio Espinosa-Lavenant, Juan Gerardo Flores-Iga, Fernando Hernández Teran, María de Lourdes Froto Madariaga, and Nagamani Balagurusamy

8.1	Introduction *157*	
8.2	Microbial Surfactant Production *158*	
8.3	Co-production of Biosurfactants in a Biorefinery *160*	
8.3.1	Co-production of Biosurfactants and Polyhydroxyalkanoates *161*	
8.3.2	Co-production of Biosurfactants and Enzymes *162*	
8.3.3	Co-production of Biosurfactants and Lipids *164*	
8.3.4	Co-production of Biosurfactants and Ethanol *165*	
8.4	Conclusions *166*	
	References *166*	

9	**Biosurfactants in Nanotechnology: Recent Advances and Applications** *173*	
	Avinash P. Ingle, Shreshtha Saxena, Mangesh Moharil, Mahendra Rai, and Silvio S. Da Silva	
9.1	Introduction *173*	
9.2	Biosurfactants and their Types *174*	
9.2.1	Glycolipid Biosurfactants *174*	
9.2.2	Rhamnolipids *174*	
9.2.3	Trehalolipids *175*	
9.2.4	Sophorolipids *175*	
9.2.5	Mannosylerythritol Lipids *175*	
9.2.6	Lipopeptide Biosurfactants *175*	
9.2.7	Phospholipid Biosurfactants *176*	
9.2.8	Polymeric Biosurfactants *176*	
9.3	Properties of Biosurfactants *178*	
9.3.1	Surface and Interface Activity *178*	
9.3.2	Efficiency *179*	
9.3.3	Foaming Capacity *179*	
9.3.4	Emulsification/Emulsion Forming and Emulsion Breaking *179*	
9.3.5	Tolerance for Temperature and pH Tolerance *180*	
9.3.6	Low Toxicity *180*	
9.3.7	Biodegradability *180*	
9.4	Conventional Methods for Biosurfactant Production *180*	
9.5	Commercial Applications of Biosurfactants *182*	
9.5.1	Application of Biosurfactants in Agriculture *182*	
9.5.2	Application of Biosurfactants in Nanotechnology *183*	
9.5.3	Applications of Biosurfactants in Commercial Laundry Detergents *184*	
9.5.4	Application of Biosurfactants in Medicine *184*	
9.5.5	Application of Biosurfactants in the Food Processing Industry *185*	
9.5.6	Application of Biosurfactants in the Cosmetic Industry *185*	
9.5.7	Application of Biosurfactants in Petroleum *185*	
9.5.8	Application of Biosurfactant in Microbial-enhanced Oil Recovery *186*	
9.6	Biosurfactants in Nanotechnology (Biosurfactant Mediated Synthesis of Nanoparticles) *186*	
9.6.1	Glycolipids Biosurfactants Produced Nanoparticles *186*	
9.6.2	Lipopeptides Biosurfactants Produced Nanoparticles *187*	
9.7	Conclusions *188*	
	References *188*	
10	**Interaction of Glycolipid Biosurfactants with Model Membranes and Proteins** *195*	
	Francisco J. Aranda, Antonio Ortiz, and José A. Teruel	
10.1	Introduction *195*	
10.2	Interaction of Glycolipid Biosurfactants with Model Membranes *196*	
10.2.1	Rhamnolipids *197*	
10.2.2	Trehalose *Lipids* *206*	
10.2.3	Other Glycolipids *209*	

10.3 Interaction of Glycolipid Biosurfactants with Proteins *211*
10.3.1 Rhamnolipids *211*
10.3.2 Trehalose Lipids *211*
10.3.3 Mannosylerythritol Lipids *212*
10.4 Conclusions *212*
References *213*

11 Biosurfactants: Properties and Current Therapeutic Applications *221*
Cristiani Baldo, Maria Ines Rezende, and Fabiana Guillen Moreira Gasparin
11.1 Production of Microbial Biosurfactants *221*
11.2 Anti-tumoral Activity of Biosurfactants *223*
11.3 Anti-inflammatory Activity of Biosurfactants *226*
11.4 Anti-microbial Activity of Biosurfactant *228*
11.4.1 Biosurfactants as Anti-bacterial Agents *229*
11.4.2 Biosurfactants as Anti-viral Agents *231*
11.4.3 Biosurfactants as Anti-fungal Agents *232*
11.5 Other Therapeutic Applications of Biosurfactants *233*
11.6 Concluding Remarks *234*
References *234*

12 Fungal Biosurfactants: Applications in Agriculture and Environmental Bioremediation Processes *243*
Láuren Machado Drumond de Souza, Débora Luiza Costa Barreto, Lívia da Costa Coelho, Elisa Amorim Amâncio Teixeira, Vívian Nicolau Gonçalves, Júlia de Paula Muzetti Ribeiro, Natana Gontijo Rabelo, Stephanie Evelinde Oliveira Alves, Mayanne Karla da Silva, Laura Beatriz Miranda Martins, Charles Lowell Cantrell, Stephen Oscar Duke, and Luiz Henrique Rosa
12.1 Biosurfactants as Agrochemicals *243*
12.1.1 Biosurfactants as Herbicide Adjuvants *244*
12.1.2 Biosurfactants and Antifungal Activity *245*
12.1.3 Biosurfactants as Insecticidal Adjuvants *246*
12.2 Insecticidal Biosurfactants for Use against Disease Vector Insects *246*
12.3 Fungal Biosurfactants in Bioremediation Processes *248*
References *249*

13 New Formulations Based on Biosurfactants and Their Potential Applications *255*
Maria Jose Castro-Alonso, Fernanda G. Barbosa, Thiago A. Vieira, Diana A. Sanchez, Monica C. Santos, Thércia R. Balbino, Salvador S. Muñoz, and Talita M. Lacerda
13.1 Introduction *255*
13.2 General Chemical and Biochemical Aspects *258*
13.3 Downstream Processing *259*
13.4 Biosurfactants in Cosmetics and Personal Care *259*
13.5 Biosurfactants in Medicine and Pharmaceutics *261*
13.6 Biosurfactants in Food and Feed *262*

13.7 Biosurfactants in Pesticides, Insecticides, and Herbicide Formulations *264*
13.8 Biosurfactants in Civil Engineering *265*
13.9 Miscellaneous *266*
13.9.1 Detergent Formulations *266*
13.9.2 Bioremediation Purposes *267*
13.9.3 Nanoparticle Synthesis *267*
13.9.4 Polymer Synthesis *268*
13.10 Overview of the Biosurfactant Market *268*
13.11 Conclusions and Future Perspectives *270*
References *270*

14 Techno-economic-environmental Analysis of the Production of Biosurfactants in the Context of Biorefineries *281*
Andreza Aparecida Longati, Andrew Milli Elias, Felipe Fernando, Furlan Everson Alves Miranda, and Roberto de Campos Giordano
14.1 Introduction *281*
14.1.1 Background *281*
14.1.2 Surfactant Versus Biosurfactant *282*
14.1.3 Biosurfactant Market, Producers, and Patents *282*
14.1.4 Biosurfactant Production Routes *283*
14.2 Economic Aspects of the BS Production *286*
14.3 Environmental Aspects *288*
14.4 Biosurfactant Production Synergies in the Brazilian Biorefineries Context *290*
14.5 Conclusion *293*
References *294*

Index *301*

List of Contributors

Elisa Amorim Amâncio Teixeira
Departamento de Microbiologia
Universidade Federal de Minas Gerais
Minas Gerais, Brazil

Francisco J. Aranda
Department of Biochemistry and Molecular Biology-A
University of Murcia
Murcia, Spain

Nagamani Balagurusamy
Laboratorio de Biorremediación
Ciudad Universitaria de la Universidad Autónoma de Coahuila
Torreón, México

Thércia R. Balbino
Biotechnology Department
University of São Paulo
São Paulo, Brazil

Cristiani Baldo
Department of Biochemistry and Biotechnology
Londrina State University
Londrina, Brazil

F. G. Barbosa
Department of Biotechnology
University of São Paulo
São Paulo, Brazil

Fernanda G. Barbosa
Biotechnology Department
University of São Paulo
São Paulo, Brazil

Heitor B. S. Bento
Department of Bioprocess Engineering and Biotechnology
São Paulo State University
São Paulo, Brazil

Charles Lowell Cantrell
Natural Products Utilization Research Unit
United States Department of Agriculture
Oxford MS, USA

Felipe R. Carlos
Postgraduate Program in Biotechnology
Federal University of Alfenas
Alfenas, Brazil

M. J. Castro-Alonso
Department of Biotechnology
University of São Paulo
São Paulo, Brazil

Maria Jose Castro-Alonso
Biotechnology Department
University of São Paulo
São Paulo, Brazil

Jonas Contiero
Universidade Estadual Paulista
"Julio de Mesquita Filho"
Instituto de Biociências
São Paulo, Brazil

Débora Luiza Costa Barreto
Departamento de Microbiologia
Universidade Federal de Minas Gerais
Minas Gerais, Brazil

Lívia da Costa Coelho
Departamento de Microbiologia
Universidade Federal de Minas Gerais
Minas Gerais, Brazil

Patrícia C. M. Da Rós
Department of Chemical Engineering,
University of São Paulo, São Paulo, Brazil.

List of Contributors

Mayanne Karla da Silva
Departamento de Microbiologia
Universidade Federal de da Silva
Minas Gerais
Minas Gerais, Brazil

Silvio S. Da Silva
Department of Biotechnology
University of São Paulo
São Paulo, Brazil

Vinícius Luiz da Silva
MicroGreen – Soluções Biotecnológicas
Piracicaba
São Paulo, Brazil

G. L. de Arruda
Department of Biotechnology
University of São Paulo
São Paulo, Brazil

Ana Karine F. de Carvalho
Postgraduate Program in Biotechnology
Federal University of Alfenas
Alfenas, Brazil

María de Lourdes Froto Madariaga
Laboratorio de Biorremediación
Ciudad Universitaria de la Universidad
Autónoma de Coahuila
Torreón, México

Stephanie Evelin de Oliveira Alves
Departamento de Microbiologia
Universidade Federal de Minas Gerais
Minas Gerais, Brazil

Yasmim Senden, dos Santos
Departamento de Microbiologia
Universidade Federal de Minas Gerais
Minas Gerais, Brazil

Láuren Machado Drumond de Souza
Departamento de Microbiologia
Universidade Federal de Minas Gerais
Minas Gerais, Brazil

Stephen Oscar Duke
National Center for Natural
Products Research
School of Pharmacy Oxford MS, USA

Embrapa Andrew Milli, Elias
Instrumentação
São Paulo, Brazil

Carlos Antonio Espinosa-Lavenant
Laboratorio de Biorremediación
Ciudad Universitaria de la Universidad
Autónoma de Coahuila, Torreón, México

Juan Gerardo Flores-Iga
Laboratorio de Biorremediación
Ciudad Universitaria de la Universidad
Autónoma de Coahuila
Torreón, México

Felipe Fernando Furlan
Department of Chemical Engineering Federal
University of São Carlos
São Paulo, Brazil

Bruno C. Gambarato
Department of Material Science
University Center of Volta Redonda
Volta Redonda, Brazil

Roberto de Campos Giordano
Department of Chemical Engineering
Federal University of São Carlos
São Paulo, Brazil

Vívian Nicolau Gonçalves
Departamento de Microbiologia
Universidade Federal de Minas Gerais
Minas Gerais, Brazil

Teran Hernández
Fernando, Laboratorio de Biorremediación
Ciudad Universitaria de la Universidad
Autónoma de Coahuila
Torreón, México

Vitor B. Hidalgo
Postgraduate Program in Biotechnology
Federal University of Alfenas
Alfenas, Brazil

Avinash P. Ingle
Department of Agricultural Botany
Dr. Panjabrao Deshmukh Krishi
Akola, India

Julio Bonilla Jaime
Escuela Superior Politécnica del Litoral
ESPOL, Centro de Investigaciones
Biotecnológicas del Ecuador (CIBE)
Guayaquil, Ecuador

Talita M. Lacerda
Biotechnology Department
University of São Paulo
São Paulo, Brazil

Jonathan Coronel León
Escuela Superior Politécnica del Litoral
ESPOL, Centro de Investigaciones
Biotecnológicas del Ecuador (CIBE)
Guayaquil, Ecuador

Andreza Aparecida Longati
Department of Materials and Bioprocess Engineering
State University of Campinas
São Paulo, Brazil

Roberta Barros Lovaglio
Universidade Federal de São Carlos
São Carlos, Brazil

P.R.F. Marcelino
Department of Biotechnology
University of São Paulo
São Paulo, Brazil

Nathália Sá Alencar do Amaral Marques
Catholic University of Pernambuco (UNICAP)
Boa Vista, Brazil

Rogger Alessandro Mata da Costa
Department of Biotechnology
University of São Paulo
São Paulo, Brazil

Laura Beatriz Miranda Martins
Departamento de Microbiologia
Universidade Federal de Minas Gerais
Minas Gerais, Brazil

Everson Alves Miranda
Department of Materials and Bioprocess Engineering, State University of Campinas
São Paulo, Brazil

Mangesh Moharil
Department of Agricultural Botany
Dr. Panjabrao Deshmukh Krishi
Akola, India

Dayana Montero Rodríguez
Catholic University of Pernambuco (UNICAP)
Boa Vista, Brazil

Fabiana Guillen Moreira Gasparin
Department of Biochemistry and Biotechnology
Londrina State University
Londrina, Brazil

Salvador S. Muñoz
Biotechnology Department
University of São Paulo
São Paulo, Brazil

Júlia de Paula Muzetti Ribeiro
Departamento de Microbiologia
Universidade Federal de Minas Gerais
Minas Gerais, Brazil

Antonio Ortiz
Department of Biochemistry and Molecular Biology-A
University of Murcia
Murcia, Spain

Fernanda Palladino
Institute of Biological Sciences
Federal University of Minas Gerais
Belo Horizonte-MG, Brazil

C.A. Prado
Department of Biotechnology
University of São Paulo
São Paulo, Brazil

Natana Gontijo Rabelo
Departamento de Microbiologia
Universidade Federal de Minas Gerais
Minas Gerais, Brazil

Mahendra Rai
Department of Biotechnology
Sant Gadge Baba Amravati University
Maharashtra, India

Maria Ines Rezende
Department of Biochemistry and
Biotechnology
Londrina State University
Londrina, Brazil

T.M. Rocha
Department of Biotechnology
University of São Paulo
São Paulo, Brazil

Rita C.L.B. Rodrigues
Lorena Engineering School
University of São Paulo
Lorena-SP, Brazil

Cintia M. Romero
Planta Piloto de Procesos Industriales
Microbiológicos, Consejo Nacional de
Investigaciones Científicas y Técnicas
Tucuman, Argentina

Luis Galarza Romero
Escuela Superior Politécnica del Litoral
ESPOL, Centro de Investigaciones
Biotecnológicas del Ecuador (CIBE)
Guayaquil, Ecuador

Carlos A. Rosa
Microbiology Department
Federal University of Minas Gerais
Belo Horizonte-MG, Brazil

Luiz Henrique Rosa
Departamento de Microbiologia
Universidade Federal de Minas Gerais
Belo Horizonte, Brazil

Daylin Rubio-Ribeaux
Department of Biotechnology
University of São Paulo,
São Paulo, Brazi

Diana A. Sanchez
Biotechnology Department
University of São Paulo
São Paulo, Brazil

S. Sánchez-Muñoz
Department of Biotechnology
University of São Paulo
São Paulo, Brazil

J.C. Santos
Department of Biotechnology
University of São Paulo
São Paulo, Brazil

Monica C. Santos
Biotechnology Department
University of São Paulo
São Paulo, Brazil

Shreshtha Saxena
Department of Agricultural Botany
Dr. Panjabrao Deshmukh Krishi
Akola, India

Gilda Mariano Silva
Department of Biotechnology
University of São Paulo
São Paulo, Brazil

José A. Teruel
Department of Biochemistry and Molecular
Biology-A
University of Murcia
Murcia, Spain

Martha Inés Vélez-Mercado
Laboratorio de Biorremediación
Ciudad Universitaria de la Universidad
Autónoma de Coahuila
Torreón, México

M.C.A. Viana
Department of Biotechnology
University of São Paulo
São Paulo, Brazil

Thiago A. Vieira
Biotechnology Department
University of São Paulo
São Paulo, Brazil

Foreword

Several famous universities and research institutes have come together in this book to offer the principles and fundamentals of biosurfactants, ecologically versatile compounds, their diverse sources and production mechanisms, as well as a vast field of applications in industry, agriculture, health, and in different environmental areas. Gathering knowledge on the basis of a solid theoretical background and bold applications is the biggest challenge facing Academia today, whose mission is to educate new generations and produce knowledge based on research, and integrate it into innovation for industrial partners and society. This book makes it accessible to undergraduate and graduate students, researchers, and professionals in the field, and demonstrates how the science of biosurfactants and the technology developed for their production and use can be applied to mission-oriented research to improve processes, improve materials and products, to meet the well-being of the population and promote sustainable development. May new disruptive ideas be born from this book, so that the frontier of knowledge is always extended for the benefit of society, with respect for the environment and the preservation of the planet. I wish you good reading, good study, and a good source of ideas!

Liedi Bernucci
Professor at Escola Politecnica of the University of São Paulo
President of Technological Research Institute of the State of São Paulo

Introduction

Paulo Ricardo Franco Marcelino[1,], Carlos Augusto Ramos[1], Guilherme de Oliveira Silva[1], Ramiro Reyes Guzman[1], Silvio Silverio da Silva[1], and Antonio Ortiz Lopez[2]*

[1] *Laboratório de Bioprocessos e Produtos Sustentáveis (LBios), Escola de Engenharia de Lorena (EEL), Universidade de São Paulo (USP), Brazil*
[2] *Departamento de Bioquímica y Biología Molecular-A, Facultad de Veterinaria, Universidad de Murcia, Spain*
* *Corresponding author*

Biosurfactants: Concept, Biological Functions, Classification, General Properties and Applications

Biosurfactants are compounds with amphipathic structures obtained from natural sources such as plants (saponins), animals (bile salts) and microorganisms (glycolipids, lipopeptides, lipoproteins, polymerics and others) (Figure I.1) [1]. They present highlighted physicochemical and biological properties, such as surfactant and/or emulsifier, antimicrobial, antitumor, larvicide, mosquitocide, anti-inflammatory, immunomodulatory, which give them numerous applications [2–7]. For a long time the concept of biosurfactants was restricted only to microbial surfactants, but more recently Wim Soetaert, in a lecture at the *Workshop on biosurfactants – Berlin, 2014*, began to divide biosurfactants into first- and second-generation compounds, according to the origin of these substances.

First-generation biosurfactants or "green surfactants" are conceptualized as compounds extracted and purified from vegetable and animal raw materials or produced entirely from renewable resources through chemical syntheses. The main examples include saponins, sugar esters, alkyl polyglycosides, and alkanolamides. Second-generation biosurfactants are compounds entirely produced from renewable resources and through a biological process (biocatalysis or fermentation), being exemplified mainly by microbial surfactants of the glycolipid and lipopeptide type [8, 9].

Microbial surfactants have several physiological functions such as emulsification, solubilization and intracellular transport of insoluble compounds in aqueous media, cell release in biofilms, antimicrobial activity, and *quorum sensing* [1, 4].

The need for some microorganisms to emulsify, solubilize, and transport compounds insoluble in aqueous medium to intracellular compartments, facilitating the consumption of these substrates for energy generation, has already been studied in depth. In addition to energy production, this biological function is of fundamental importance for technologies for the bioremediation of insoluble organic compounds in aqueous media [10–16].

The production of biosurfactants by microorganisms can also be considered a regulatory mechanism for cell adherence and release on surfaces, during a signaling/communication process dependent on population density, the so-called *quorum sensing* [17–19]. Adherence and cell release may occur due to surface properties (surface tension, surface enthalpy per

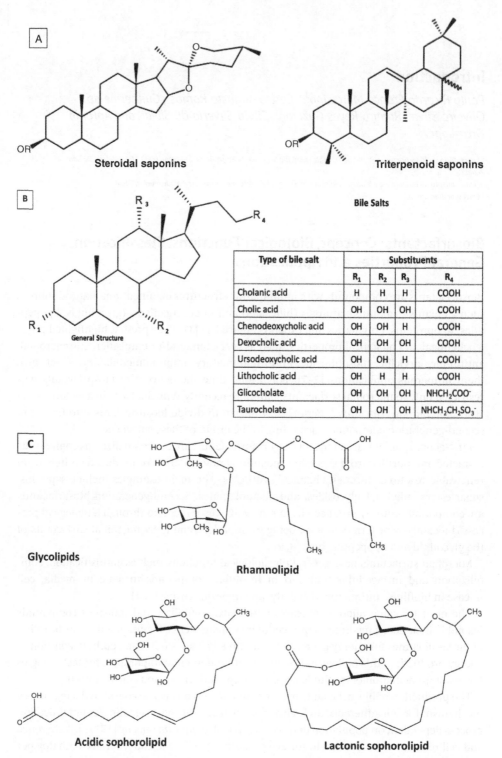

Figure I.1 Types of biosurfactant: (A) plant biosurfactants: saponins, (B) animal biosurfactants: bile salts, (C) microbial biosurfactants: glycolipids, lipopeptides, lipoproteins and polymerics.

unit area, surface composition, among others), the interfaces of supports, and microbial membranes. In addition, the microbiological properties and characteristics of the surrounding medium, such as temperature, pH, ionic strength, and availability of nutrients, are also determinant in the reported process [20–22]. Biosurfactants are molecules of fundamental importance in cell adherence and release processes. Microorganisms can use cell wall surfactant substances to regulate surface properties, aiming to adhere to or detach from a given location according to their need to find new habitats with greater availability of nutrients or to get rid of environments with unfavorable conditions [23, 24].

Biosurfactant compounds are also important for microorganisms because they have antimicrobial activity, becoming a defense mechanism in a competitive environment. Lipopeptides and glycolipids are the main biosurfactants with antimicrobial properties [25–28]. The antimicrobial action of biosurfactants is due to the interaction of these compounds with short-chain phospholipids found in the plasma membrane, causing an increase in membrane permeability through its solubilization, rupture, and disruption [29]. Among the biosurfactants with antimicrobial action, the class of lipopeptides is the most reported and studied [28]. With the recent pandemic caused by the SARS-COV2 coronavirus, some works have also highlighted the antiviral potential of biosurfactants [30–32].

The physical-chemical properties and classifications of biosurfactants are due to their structural characteristics. According to Bognolo [33] and Banat et al. [34], in biosurfactants, hydrophobic portions consisting of hydrocarbon chains with one or more fatty acids are observed, which can be saturated, unsaturated, hydroxylated or branched, linked to a hydrophilic portion, which it can be an ester, a hydroxyl group, a phosphate, a carboxylate, a carbohydrate, an amino acid, or a peptide. Most biosurfactants are neutral or anionic (present negatively charged polar groups), ranging from small fatty acids to large polymers. Based on the nature of their hydrophilic groups, biosurfactants are also commonly classified according to their chemical composition, as can be seen in Table I.1 [35, 36].

Although the majority of biosurfactants reported in the literature are microbial metabolites associated, partially or not associated with cell growth, Kappeli and Finnerty [37] and Santos et al. [38] reported that some microbial cells, due to their high surface hydrophobicity, can also be considered surfactants by themselves. As an example, we have some hydrocarbonclastic microorganisms and species of *Cyanobacteria, Staphylococcus, Serratia, Acinetobacter calcoaceticus*, and *Pseudomonas marginalis*. There are also reports of some microbial extracellular vesicles that have high surfactant activity being classified as particulate biosurfactants.

In addition to natural and synthetic biosurfactants, others can be obtained through enzymatic/chemical transformations in a pre-existing structure, the so-called semi-synthetic or modified biosurfactants [39, 40]. Such modification mechanisms allow not only the structural change of the molecule but also the physical-chemical and biological properties. Zinjarde and Ghosh [41] reported the use of enzymatic systems, such as lipases and glycosidases, in the structural modification of precursor molecules of biosurfactants, such as carbohydrates and lipids. Delbeke and collaborators [42] synthesized a quaternary ammonium sophorolipid. From a sophorolactone produced by *Starmerella bombicola*, organic reactions were carried out in order to insert an amine group in the biosurfactant molecule. The modified sophorolipid obtained was tested against pathogenic bacteria, proving to be an excellent antimicrobial in the treatment of gram-positive strains, with

Table I.1 Types of biosurfactants and their main representatives.

Type of biosurfactant	Main representatives
Saponins or saponosides	Steroidal saponins
	Triterpene saponins
Glicolipids	Rhamnolipids
	Sophorolipids
	Mannosylerythritol lipids (MEL)
	Cellobiose lipids/ustilagic acid
	Xylolipids
	Trealolipids
	Mycolates/trealomycolates
	Oligosaccharide lipids
	Lipid polyols
Fatty acids, neutral lipids and phospholipids	Fatty acids
	Neutral lipids
	Phospholipids
Polymeric surfactants	Emulsan
	Biodispersant
	Liposan
	Carbohydrate-lipid-protein complex
	Lipo heteropolysaccharides
	Mannoproteins
	Mannan-lipid
Lipopeptides and lipoproteins	Surfactin
	Iturin
	Serrawettin
	Subtilisin
	Cerelipin
	Gramicidin
	Viscosine
	Amphisine
	Tolaasin
	Syringomycin
	Peptide-lipids (lysine lipid and ornithine lipid)
Particulates	Membrane vesicles, fimbriae, and whole cells

greater potential than gentamicin sulfates already used in standard antibiotic-therapies, proving to have great promise in the clinical area. Recently, rhamnolipids were also functionalized with the amino acids arginine and lysine, obtaining cationic rhamnolipids (RLs) derivatives (Figure I.2) with outstanding physicochemical properties, in addition to

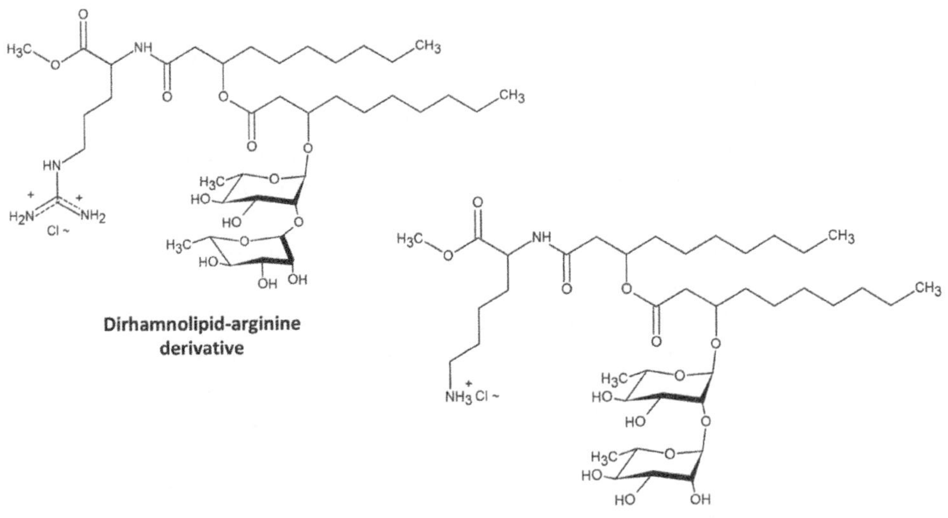

Biosurfactants: Concept, Biological Functions, Classification, General Properties and Applications | 5

Figure I.2 Chemical structures of new semi-synthetic or modified surfactants based on rhamnolipids and amino acids: dirhamnolipid-arginine derivative and dirhamnolipid-lysine derivative.

presenting notable DNA binding affinity and good antimicrobial activity against gram-positive bacteria, including methicillin-resistant *Staphylococcus aureus* [40].

The classification of biosurfactants can also be given according to their molecular weights. Some authors commonly classify them as high molar mass compounds (polysaccharides, lipopolysaccharides, lipoproteins, or complex mixtures of these biopolymers) or low molar mass (glycolipids and lipopeptides) [43]. This classification, despite still being widely used, has received criticism, since low molar mass molecules tend to have more surfactant characteristics, while high molecular mass molecules have emulsifying characteristics. Because of this, these molecules have different physicochemical and biological characteristics, and it is not correct to classify them all as biological surfactants. According to some authors, the correct way to classify these molecules would be to name molecules with a high molar mass of bioemulsifiers, while those with low molar mass would be called biosurfactants [44].

Biosurfactants are of fundamental importance in the current scenario because they are considered ecologically correct products, due to their low-/non-toxicity and high biodegradability. In addition, when compared to synthetic surfactants, they are more efficient, as they exhibit reduced surface and interfacial activities; are tolerant of high temperatures and extremes in pH and ionic strength; are specific and biocompatible [3, 33, 38, 45–48].

Biosurfactants are considered versatile molecules due to their wide possibilities in industrial applications. They can be used in the oil, chemical, food, and pharmaceutical industries and also in environmental and agricultural applications [4, 6, 38, 49–55]. However, these biomolecules are still not widely used due to high production costs, associated with inefficient product recovery methods and the use of expensive substrates [4]. In addition, several research groups have been seeking a better understanding of the biochemical production pathways for these products and thus developing new strategies to increase production.

The production of biosurfactants in the context of biorefineries can be a way to reduce the production costs of these molecules. In recent years, studies have begun on the production of biosurfactants in these biofactories and the technical and economic viability of these processes and products.

This book intends to present an overview of trends in the production of biosurfactants in biorefineries using fermentative and enzymatic processes and the versatility of applications of these bioproducts in the contemporary world that seeks sustainable development.

In Chapter 1, the main microorganisms produced by biosurfactants will be shown and how they can have their potential explored in different biorefineries. In addition, it will be briefly presented how extremophile microorganisms can be used in the biosurfactant production.

In Chapter 2, topics related to the selection of microorganisms that produce biosurfactants and lipase enzymes will be addressed, from conventional microorganisms to extremophiles.

In Chapter 3 some concepts related to the use of metabolic engineering tools in the microbial production of biosurfactants will be presented.

In Chapter 4, the production of biosurfactants in the context of biorefineries will be discussed. The possibility of biosurfactant production as a value-added product in biorefineries will be shown. In addition, the main biomasses used and their pre-treatments and important aspects of fermentation processes in this context will also be presented.

In Chapter 5, the production of biosurfactants by solid-state fermentation will be discussed, which is of fundamental importance in lignocellulosic biorefineries, but still little explored in the literature.

In Chapter 6, the development and challenges in the production of biosurfactants will be discussed. Some of the main production problems and how they can be overcome will be highlighted. It is noteworthy that this chapter will address some important bioprocess engineering concepts for the production of biosurfactants, mainly in submerged fermentations.

In Chapter 7, important concepts in the enzymatic production of biosurfactants will be discussed. The enzymes and conditions used in the enzymatic synthesis of biosurfactants, the bioreactors and the advantages and disadvantages of this form of sustainable production of biosurfactants will be discussed.

In Chapter 8, the production of biosurfactants will be discussed concomitantly with other important bioproducts for biorefineries, such as microbial polyesters, enzymes, lipids, and ethanol.

In Chapter 9, the use of biosurfactants in nanotechnological processes will be discussed, a topic of relevance in the current scenario.

In Chapter 10, the interaction of biosurfactants with models of membranes and proteins will be discussed. This theme is extremely important to understand future applications of biosurfactants, mainly in the pharmaceutical area, as antimicrobials.

In Chapter 11, the properties and therapeutic applications of biosurfactants will be discussed. It will be shown briefly from the purification processes necessary for the application of these metabolites in the pharmaceutical area to their antimicrobial, antitumor and anti-inflammatory potential.

In Chapter 12, the production of biosurfactants by fungi and the application of these metabolites in agriculture and bioremediation processes will be discussed.

In Chapter 13, the application of biosurfactants in the development of new products will be discussed. It will be shown how important biosurfactants are for the pharmaceutical, food and even civil engineering industries.

In Chapter 14, a brief discussion of the aspects necessary for techno-economic-environmental analysis of the production of biosurfactants in the context of biorefineries will be presented.

References

1 Barbosa FG, Ribeaux DR, Rocha T, Costa RA, Guzmán RR, Marcelino PR, da Silva SS. Biosurfactants: sustainable and Versatile Molecules. *J Braz Chem Soc* 2022;33:870–893.
2 Desai JD, Banat IM. Microbial Production of surfactants and their commercial potential. *Microbiol Mol Biol Rev.* 1997;6(1):47–64.
3 Cameotra SS, Makkar RS. Synthesis of biosurfactants in extreme conditions. *Appl Microbiol Biotechnol.* 1998;50(5):520: 9.
4 Nitschke M, Pastore GM. Biossurfactantes: propriedades e aplicações. *Química nova* 2002;25:772–776.
5 Fontes GC, Amaral PFF, Coelho MAZ. Produção de biossurfactante por levedura. *Química Nova.* 2008;31(8):2091–2099.
6 Silva VL, Lovaglio RB, Von Zuben CJ, Contiero J. Rhamnolipids: solution against Aedes aegypti? *Front Microbiol* 2015;6:88.
7 Franco Marcelino PR, da Silva VL, Rodrigues Philippini R, Von Zuben CJ, Contiero J, Dos Santos JC, da Silva SS. Biosurfactants produced by *Scheffersomyces stipitis* cultured in sugarcane bagasse hydrolysate as new green larvicides for the control of Aedes aegypti, a vector of neglected tropical diseases. *PLoS One.* 2017;12(11):e0187125.
8 Albano TJS Development of production processes for new-to-nature biosurfactants. Lisbon: Technical Institute of Lisbon; 2014. Avaibable from: https://www.semanticscholar.org/paper/Development-of-production-processes-for-Albano/d2fd3f5dcdb8dad549dced52e776453dc78008ab (Accessed 23 January 2023)
9 Soetaert W Introduction to Biosurfactants. Berlim, Biosurfactants workshop – BioTic, 2014. (Comunicação Oral/Palestra).
10 Rosenberg E, Zuckerberg A, Rubinovitz C, Gutnick D. Emulsifier of Arthrobacter RAG-1: isolation and emulsifying properties. *Appl Environ Microbiol.* 1979;37(3):402–408.
11 Reddy PG, Singh HD, Roy PK, Baruah JN. Predominant role of hydrocarbon solubilization in the microbial uptake of hydrocarbons. *Biotechnol Bioengineering.* 1982;24(6):1241–1269.
12 Zhang YIMIN, Miller R. Enhanced octadecane dispersion and biodegradation by a Pseudomonas rhamnolipid surfactant (biosurfactant). *Appl Environ Microbiol.* 1992;58(10):3276–3282.
13 Patricia B, Jean-Claude B. Involvement of bioemulsifier in heptadecane uptake in Pseudomonas nautica. *Chemosphere.* 1999;38(5):1157–1164.

14. Noordman WH, Wachter JH, De Boer GJ, Janssen DB. The enhancement by surfactants of hexadecane degradation by Pseudomonas aeruginosa varies with substrate availability. *J Biotechnol*. 2002;94(2):195–212.
15. Cameotra SS, Singh P. Synthesis of rhamnolipid biosurfactant and mode of hexadecane uptake by Pseudomonas species. *Microbial Cell Factories*. 2009;8(1):1–7.
16. Karlapudi AP, Venkateswarulu TC, Tammineedi J, Kanumuri L, Ravuru BK, ramu Dirisala V, et al. Role of biosurfactants in bioremediation of oil pollution-a review. *Petroleum* 2018;4(3):241–249.
17. Van Hamme JD, Singh A, Ward OP. Physiological aspects: part 1 in a series of papers devoted to surfactants in microbiology and biotechnology. *Biotechnol Adv*. 2006;24(6):604–620.
18. Victor IU, Kwiencien M, Tripathi L, Cobice D, McClean S, Marchant R et al. Quorum sensing as a potential target for increased production of rhamnolipid biosurfactant in Burkholderia thailandensis E264. *Appl Microbiol Biotechnol*. 2019;103(16):6505–6517.
19. Saadati F, Shahryari S, Sani NM, Farajzadeh D, Zahiri HS, Vali H et al. Effect of MA01 rhamnolipid on cell viability and expression of quorum-sensing (QS) genes involved in biofilm formation by methicillin-resistant Staphylococcus aureus. *Sci Rep*. 2022;12(1):1–12.
20. Ubbink J, Schär-Zammaretti P. Colloidal properties and specific interactions of bacterial surfaces. *Curr Opin Colloid Interface Sci*. 2007;12(4–5):263–270.
21. Jimoh AA, Lin J. Biosurfactant: a new frontier for greener technology and environmental sustainability. *Ecotoxicol Environm Safety* 2019;184:109607.
22. Sharma J, Sundar D, Srivastava P. Biosurfactants: potential agents for controlling cellular communication, motility, and antagonism. *Front Mol Biosci*. 2021>;8:727070.
23. Rosenberg E, Ron EZ. High-and low-molecular-mass microbial surfactants. *Appl Microbiol Biotechnol*. 1999;52(2):154–162.
24. Nickzad A, Déziel E. The involvement of rhamnolipids in microbial cell adhesion and biofilm development–an approach for control? *Letters Appl Microbiol*. 2014;58(5):447–453.
25. Araujo LVD, Freire DMG, Nitschke M. Biossurfactantes: propriedades anticorrosivas, antibiofilmes e antimicrobianas. *Química Nova* 2013;36:848–858.
26. Cortés-Sánchez AJ, Hernández-Sánchez H, Jaramillo-Flores ME. Biological activity of glycolipids produced by microorganisms: new trends and possible therapeutic alternatives. *Microbiol Res*. 2013;168(1):22–32.
27. Meena KR, Kanwar SS Lipopeptides as the antifungal and antibacterial agents: applications in food safety and therapeutics. *BioMed Res International*. 2015; 2015.
28. De Giani A, Zampolli J, Di Gennaro P. Recent trends on biosurfactants with antimicrobial activity produced by bacteria associated with human health: different perspectives on their properties, challenges, and potential applications. *Front Microbiol*. 2021;12:655150.
29. Bouffioux O, Berquand A, Eeman M, Paquot M, Dufrêne YF, Brasseur R et al. Molecular organization of surfactin–phospholipid monolayers: effect of phospholipid chain length and polar head. *Biochim et Biophys Acta (BBA)-Biomembranes*. 2007;1768(7):1758–1768.
30. Smith ML, Gandolfi S, Coshall PM, Rahman PK. Biosurfactants: a Covid-19 perspective. *Front Microbiol* 2020; 11:1341.
31. Çelik PA, Manga EB, Çabuk A, Banat IM. Biosurfactants' potential role in combating COVID-19 and similar future microbial threats. *Appl Sci*. 2021;11(1):334.

32. Raza ZA, Shahzad Q, Rehman A, Taqi M, Ayub A Biosurfactants in the sustainable eradication of SARS COV-2 from the environmental surfaces. 2022;12(10):1–12.
33. Bognolo G. Biosurfactants as emulsifying agents for hydrocarbons. *Colloids Surf A: Physicochem Engineering Aspects*. 1999;152(1–2):41–52.
34. Banat IM, Franzetti A, Gandolfi I, Bestetti G, Martinotti MG, Fracchia L et al. Microbial biosurfactants production, applications and future potential. *Appl Microbiol Biotechnol*. 2010;87(2):427–444.
35. Desai JD, Desai AJ. Production of biosurfactants. In: Kosaric N editors. *Biosurfactants: Production, Properties, Applications*, New York: CRC Press. 1993; 65–92.
36. Gayathiri E, Prakash P, Karmegam N, Varjani S, Awasthi MK, Ravindran B. Biosurfactants: potential and eco-friendly material for sustainable agriculture and environmental safety - a review. *Agronomy*. 2022;12(3):662.
37. Kappeli O, Finnerty WR. Partition of alkane by an extracellular vesicle derived from hexadecane-grown Acinetobacter. *J Bacteriology*. 1979;140(2):707–712.
38. Santos DK, Rufino RD, Luna JM, Santos VA, Sarubbo LA. Biosurfactants: multifunctional biomolecules of the 21st century. *Int J Mol Sci*. 2016;17(3):401.
39. Recke VK, Gerlitzki M, Hausmann R, Syldatk C, Wray V, Tokuda H et al. Enzymatic production of modified 2-dodecyl-sophorosides (biosurfactants) and their characterization. *Eur J Lipid Sci Technol*. 2013;115(4):452–463.
40. da Silva AR, Má M, Pinazo A, García MT, Pérez L. Rhamnolipids functionalized with basic amino acids: synthesis, aggregation behavior, antibacterial activity and biodegradation studies. *Colloids Surf B: Biointerfaces* 2019;181:234–243.
41. Zinjarde SS, Ghosh M. Production of surface active compounds by biocatalyst technology. In: Sen R editors. *Biosurfactants*, New York: Springer. 2010; 289–303.
42. Delbeke EIP, Roman BI, Marin GB, Van Geem KM., Stevens CV. A new class of antimicrobial biosurfactants: quaternary ammonium sophorolipids. *Green Chem*. 2015;17(6):3373–3377.
43. Ron EZ, Rosenberg E. Biosurfactants and oil bioremediation. *Current Opinion in Biotechnology*. 2002;13(3):249–252.
44. Uzoigwe C, Burgess JG, Ennis CJ, Rahman PK. Bioemulsifiers are not biosurfactants and require different screening approaches. *Front Microbiol*. 2015;6:245.
45. Cooper DG, Paddock DA. Production of a biosurfactant from Torulopsis bombicola. *Appl Environ Microbiol*. 1984;47(1):173–176.
46. Horowitz S, Gilbert JN, Griffin WM. Isolation and characterization of a surfactant produced byBacillus licheniformis 86. *J Ind Microbiol*. 1990;6(4):243–248.
47. Mulligan CN, Gibbs BF. Factors influencing the economics of biosurfactants. In: Kosaric N editors. *Biosurfactants: Production, Properties, Applications*, New York: CRC Press. 1993; 65–92.
48. Flasz A, Rocha CA, Mosquera B, Sajo C. A comparative study of the toxicity of a synthetic surfactant and one produced by Pseudomonas aeruginosa ATCC 55925. *Med Sci Res*. 1998;26(3):181–185.
49. Van Dyke MI, Lee H, Trevors JT. Applications of microbial surfactants. *Biotechnol Adv*. 1991;9(2):241–252.
50. Nitschke M, Costa SGVAO. Biosurfactants in food industry. *Trends Food Sci Technol*. 2007;18(5):252–259.

51 Lourith N, Kanlayavattanakul M. Natural surfactants used in cosmetics: glycolipids. *International Journal of Cosmetic Science.* 2009;31(4):255–261.

52 Gharaei-Fathabad E. Biosurfactants in pharmaceutical industry: a mini-review. *Am J Drug Disc Develop.* 2011;1(1):58–69.

53 Campos JM, Montenegro Stamford TL, Sarubbo LA, de Luna JM, Rufino RD, Banat IM. Microbial biosurfactants as additives for food industries. *Biotechnol Prog.* 2013;29(5):1097–1108.

54 Naughton PJ, Marchant R, Naughton V, Banat IM. Microbial. Microbial biosurfactants: current trends and applications in agricultural and biomedical industries. *J Appl Microbiol.* 2019;127(1):12–28.

55 Eras-Muñoz E, Farré A, Sánchez A, Font X, Gea T. Microbial biosurfactants: a review of recent environmental applications. *Bioengineered.* 2022;13(5):12365–12391.

1

Microorganisms Producing Biosurfactants in the Current Scenario

Fernanda Palladino[1], Rita C.L.B. Rodrigues[2], Yasmim Senden dos Santos[1], and Carlos A. Rosa[1,*]

[1] *Microbiology Department, Institute of Biological Sciences, Federal University of Minas Gerais, 31270-901, Belo Horizonte-MG, Brazil*
[2] *Biotechnology Department, Lorena Engineering School, University of São Paulo, 12602-810, Lorena-SP, Brazil*
* *Corresponding author*

1.1 Introduction

Surfactants are compounds used in different industry segments (oil, detergent soap, food, and beverages) [1, 2]. They are amphipathic molecules that reduce the surface tension at the oil–water interface, increasing the solubility of immiscible substances in water. The global demand for surfactants is overgrowing, and according to market surveys, the outlook is for more than 520 billion tonnes of surfactants to be on the global market by 2022 [3]. The production of synthetic surfactants began in the first half of the 20th century. However, with a more expressive development of the petrochemical industry after World War II, the production of these compounds became more significant since the petrochemical industry provides raw materials to produce synthetic surfactants [4]. In the last decades, surfactants have attracted attention for their wide range of applications in modern society and their harmful effects on the environment. They have low biodegradability because they are derived from petroleum ecotoxicity and bioaccumulation [5].

In recent years, there has been an increase in the search for biodegradable compounds of natural origin that can be a more sustainable alternative to synthetic surfactants, corresponding to environmentally friendly and versatile products with extensive industrial applications [6, 7]. Natural surfactants are synthesized from living organisms, such as microorganisms (biosurfactants), plants (saponins), and vertebrates (bile salts) [8]. However, the best alternative for the production of surfactants that are less harmful to the environment is those produced by microorganisms, as they offer advantages over synthetic surfactants, such as lower toxicity, more significant foaming potential, biodegradability, thermal stability, selectivity and specificity, resistance to extreme values of pH and ionic strength, and solubility in alkaline medium [9, 10]. Although several studies involve bacteria in producing biosurfactants, yeasts are being studied because some bacteria are not GRAS (generally regarded as safe). Thus, this chapter will address microorganisms classified as the best biosurfactant producers and their applications.

1.2 Microbial Biosurfactants

Biosurfactants are a structurally diverse group of surface-active molecules produced by microorganisms [11]. These compounds can be synthesized by bacteria, yeasts, and filamentous fungi from various substrates, such as oils, alkanes, sugars, and industrial by-products [12]. Its production occurs when microorganisms are growing aerobically. During growth, they are secreted into the culture medium and help translocate and transport hydrophobic substrates across cell membranes, making them more available for metabolism and uptake and allowing microorganisms to develop on immiscible substrates [13]. Due to their characteristics, biosurfactant molecules have several properties, such as emulsification, detergency, wetting, foaming, dispersion, solubilization, and antimicrobial activity [12].

Several microorganisms can produce molecules with interfacial activity. There has been increasing interest in identifying and isolating new microorganisms that produce surfactant molecules with good characteristics, such as low critical micelle concentration (CMC), low toxicity, reduced surface tension, high emulsification activity,and ion complexation metallic [14, 15]. Studies approach biosurfactants of bacterial origin, produced by *Pseudomonas*, *Acinetobacter*, *Bacillus*, and *Arthrobacter*; however, the yeast species capable of producing these compounds are gaining more and more attention from the world scientific community probably due to these bacterial genera having a pathogenic nature, making their application in the food industry restricted. On the other hand, yeast has the advantage that some species have GRAS status, which means that microorganisms with this status are not pathogenic or toxic and can be applied in medicinal and food products [8]. Some yeasts of the genus *Candida*, *Saccharomyces*, *Starmerella*, *Rhodotorula*, *Pseudozyma*, *Yarrowia*, *Pichia*, *Ustilago*, *Schizonella*, *Kluyveromyces*, *Wickerhamiella*, *Kurtzmanomyces*, *Debaryomyces*, *Cutaneotrichosporon*, *Spathaspora*, *and Scheffersomyces e Meyerozyma* have been studied and reported in the literature as good biosurfactants producers [13, 16–18].

1.2.1 Structure and Classification of Biosurfactants

Biosurfactants are surface-active compounds produced by microorganisms, which have different structures obtained from renewable resources, possessing both hydrophilic and hydrophobic moieties [19]. Its amphiphilic structure modifies the surface properties or interfaces of complex systems. Biosurfactants are being studied as alternatives to cationic surfactants (quaternary ammonium), anionic surfactants (sodium lauryl sulfate, alkylbenzenesulfonate, alkyl ether phosphate), and amphoteric surfactants (phosphatidylcholine), mainly in the pharmaceutical industry [20].

Biosurfactants can be divided into two categories according to molecular weight [15]. One of these categories comprises low molecular weight, effective surface, and interfacial tension reducers. The other category is high molecular weight, called bioemulsifiers, which can form and stabilize emulsions but not necessarily reduce surface tension [21, 2, 22, 23]. Table 1.1 presents a classification of biosurfactants and producing microorganisms.

The main parameter used to identify the presence of biosurfactants in a system is the reduction of the surface tension of the medium. This substance in a system forms micelles (amphipathic molecules), which aggregate into hydrophilic and hydrophobic portions. More micelles lead to lower surface tension and are directly influenced by the increased concentration of biosurfactants [43, 44]. However, screening methods to identify these compounds based solely on surface tension reduction may eliminate bioemulsifiers as they will not always significantly reduce surface tension. Therefore, different tests must be performed, not just surface tension measurements, which are often used as preliminary tests [45].

There are many types of biosurfactants based on their chemical nature, such as glycolipids, lipopolysaccharides, oligosaccharides, and lipopeptides reported as being produced by various bacterial genera [46, 47]. Among the most studied is the genus *Pseudomonas*

Table 1.1 Classification of biosurfactants and producing microorganisms.

Molecular weight	Biosurfactant	Microorganism	References
Low molecular weight	Rhamnolipids	*Pseudomonas* sp.[1]	Phulpoto et al. [24]
		Aspergillus sp.[3]	Kiran et al. [25]
		Planococcus spp.[1]	Gaur et al. [26]
	Trehaloselipids	*Rhodococcus* sp.[1]	Bages-Estopa et al. [27]
	Sophorolipids	*Starmerella bombicola*[2]	Ceresa et al. [28]
	Mannosylerythritol lipids (MELs)	*Pseudozyma tsukubaensis*[2]	Andrade et al. [29]
		Ustilago maydis[3]	Becker et al. [30]
	Glucose lipids	*Alcanivorax borkurnensis*[1]	Yakimov et al. [31]
	Cellobiose lipids	*Pseudozyma aphidis*[2], *P. hubeiensis*[2]	Morita et al. [32]
	Fatty acids	*Corynebacterium lepus*[1]	Cooper et al. [33]
High molecular weight	Phospholipid	*Sphingobacterium* sp.[1]	Burgos-Díaz et al. [34]
	Surfactin/iturin/fengycin	*Bacillus subtilis*[1]	Arima et al. [35]
	Surfactin/fengycin	*Alcaligenes aquatilis*[1]	Yalaoui-Guellal et al.[36]
	Polymyxines	*Paenibacillus polymyxa*[1]	Deng et al. [37]
	Viscosin	*Pseudomonas fluorescens*[1]	Bonnichsen et al. [38]
	Serrawettin	*Serratia marcescens*[1]	Clements et al. [39]
	Emulsan	*Acinetobacter venetianus*[1]	Castro et al. [40]
	Liposan	*Candida lipolytica*[2]	Cirigliano et al. [41]
	Vesicle	*Acinetobacter* sp.[1]	Kappeli et al. [42]

[1] Bacteria
[2] Yeast
[3] Filamentous fungus. Vieira et al., 2021 / with permission of Elsevier

and *Bacillus*. Silva et al. [48] evaluated the commercial production of a biosurfactant from *Pseudomonas cepacia* CCT6659 grown in industrial by-products in a 50 L semi-industrial bioreactor for use in removing hydrocarbons from oily effluents. A concentration of 40.5 gL^{-1} was achieved on scale-up, and the surface tension was reduced to 29 mN m^{-1}. Huang [49] isolated *Serratia marcescens* ZCF25 from the sludge of an oil tanker. This author carried out tests for biosurfactants production and observed that the biosurfactants produced was classified as a lipopeptide that managed to reduce the water surface tension from 72–29.50 mN m^{-1}, showed high tolerance in a wide range of pH (2–12), high temperatures (50–100 °C) and salinity (10–100 gL^{-1}). This microorganism was considered promising to be used in industrial applications and bioremediation. Hu et al. [50] evaluated the biosurfactant production by *Bacillus subtilis* ATCC 21332 in a 7 L and 100 L bioreactor using fish peptone generated by enzymatic hydrolysis of crushed tuna (*Katsuwonus pelamis*). Results showed that *Bacillus subtilis* ATCC 21332 could effectively use fish waste peptones for surfactin production. The highest surfactin productivity achieved in pilot-scale experiments was 274 mg L^{-1}. Phulpoto et al. [24] evaluated several microorganisms isolated from freshwater lakes to produce biosurfactants. Among them, bacteria of the Pseudomonas and Bacillus species stood out, presenting an emulsification index of 32.70% and 55% and surface tension of 33.15 mN m^{-1} and 34.15 mN m^{-1}, respectively. Liu et al. [51] evaluated the biosurfactant production by Bacillus licheniformis. The highest emulsification index was obtained using lactose-based mineral salt solution. The samples were characterized, and the biosurfactant found was a series of lipopeptides containing C13-, C14-, C15-, C16- surfactin, and other types of lipopeptides. They also observed that the biosurfactant altered the wettability of the hydrophobic surface of core slices with different permeability. This biosurfactant can maintain emulsification activity at pH 4–11, 85 °C, 25% NaCl solution, or 17.5% CaCl$_2$ solution. Although there are several studies involving bacteria in the production of biosurfactants, yeasts are being studied.

1.2.2 Biosurfactants Producing Yeasts

The attention to the use of yeasts as biofactories in the production of biosurfactants is because some species are GRAS, making their application in the food, pharmaceutical, and cosmetic industries safer. For this reason, research involving the production of biosurfactants by yeasts has increased, and several strains of non-pathogenic species have been reported as good producers [52]. The genus Candida is a well-recognized producer of biosurfactants [53]. Some examples of *Candida* species able to produce biosurfactants are lipopeptide produced by *Candida lipolytica* [54], sophorolipids produced by *Candida bombicola* [55] or by *Candida apicola* [56] and mannosylerythritol lipid produced by *Candida antarctica* [57].

Some microorganisms can produce biosurfactants when they grow on different substrates, ranging from carbohydrates to hydrocarbons. Different carbon sources alter the structure of the biosurfactants produced and, consequently, their emulsifying properties. The biosurfactants production by yeasts is influenced by carbon and nitrogen sources and the presence or absence of phosphorus, iron, magnesium, and manganese in the substrate. Carbon sources can be alone or combined with organic or inorganic nitrogen sources such as yeast extract, ammonium salts, and nitrate salts [58]. Other parameters can change the yield of the process, such as temperature, pH, speed of agitation, aeration of the medium, and fermentation time. Table 1.2 shows microorganisms, substrates, and parameters used to produce biosurfactants [58, 8].

Sophorolipids are among the best-known and most studied biosurfactants, offering several advantages over synthetic surfactants [66]. Its production has been reported in the literature by the yeasts *Starmerella bombicola, S. batistae, S. apicola, Rhodotorula bogoriensis, Wickerhamiella domercqiae, S. riodocensis, S. stellata, e Candida* sp.NRR Y-27208 [67, 68].

Most biosurfactants are generally produced when cultures reach the stationary phase of growth. However, some species may show small production during the exponential growth

Table 1.2 Microorganisms and substrates for the production of biosurfactants.

Microorganisms	Carbon source	Surface tension (mNm^{-1})	E24 (%)	Cultivation Method	References
Yarrowia lipolytica	Glucose (10 g L^{-1})	34.7	68	Flasks	Yalçin et al. [18]
Apiotrichum loubieri		35.3	67	Flasks	
Pseudomonas aphidis	Glucose (40 g L^{-1})	32.83	ns*	Flasks	Niu et al. [59]
Candida glaebosa	Glycerol (40 g L^{-1})	ns*	30	Flasks	Bueno et al. [53]
Bacillus Subitilis	Glucose (4% w v^{-1})	25.9	ns*	Flasks	Queiroga et al. [60]
Bacillus licheniformis	Lactose (10 g L^{-1})	ns*	96	Flasks	Liu et al. [51]
Naganishia adellienses	Sugarcane straw hemicellulosic hydrolysate (40 g L^{-1} Xylose)	ns*	52	Flasks	Chaves et al. [9]
Cutaneotrichosporon mucoides	Sugarcane bagasse hemicellulosic hydrolysate (40 g L^{-1} Xylose)	ns*	70	Flasks	Marcelino, et al. [17]
Meyerozyma guilliermondii		ns*	70	Flasks	
Trichosporon montevideense	Sunflower oil (20 g L^{-1})	44.9	75.80	Flasks	Monteiro et al. [61]
Aspergillus niger	Banana stalks powder	ns*	57	Solid State	Asgher et al. [62]
Candida sphaerica	Refinery residue of soybean oil (90 g L^{-1}) and corn steep liquor (90 g L^{-1})	25.00	ns*	Flasks	Luna et al. [63]
		27.48	ns*	Bioreactor	
Candida tropicalis	Sugarcane molasses (25 g L^{-1}), Corn steep liquor (25 g L^{-1}), Waste frying oil (25 g L^{-1}),	29.98	ns*	Flasks	Almeida et al. [64]
		34.12	ns*	Bioreactor	
Pseudomonas aeruginosa	Corn steep liquor (10% v v^{-1}), sugarcane molasses (10% v v^{-1})	31.4	59.0	Flasks	Gudiña et al. [65]

*ns (Not specified)

phase. Sophorolipid biosynthesis is stimulated under nitrogen-limiting conditions since the enzymatic activity is high under these conditions. Furthermore, it occurs at the end of the exponential phase and the beginning of the microorganism's stationary growth phase [69]. The proportion, type, and production of acidic or lactone forms of sophorolipids depend on some variables, such as the composition of the medium (carbon, nitrogen, and salt sources), production lineage, environmental conditions (pH, temperature, agitation, time, and aeration) and the type of cultivation process (batch or continuous) [70, 66]. The sophorolipid production is highly stimulated when two carbon sources of lipophilic origin (alkanes, saturated and unsaturated fatty acids, alcohols, oils, and fats) and hydrophilic (glucose) are present in the culture medium. Production can be lower when there are no hydrophobic sources in the middle.

The amphiphilic nature of biosurfactant creates surface tension at the interphase of two mediums and thus, helps microorganisms utilize hydrophobic carbon sources as nutrients and protects microorganisms from adverse environmental conditions [71]. Furthermore, the type of hydrophobic source will influence the sophorolipid final composition [14]. evaluated the production of sophorolipid by *Starmerella bombicola* using biodiesel as a hydrophobic substrate. They used a fed-batch fermentation system and obtained higher yields than flask cultivation (58.1–224.2 gL^{-1}). It showed good surfactant properties, such as surface tension of 34.2 $nM\ m^{-1}$ and a critical micellar concentration of 25.1 $mg\ L^{-1}$.

Two basic approaches are adopted globally for cost-efficient production of biosurfactants: (1) utilization of abundant, inexpensive, and waste biomass as a substrate for the production media resulting in low initial raw material costs required for the process, and (2) development and optimization of bioprocesses for maximizing biosurfactant production and recovery, leading to reduced operating costs [72, 73]. By-products from different industrial segments have been used for the biotechnological production of biosurfactants. Among them, we can have glycerol [74]; animal fat [75]; soy oil refinery by-product [76]; a by-product of a peanut oil refinery [77]; residual frying oil [78]; clarified cashew apple juice [79]; corn steep liquor [80], cassava wastewater [81], vinasse [79]; molasses [82]; sugarcane bagasse hemicellulosic hydrolysate [83]. A single carbon source is used in biosurfactants produced via microbial fermentation (carbohydrate or a combination of two substrates – sugars and lipids) [22].

The yeast produces fatty acids in its growth phase to form new lipid membranes, organelles, and compounds such as phospholipids, triacylglycerols, and neutral lipids. For this to occur, it is necessary to use precursors derived from the glycolytic pathway (acetyl-CoA, malonyl-CoA) and energy (ATP, NADPH), and it is essential to maintain controls over the fermentative parameters during the process so that the formation of these chains and the accumulation of these molecules by the yeast occurs satisfactorily.

The synthesis of a fatty acid chain from glucose (Figure 1.1) begins with the oxidation of this molecule to pyruvate and its conversion to acetyl-CoA, a compound with several cellular fates and functions. This compound is synthesized in mitochondria, nucleus, peroxisome, and cytoplasm, where oxidation reactions occur [84]. In the case of fatty acid biosynthesis, it occurs in the cytosol, within the mitochondria, and from acetyl-Coa, using the enzyme pyruvate decarboxylase, it converts pyruvate into acetaldehyde, forming acetyl-CoA. This molecule is converted into malonyl-CoA through acetyl-CoA carboxylase, using biotin-containing and sodium bicarbonate as cofactors [84]. This chain is extended

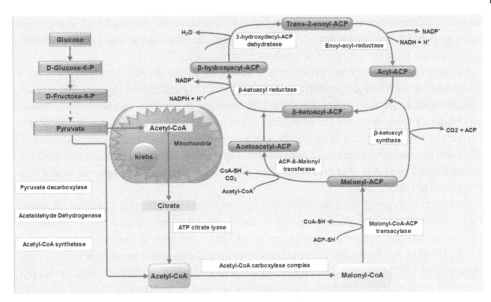

Figure 1.1 Fatty acid synthesis in yeast [Adapted 86].

by involving five reactions with the following enzymatic activities: malonyl transferase, ketoacyl synthase, ketoacyl reductase, 3-hydroxyacyl dehydratase, and enoyl-acyl-reductase. By yeast, fatty acid biosynthesis is terminated with the release of fatty acyl-CoA. These are converted into products such as free fatty acids, fatty acid ethyl esters, palmitic acids, and oleic acids, among others [85]. Figure 1.1 presents a description of fatty acid synthesis from glucose. First, the cytosolic acetyl-CoA compound resulting from glycolysis is converted into malonyl-CoA by the acetyl-CoA carboxylase complex. This compound is then converted by the enzymatic activity of malonyl-CoA transacylase, which converts this compound to malonyl-ACP. Then, using an acetyl-CoA and a malonyl-ACP as substrate, the enzyme ACP-S-malonyl transferase synthesizes the compound acetoacetyl-ACP that will be used by the enzyme β-ketoacyl reductase, together with the compound β-ketoacyl-ACP to synthesize a β-hydroxyacyl-ACP. Continuing the cycle, the 3-hydroxyacyl dehydratase enzyme will use the β-hydroxyacyl-ACP and convert it into trans-2-enoyl-ACP, generating an acyl-ACP chain through the action of the enoyl-acyl-reductase enzyme[86].

1.2.3 Biosurfactants Produced by Extremophile Microorganisms

Antarctica is a remote and inhospitable continent with the coldest and driest climate on Earth, with low temperature, low water availability, frequent freeze–thaw cycles, low annual precipitation, strong winds, high sublimation, and evaporation, and high incidence of solar radiation. Despite this, it is dominated by microorganisms with a high level of adaptation and the ability to resist extreme conditions [53, 87]. The ability to adapt to extreme environments has led to the selection of extremotolerant/extremophilic strains [88] which are highly specialized microorganisms such as Archaea, bacteria, cyanobacteria, algae, protozoa, filamentous fungi, and yeasts [89; 53]. Studies have reported that some yeasts such as *Pseudozyma antarctica*, *Candida lipolytica*, *Candida sphaerica*, and *Rhodotorula mucilaginosa* have the potential to produce biosurfactants [90–93].

Extreme environments and extremophile microorganisms are excellent sources of novel biomolecules with unique properties. Biosurfactants play critical ecological roles in many cold habitats, improving their habitability, for example, in catalysis and recovery of natural gas hydrates from deep-sea environments. They participate in carbon-cycling processes, increasing the bioavailability of poorly soluble compounds, including pollutants, in cold soils and marine environments. These natural capabilities can be used to develop new biotechnological products and processes with low energy demand and operate under low-temperature regimes [94].

Chaves et al. [9] evaluated the biosurfactants produced by Antarctic yeasts using detoxified sugarcane straw hemicellulose hydrolyzate (DSSHH). They observed that *Naganishia adelienses* had the highest emulsification index (52%) and total xylose consumption (40 g L^{-1}) in DSSHH. Furthermore, the emulsion remained stable under low-temperature conditions (0 and 4 °C), high salt concentration (10%), and alkaline conditions. Bueno et al. [53] screened biosurfactants-producing yeast strains isolated from Antarctic soil and evaluated the fermentation process kinetics. From this screening using 68 isolated yeast strains, 11 strains produced biosurfactants in a medium containing glycerol as a carbon source. The most promising yeast was *Candida glaebosa*, presenting an E.I of 30% at the end of fermentation. These authors concluded that the Maritime Antarctic environment is a promising source for isolating biosurfactants-producing microorganisms.

1.3 Industrial Applications of Biosurfactants

Biosurfactants have a wide range of applications due to their low toxicity, wide working range, biocompatibility, and biodigestibility. It can be used in agriculture [95], microbial enhanced oil recovery [58], soil and water remediation [96, 97], biomedical science [98], food industry [99], nanotechnology [100], cleaning products [101]. Figure 1.2 shows several applications of biosurfactants in the industrial segment.

Commercial production and development of biosurfactants present safety problems and high production costs. Research has been carried out with several microorganisms for biosurfactant production using renewable substrates such as sugars, oils, alkanes, and agro-industrial residues, including molasses, potato processing residues, olive oil effluent, vegetable oil extracts and residues, distillery, and residues of whey, by-products from vegetable industries, dairy and residues from the sugar industry and cassava wastewater [6, 73]. In addition to using renewable raw materials, it is essential to improve the stability of biosurfactants and increase productivity so that they can be produced on a large scale [73].

Studies have been carried out on biosurfactants in the synthesis of nanoparticles, as this product reduces the formation of aggregates, facilitating the homogeneous and uniform morphology of the nanoparticles during synthesis [102]. For this application, surfactin has been studied as a biodegradable stabilizing agent and less toxic in nanoparticles [103].

Studies have been carried out using biosurfactants to convert agricultural biomass into sugars, during pretreatment, for the biotechnological production of several metabolites.

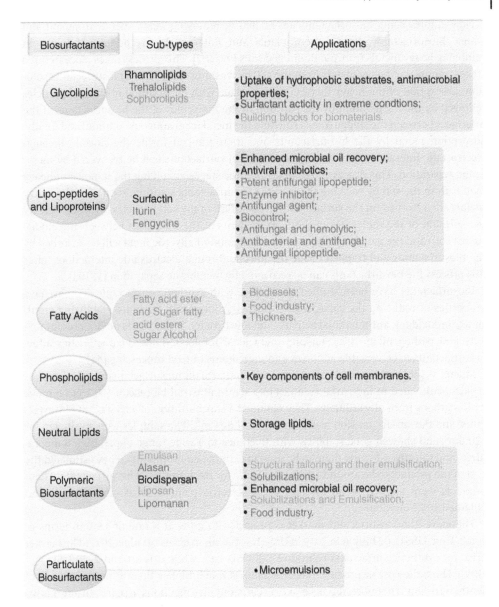

Figure 1.2 Industrial applications of biosurfactants.

This strategy will assist in producing second-generation bioethanol [104]. Recently, several studies have shown the antimicrobial, antiviral and anticancer potentials of biosurfactant molecules. Biosurfactants destroy bacterial cells, directly disrupting the integrity of the plasma membrane or cell wall. Antiviral activity occurs because of the physicochemical interaction with the virus's lipid envelope. Studies related to the anticancer potential revealed that the biosurfactant could induce cell cycle arrest, apoptosis, and metastasis arrest in tumor cells, without influencing non-tumor cells [105, 106].

New studies using biosurfactants are related to removing heavy metals from soil and water. Biosurfactants derived from plants and microorganisms performed better in removing heavy metals from contaminated soils [92, 107]. This process takes place in three stages: removal of heavy metals from the soil by washing them with a biosurfactants solution; absorption of the heavy metals on the surface with the solution of the soil particles followed by their separation by the sorption of the molecules of the biosurfactants at the interfaces between the sludge (moist soil) and the metal in an aqueous solution; and finally, adsorption metal by the biosurfactants and their trapped inside the micelle through electrostatic interactions, after this process the biosurfactants can be recovered by membrane separation. This process takes place in three stages, involving the removal of heavy metals from the soil by washing them with a biosurfactants solution; soon after the heavy metals are adsorbed on the surface with the solution of the soil particles, they separate by the sorption of the molecules of the biosurfactants at the interfaces between the sludge (moist soil) and the metal in an aqueous solution; and finally, the meta will be adsorbed by the biosurfactants and trapped inside the micelle through electrostatic interactions, after this process the biosurfactants can be recovered by membrane separation [1, 108].

Biosurfactants have been studied to compete with synthetic surfactants, as they have properties of reducing the capacity of surface and interfacial tensions, stabilizing emulsions, antioxidant, antimicrobial activity, and low toxicity. Some yeasts do not present toxicity and pathogenicity risks, making them ideal for food. These characteristics allow biosurfactants to be versatile additives and ingredients in food processing [44].

Silva et al. [48] evaluated the commercial production of the biosurfactants by *Pseudomonas cepacia*, cultivated in industrial residues in a semi-industrial bioreactor (50 L) to remove hydrocarbons from oily effluents. They obtained a concentration of 40.5 g L^{-1} of biosurfactants, and the surface tension was reduced to 29 mN m^{-1}. The samples were evaluated for 120 days and showed good stability, with tolerance to a wide range of pH, high temperatures, and salinity, allowing them to be used in extreme environments. They estimated the cost of the product to be around US$0.14–0.15 L^{-1} for the formulated biosurfactants. These products were applied in an oil-fired thermoelectric plant to treat oily effluents and obtained 100% oil removal.

Therefore, the biosurfactant market is projected to grow at a rate of 4.9% in terms of value, from US$15.02 billion in 2019 to US$21.84 billion in terms of value. 2027. The market price for synthetic surfactants is around 1–3 USD kg^{-1}, while it is around 2–20 USD kg^{-1} [109]. Thus, the cost to produce biosurfactants is much higher than to produce those of synthetic origin. Therefore, it is necessary to carry out investigations in search of alternative substrates that reduce the cost of production of these compounds and also microorganisms that are promising for this process.

References

1 Akbari S, Abdurahman NH, Yunus RM, Fayaz F, Alara OR. *Biosurfactants-a New Frontier for Social and Environmental Safety: A Mini Review*. Biotechnology Research and Innovation, 2018.

2. Drakontis CE, Amin S. Biosurfactants: formulations, properties, and applications. *Curr Opin Colloid Interface Sci.* 2020;48:77–90.
3. Singh P, Patil Y, Rale V. Biosurfactant production: emerging trends and promising strategies. *J Appl Microbiol.* 2019;126:2–13.
4. Karsa DR, Houston J. Chapter 1: What are surfactants. In: Farn RJ, editor. *Chemistry and technology of surfactants.* Blackwell Publishing Ltd., Oxford; 2006. pp. 1–23. ISBN: 978- 14051-2696-0.
5. Vieira IMM, Santos BLP, Ruzene DS, Silva DP. An overview of current research and developments in biosurfactants. *J Ind Eng Chem.* 2021;100:1–18.
6. Makkar RS, Banat IM, Cameotra SS. Advances in utilization of renewable substrates for biosurfactant production. *AMB Express.* 2011;1:5–17.
7. Miao S, Dashtborozorg SS, Callow NV, Ju LK. Rhamnolipids as platform molecules for production of potential antizoospore agrochemicals. *J Agr Food Chem.* 2015;63:3367–3376.
8. Marcelino PRF, Gonçalves F, Jimenez IM, Carneiro BC, Santos BB, Silva SS. Sustainable production of biosurfactants and their applications. *Lignocellulosic Biorefining Technol.* 2020:159–183.
9. Chaves FDS, Brumano LP, Franco Marcelino PR, da Silva SS, Sette LD, Felipe MDGDA. Biosurfactant production by Antarctic-derived yeasts in sugarcane straw hemicellulosic hydrolysate. *Biomass Convers Biorefin.* 2021.
10. Saravanakumari P, Mani K. Structural characterization of a novel xylolipid biosurfactant from *Lactococcus lactis* and analysis of antibacterial activity against multi-drug resistant pathogens. *Biores Technol.* 2010;101(22):8851–8854.
11. Mir S, Jamal P, Alamal MZ, Mir AB, Ansari AH. Microbial surface tensio-active compounds: production and industrial application perspectives: a review. *Int J Biotechnol Bioeng.* 2017;3:282–301.
12. Sourav D, et al. Review on natural surfactants. *R Soc Chem.* 2015;5:65757–65767.
13. Fenibo E, Dlouglas SI, Stanley HO. A review on microbial surfactants: production, classifications, properties and characterization. *J Adv Microbiol.* 2019;18:1–22.
14. Kim J-H, Oh Y-R, Hwang J, Jang Y-A, Lee SS, Hong SH, Eom GT. Value-added conversion of biodiesel into the versatile biosurfactant sophorolipid using *Starmerella bombicola*. *Cleaner Eng Technol.* 2020;1:100027.
15. Rosenberg E, Ron EZ. High- and low-molecular-mass microbial surfactants. *Appl Microbiol Biotechnol.* 1999;52(2):154–162.
16. Jezierska S, Claus S, Van Bogaert I. Yeast glycolipid biosurfactants. *FEBS Lett.* 2017;592(8):1312–1329.
17. Marcelino PRF, Peres GFD, Terán-Hilares R, Pagnocc FC, Rosa CA, Lacerda TM, Santos JC, Silva SS. Biosurfactants production by yeasts using sugarcane bagasse hemicellulosic hydrolysate as new sustainable alternative for lignocellulosic biorefineries. *Ind Crops Prod.* 2019;129(2019):212–223.
18. Yalçin HT, Ergin-Tepebaşı G, Uyar E. Isolation and molecular characterization of biosurfactant producing yeasts from the soil samples contaminated with petroleum derivatives. *J Basic Microbiol.* 2018.
19. Kugler JH, Le, Roes-Hill M, Syldatk C, Hausmann R. Surfactants tailored by the class Actinobacteria. *Front Microbiol.* 2015;6:212–219.

20. Merchant R, Banat IM. Biosurfactants: a sustainable replacement for chemical surfactants? *Biotechnol Lett*. 2012;34(9):1597–1605.
21. Bento MF, et al. Biossurfactantes. In: Melo IS, Azevedo JL, (org.). *Microbiologia Ambiental*. Embrapa Meio Ambiente, Jaguariúna; 2008. pp. 152–184.
22. Moutinho LF, Moura FR, Silvestre RC, Romão-Dumaresq AS. Microbial biosurfactants: a broad analysis of properties, applications, biosynthesis, and techno-economical assessment of rhamnolipid production. *Biotechnol Prog*. 2020;37(2).
23. Varjani SJ, Upasani VN. Critical review on biosurfactant analysis, purification and characterization using rhamnolipid as a model biosurfactant. *Biores Technol*. 2017;232:389–397.
24. Phulpoto IA, Hu B, Wang Y, Ndayisenga F, Li J, Yu Z. Effect of natural microbiome and culturable biosurfactants-producing bacterial consortia of freshwater lake on petroleum-hydrocarbon degradation. *Sci Total Environ*. 2021:141720.
25. Kiran GS, Thajuddin N, Hemma TA, Idhayadhulla A, Kumar RS, Selvin J. Optimization and characterization of rhamnolipid biosurfactant from sponge associated marine fungi *Aspergillus* sp. MSF1. *Desalination Water Treat*. 2010;24.
26. Gaur VK, Tripathi V, Gupta P, Dhiman N, Regar RK, Gautam K, Srivastava JK, Patnaik S, Patel DK, Manickam N. Rhamnolipids from *Planococcus* spp. and their mechanism of action against pathogenic bacteria. *Biores Technol*. 2020;https://www.sciencedirect.com/journal/bioresource-technology/vol/307/suppl/C307:123206.
27. Bages-Estopa S, White DA, Winterburn JB, Webb C, Martin PJ. Production and separation of a trehalolipid biosurfactant. *Biochem Eng J*. 2018;139(15):85–94.
28. Ceresa C, Rinaldi M, Tessarolo F, Maniglio D, Fedeli E, Tambone E, Caciagli P, Banat IM, Rienzo MAD, Fracchia L. Inhibitory Effects of Lipopeptides and Glycolipids on *C. albicans*–*Staphylococcus* spp. Dual-Species Biofilms. *Front Microbiol*. 2020;11:545654.
29. Andrade CJ, Andrade LM, Bution ML, Heididolder MA, Cavalcante Barros FF, Pastore GM. Optimizing alternative substrate for simultaneous production of surfactin and 2, 3-butanediol by *Bacillus subtilis* LB5a. *Biocatal Agric Biotechnol*. 2016;6:209–218.
30. Becker J, Tehrani HH, Ernst P, Blank LM, Wierckx N. An Optimized *Ustilago maydis* for Itaconic Acid Production at Maximal Theoretical Yield. *J Fungi*. 2021;7(1):20.
31. Yakimov MM, Golyshin PN, Lang S, Moore ER, Abraham WR, Lünsdorf H, Timmis KN. Alcanivorax borkumensis gen. nov., sp. nov., a new, hydrocarbon-degrading and surfactant-producing marine bacterium. *Int J Syst Bacteriol*. 1998;48(2):339–348.
32. Morita T, Fukuoka T, Imura T, Kitamoto D. Accumulation of cellobiose lipids under nitrogen-limiting conditions by two ustilaginomycetous yeasts, *Pseudozyma aphidis* and *Pseudozyma hubeiensis*. *FEMS Yeast Res*. 2013;13(1):44–49.
33. Cooper DG, Zajic JE, Gerson DF. Production of surface-active lipids by *Corynebacterium lepus*. *Appl Environ Microbiol*. 1979;37(1):4–10.
34. Burgos-Díaz C, Pons R, Espuny MJ, Aranda FJ, Teruel JA, Manresa A, Ortiz A, Marqués AM. Isolation and partial characterization of a biosurfactant mixture produced by *Sphingobacterium* sp. isolated from soil. *J Colloid Interface Sci*. 2011;361:195–204.
35. Arima K, Kakinuma A, Tamura G. Surfactin, a crystalline peptidelipid surfactant produced by *Bacillus subtilis*: isolation, characterization and its inhibition of fibrin clot formation. *Biochem Biophys Res Commun*. 1968;10;31(3):488–494.

36. Yalaoui-Guellal D, Fella-Temzi S, Djafri-Dib S, Sahu SK, Irorere VU, Banat IM, Madani K. The petroleum-degrading bacteria *Alcaligenes aquatilis* strain YGD 2906 as a potential source of lipopeptide biosurfactant. *Fuel*. 2021;285:1–7.
37. Deng Y, Lu Z, Lu F, Zhang C, Wang Y, Zhao H, Bie X. Identification of LI-F type antibiotics and di-*n*-butyl phthalate produced by *Paenibacillus polymyxa*. *J Microbiol Methods*. 2011;85(3):175–182.
38. Bonnichsen L, Svenningsen NB, Rybtke M, Bruijn I, Raaijmakers JM, Tolker-Nielsen T, Nybroe O. Lipopeptide biosurfactant viscosin enhances dispersal of *Pseudomonas fluorescens* SBW25 biofilms. *Microbiology (Reading)*. 2015;161(12):2289–2297.
39. Clements T, Ndlovu T, Khan W. Broad-spectrum antimicrobial activity of secondary metabolites produced by *Serratia marcescens* strains. *Microbiol Res*. 2019;229:126329.
40. Castro GR, Panilaitis B, Kaplan DL. Emulsan, a tailorable biopolymer for controlled release. *Biores Technol*. 2008;99:4566–4571.
41. Cirigliano MC, Carman GM. Purification and Characterization of Liposan, a Bioemulsifier from *Candida lipolytica*. *Appl Environ Microbiol*. 1985;50(4).
42. Käppeli O, Finnerty WR. Partition of alkane by an extracellular vesicle derived from hexadecane-grown Acinetobacter. *J Bacteriol*. 1979;140(2).
43. Araújo SCdS, Silva-Portela RCB, de Lima DC, Fonsêca MMB, Araújo WJ, da Silva UB, Napp AP, Pereira E, Vainstein MH, Agnez-Lima LF. MBSP1: a biosurfactant protein derived from a metagenomic library with activity in oil degradation. *Sci Rep*. 2020;10:1340.
44. Ribeiro BG, Guerra JMC, Sarubbo LA. *Biosurfactants: Production and Application Prospects in the Food Industry*. Biotechnology Progress, 2020.
45. Uzoigwe C, Burgess J, Grant ECJ, Rahman PKSM. Bioemulsifiers are not biosurfactants and require different screening approaches. *Front Microbiol*. 2015;6.
46. Banat IM, Franzetti A, Gandolfi I, Bestetti G, Martinotti MG, Fracchia L, Smyth TJ, Marchant R. Microbial biosurfactants production, applications and future potential. *Appl Microbiol Biotechnol*. 2010;87:427–444.
47. Franzetti A, Tamburini E, Banat IM. Application of biological surface active compounds in remediation technologies. In: Sen R, editor. *Biosurfactants: Advances in Experimental Medicine and Biology*, Vol. 672. Springer-Verlag, Berlin Heidelberg; 2010. pp. 121–134.
48. Silva R, de CFS, de Almeida DG, Brasileiro PPF, Rufino RD, de Luna JM, Sarubbo LA. *Production, Formulation and Cost Estimation of a Commercial Biosurfactant*. Biodegradation, 2018.
49. Huang Y, Zhou H, Zheng G, Li Y, Xie Q, You S, Zhang C. Isolation and characterization of biosurfactant-producing *Serratia marcescens* ZCF25 from oil sludge and application to bioremediation. *Environ Sci Pollut Res Int*. 2020;27(22):27762–27772.
50. Hu J, Luo J, Zhu Z, Chen B, Ye X, Zhu P, Zhang B. Multi-scale biosurfactant production by *Bacillus subtilis* using tuna fish waste as substrate. *Catalysts*. 2021;11(4):456.
51. Liu Q, Niu J, Yu Y, Wang C, Lu S, Zhang S, Ly J, Penga B. Production, characterization and application of biosurfactant produced by *Bacillus licheniformis* L20 for microbial enhanced oil recovery. *J Clean Prod*. 2021;307:127193.
52. Dailin DJ, Malek RA, Selvamani S, Nordin NZ, Tan LT, Sayyed RZ, Hisham AM, Wehbe R, Enshasy HE. Yeast biosurfactants biosynthesis, production and application. In: Sayyed RZ, El-Enshasy HA, Hameeda B, editors. *Microbial Surfactants Volume I: Production and Applications*. CRC Press; 2021. pp. 196–221.

53 Bueno JL, Santos RR, Moguel IS, Pessoa A Jr, Vianna MV, Pagnocca LD, Gurpilhares DB. Biosurfactant production by yeasts from different types of soil of the South Shetland Islands (Maritime Antarctica). *J Appl Microbiol.* 2019.

54 Rufino RD, Luna JM, Campos-Takaki GM, Sarubbo LA. Characterization and properties of the biosurfactant produced by *Candida lipolytica* UCP 0988. *Electronic J Biotech.* 2014;17:34–38.

55 Daverey A, Pakshirajan K. Production, characterization, and properties of sophorolipids from the yeast *Candida bombicola* using a low-cost fermentative medium. *Appl Biochem Biotechnol.* 2009;158(3):663–674.

56 Hommel K, Weber L, Rilke O, Himmelreich U, Kleber HP. Production of sophorose lipid by *Candida (Torulopsis) apicola* grown on glucose. *J Biotechnol.* 1994;33(2):147–155.

57 Kitamoto D, Ikegami T, Suzuki GR, Sasaki A, Takeyama Y, Idemoto Y, Koura N, Yanagishita H. Microbial conversion of n-alkanes into glycolipid biosurfactants, mannosyl erythritol lipids, by *Pseudozyma (Candida)* antarctica). *Biotecnol Lett.* 2001;23:1709–1714.

58 Geetha SJ, Banat IM, Joshi SJ. Biosurfactants: production and potential applications in Microbial Enhanced Oil Recovery (MEOR). *Biocatal Agric Biotechnol.* 2018;14:23–32.

59 Niu Y, Wu J, Wang W, Chen Q. Production and characterization of a new glycolipid, mannosylerythritol lipid, from waste cooking oil biotransformation by *Pseudozyma aphidis* ZJUDM34. *Food Sci Nutr.* 2019;7(3):937–948.

60 Queiroga CL, Nascimento LR, Serra GE. Evaluation of paraffins biodegradation and biosurfactant production by *Bacillus subtilis* in the presence of crude oil. *Braz J Microbiol.* 2003;34(4):321–324.

61 Monteiro AS, Coutinho JOPA, Júnior AC, Rosa CA, Siqueira EP, Santos VL. Characterization of new biosurfactant produced by *Trichosporon montevideense* CLOA 72 isolated from dairy industry effluents. *J Basic Microbiol.* 2009;49(6):553–563.

62 Asgher M, Arshad S, Qamar SA, Khalid N. Improved biosurfactant production from *Aspergillus niger* through chemical mutagenesis: characterization and RSM optimization. *SN Appl Sci.* 2020;2(5).

63 Luna JM, Rufino RD, Jara AMAT, Brasileiro PPF, Sarubbo LA. Environmental applications of the biosurfactant produced by *Candida sphaerica* cultivated in low-cost substrates. *Colloids Surf A Physicochem Eng Asp.* 2015;480(5):413–418.

64 Almeida DG, Soares da Silva R, de CF, Luna JM, Rufino RD, Santos VA, Sarubbo LA. Response surface methodology for optimizing the production of biosurfactant by *Candida tropicalis* on industrial waste substrates. *Front Microbiol.* 2017;8.

65 Gudiña EJ, Rodrigues AI, de Freitas V, Azevedo Z, Teixeira JA, Rodrigues LR. Valorization of agro-industrial wastes towards the production of rhamnolipids. *Biores Technol.* 2016;212:144–150.

66 Oliveira M, CAmilios-Neto D, Baldo C, Magri A, Celligoi MAPC. Biosynthesis and production of Sophorolipids. *Int J Sci Technol Res.* 2014;3:133–146.

67 Konishi M, Yoshida Y, Horiuchi J. Efficient production of sophorolipids by *Starmerella bombicola* using a corncob hydrolysate medium. *J Biosci Bioeng.* 2015;119:317–322.

68 Kurtzman CP, Price NPJ, Ray KJ, Kuo TM. Production of sophorolipid biosurfactants by multiple species of the *Starmerella (Candida) bombicola* yeast clade. *FEMS Microbiol Lett.* 2010;311:140–146.

69 Afonso L, Silveira VAI, Caretta TDO, Borsato D, Celligoi MAPC. Application of agro-industrial by-products as nitrogen sources in the production of sophorolipids by *Starmerella bombicola*. *Braz J Dev*. 2020;6(3):14875–14887.
70 Jahan R, Bodratti AM, Tsianou M, Alexandridis P. Biosurfactants, natural alternatives to synthetic surfactants: physicochemical properties and applications. *Adv Coll Interf Sci*. 2019:102061.
71 Vollbrecht E, Heckmann R, Wray V, Nimtz M, Lang S. Production and structure elucidation of di- and oligosaccharide lipids (biosurfactants) from *Tsukamurella* sp. Nov. *Appl Microbiol Biotechnol*. 1998;50:530–537.
72 Pereira BL, Francisco SM, da Silva SS. Recent advances in sustainable production and application of biosurfactants in Brazil and Latin America. *Indus Biotechnol*. 2016;12(1):31–39.
73 Sharma R, Oberoi HS. Biosurfactant-aided bioprocessing: industrial applications and environmental impact. *Recent Adv Appl Microbiol*. 2017:55–88.
74 Silva SNRL, Farias CBB, Rufino RD, Luna JM, Sarubbo LA. Glycerol as substrate for the production of biosurfactant by *Pseudomonas aeruginosa* UCP0992. *Colloids Surf B Biointerfaces*. 2010;1;79(1):174–183.
75 Santos DKF, Rufino RD, Luna JM, Santos VA, Salgueiro AA, Sarubbo LA. Synthesis and evaluation of biosurfactant produced by *Candida lipolytica* using animal fat and corn steep liquor. *J Pet Sci Eng*. 2013;105:43–50.
76 Luna JM, Rufino RD, Albuquerque CDC, Sarubbo LA, Campos-Takaki GM. Economic optimized medium for tenso-active agent production by *Candida sphaerica* UCP 0995 and application in the removal of hydrophobic contaminant from sand. *Int J Mol Sci*. 2011;12:2463–2476.
77 Sobrinho HBS, Rufino RD, Luna JM, Salgueiro AA, Campos-Takaki GM, Leite LFC, et al. Utilization of two agroindustrial by-products for the production of asurfactant by *Candida sphaerica* UCP0995. *Process Biochem*. 2008;43:912–917.
78 Batista RM, Rufino RD, Luna JM, Souza JEG, Sarubbo LA. Effect of medium components on the production of a biosurfactant from Candida tropicalis applied to the removal of hydrophobic contaminants in soil. *Water Environ Res*. 2010;82:418–425.
79 Oliveira JG, Garcia-Cruz CH. Properties of a biosurfactant produced by *Bacillus pumilus* using vinasse and waste frying oil as alternative carbon sources. *Braz Arch Biol Technol*. 2013;56:155–160.
80 Silva RCFS, Rufino RD, Luna JM, Farias CBB, Filho HJB, Santos VA et al. Enhancement of biosurfactant production from *Pseudomonas cepacia* CCT6659 through optimisation of nutritional parameters using response surface methodology. *Tenside Surf Det*. 2013;50:137–142.
81 Barros FFC, Ponezi AN, Pastore GM. Production of biosurfactant by *Bacillus subtilis* LB5a on a pilot scale using cassava wastewater as substrate. *J Ind Microbiol Biotechnol*. 2008;35(9):1071–1078.
82 Santos AC, Bezerra MS, Pereira HS, Santos ES, Macedo GR. Production and recovery of rhamnolipids using sugar cane molasses as carbon source. *J Chem Eng Chem Eng*. 2010;4:1–27.
83 Marcelino PRF, Peres GFD, Terán-Hilares R, Pagnocca FC, Rosa CA, Lacerda TM, Santos JC, Silva SS. Biosurfactants production by yeasts using sugarcane bagasse hemicellulosic

hydrolysate as new sustainable alternative for lignocellulosic biorefineries. *Ind Crops Prod.* 2019;129:212–223.

84 Hu Y, Zhu Z, Nielsen J, Siewers V. Engineering *Saccharomyces cerevisiae* cells for production of fatty acid-derived biofuels and chemicals. *Open Biol.* 2019;9(5):190049.

85 Yu T, Zhou YJ, Wenning L, Liu Q, Krivoruchko A, Siewers V, Nielsen J, David F. Metabolic engineering of Saccharomyces cerevisiae for production of very long chain fatty acid-derived chemicals. *Nat Commun.* 2017;8:15587.

86 Silva OB. Engineering *Saccharomyces cerevisiae* to increase the synthesis of genetic proteins for biosurfactant production. Thesis (Doctor molecular biology). Brasília University, 2019.

87 Ruisi S, Barreca D, Selbmann L, Zucconi L, Onofri S. Fungi in Antarctica. *Rev Environ Sci Biotechnol.* 2007;6:127–141.

88 Onofri S, Selbmann L, de Hoog GS, Grube M, Barreca D, Ruisi S, Zucconi L. Evolution and adaptation of fungi at boundaries of life. *Adv Space Res.* 2007;40:1657–1664.

89 Buzzini P, Branda E, Goretti M, Turchetti B. Psychrophilic yeasts from worldwide glacial habitats: diversity, adaptation strategies and biotechnological potential. *FEMS Microbiol Ecol.* 2012;82(2):217–241.

90 Konishi M, Morita T, Fukuoka T, Imura T, Kakugawa K, Kitamoto D. Production of different types of mannosylerythritol lipids as biosurfactants by the newly isolated yeast strains belonging to the genus Pseudozyma. *Appl Microbiol Biotechnol.* 2007;75:521–531.

91 Luna JM, Rufino RD, Sarubbo LA. Biosurfactant from *Candida sphaerica* UCP0995 exhibiting heavy metal remediation properties. *Process Saf Environ Prot.* 2016;102:558–566.

92 Rufino RD, Sarubbo LA, Campos-Takaki GM. Enhancement of stability of biosurfactant produced by Candida lipolytica using industrial residue as substrate. *World J Microbiol Biotechnol.* 2008;23:729–734.

93 Sousa TGC, Pinheiro TA, Coelho DF, Tambourgi EB, Sette LD, Pessoal A, Cardoso VL, Campos ES, Coutinho-Filho U. (2016) Evaluation of biosurfactant production by yeasts from Antarctica. In: *5th International Symposium On Industrial Biotechnology (IBIC).* 49:547–552.

94 Perfumo A, Banat IM, Marchant R. Going green and cold: biosurfactants from low-temperature environments to biotechnology applications. *Trends Biotechnol.* 2018;36(3):277–289.

95 Sachdev DP, Cameotra SS. Biosurfactants in agriculture. *Appl Microbiol Biotechnol.* 2013;97:1005–1016.

96 Mahanti P, Kumar S, Patra JK. Biosurfactants: an agent to keep environment clean. In: *Microbial Biotechnology.* Springer, Singapore; 2017. pp. 413–428.

97 Mulligan CN. Biosurfactants for the remediation of metal contamination. In: Das S, Dash HR, editors. *Handbook of Metal-Microbe Interactions and Bioremediation.* CRC Press; 2017. pp. 299–316.

98 Saha P, Nath D, Choudhury MD, Talukdar AD. Probiotic biosurfactants: a potential therapeutic exercises in biomedical sciences. In: Patra JK, Das G, Shin H, editors. *Microbial Biotechnology.* Springer, Singapore; 2018. pp. 499–514.

99 Nitschke M, Silva SSE. Recent food applications of microbial surfactants. *Crit Rev Food Sci Nutr.* 2018;58(4):631–638.

100 Singh P, Ravindran S, Suthar JK, Deshpande P, Rokade R, Rale V. Production of biosurfactant stabilized nanoparticles. *Int J Pharma Bio Sci*. 2017;8(2):701–707.

101 Kourmentza C, Freitas F, Alves V, Reis MA. Microbial conversion of waste and surplus materials into high-value added products: the case of biosurfactants. *Microb Appl*. 2017;1:29–77.

102 Mujumdar S, Bashetti S, Pardeshi S, Thombre RS. Industrial applications of biosurfactants. In: Thangadurai D, Sangeetha J, editors. *Industrial Biotechnology: Sustainable Production and Bioresource Utilization*. CRC Press, Boca Raton; 2016. pp. 61–90. ISBN 177188262X, 9781771882620.

103 Reddy AS, Chen CY, Bakerm SC, Chen CC, Jean JS, Fan CW, Chen HR, Wang JC. Synthesis of silver nanoparticles using surfactin: a biosurfactant stabilizing agent. *Mater Lett*. 2009;63:1227–1230.

104 Mesquita JF, Ferraz A, Aguiar A. Alkaline-sulfite pretreatment and use of surfactants during enzymatic hydrolysis to enhance ethanol production from sugarcane bagasse. *Bioprocess Biosyst Eng*. 2016;39:441–448.

105 Sajid M, Ahmad Khan MS, Singh Cameotra S, Safar Al-Thubiani A. *Biosurfactants: Potential Applications as Immunomodulator Drugs*. Immunology Letters, 2020.

106 Santos MS, Tavares FW, Biscaia EC Jr. Molecular thermodynamics of micellization: micelle size distributions and geometry transitions. *Braz J Chem Eng*. 2016;33:515–523.

107 Tang J, He J, Liu T, Xin X. Removal of heavy metals with sequential sludge washing techniques using saponin: optimization conditions, kinetics, removal effectiveness, binding intensity, mobility and mechanism. *RSC Adv*. 2017;7(53):33385–33401.

108 Guan R, Yuan X, Wu Z, Wang H, Jiang L, Li Y, Zeng G. Functionality of surfactants in waste-activated sludge treatment: a review. *Sci Total Environ*. 2017;609:1433–1442.

109 Sawant R, Devale A, Mujumdar S, Pardesi K, Shouche Y. Promising strategies for economical production of biosurfactants: the green molecules. In: Sayyed RZ, El-Enshasy HA, Hameeda B, editors. *Microbial Surfactants Volume I: Production and Applications*. CRC Press; 2021. pp. 266–286. ISBN 9780367521189.

2

Selection of Biosurfactant-Producing Microorganisms

Julio Bonilla Jaime[1,2],*, Luis Galarza Romero[1,2],*, and Jonathan Coronel León[1,3],*

[1] *Escuela Superior Politécnica del Litoral, ESPOL, Centro de Investigaciones Biotecnológicas del Ecuador (CIBE), Campus Gustavo Galindo, Km 30.5, Via Perimetral, P.O. Box 09-01-5863 Guayaquil, Ecuador*
[2] *Escuela Superior Politécnica del Litoral, ESPOL, Facultad de Ciencias de la Vida, Escuela Superior Politécnica del Litoral, ESPOL, Campus Gustavo Galindo, Km 30.5, Via Perimetral, P.O. Box 09-01-5863 Guayaquil, Ecuador*
[3] *Escuela Superior Politécnica del Litoral, ESPOL, Facultad de Ingeniería Mecánica y Ciencias de la Producción, Campus Gustavo Galindo, Km 30.5, Via Perimetral, P.O. Box 09-01-5863 Guayaquil, Ecuador*
* *Corresponding authors*

2.1 Introduction

New strategies to protect the environment include alternative biotechnological products. However, despite this new trend, one must be aware that chemical synthesis products are essential for daily life. Surfactants are compounds that have been produced since ancient times as far back as the Babylonian Empire. Over the years, surfactants have been used in disinfectants, detergents, and soaps [1]. The versatility of these compounds has also allowed them to be used in industrial applications either as additives in paints, fabric softeners, and even as biocides since they have antimicrobial properties [2]. A surfactant, also called a surface-active agent, is a substance that, when present at low concentration in a system, has the property of adsorbing onto surfaces or interfaces of the system and altering to a marked degree the surface or interfacial free energies of those surfaces or interfaces.

Consequently, a surfactant is a substance that at low concentration adsorbs at some boundary (interface) of a system and significantly changes the amount of work required to expand those interfaces. The use of chemical surfactants is widespread; during the last 30 years, its consumption has grown exponentially. In 1984, the industry started with 1.7 million tons, increasing to 15.93 million tons in 2014. This consumption forecasts an expected exponential increase of 24.19 tons by 2022 [3]. Due to their nature, surfactants are widely used in laundry and cosmetic products [4] as emulsifying agents and foam promoters. Their characteristics allow them to be used frequently in domestic activities, food, agricultural, pharmaceutical, and cosmetic industries. Although there is scientific evidence of the toxicity of surfactants, their importance in domestic and industrial activities is unquestionable. Therefore, the need for environmentally friendly surfactants is evident. In this sense, microbial surfactants or biosurfactants are gaining interest in the industry as an alternative

Biosurfactants and Sustainability: From Biorefineries Production to Versatile Applications, First Edition.
Edited by Paulo Ricardo Franco Marcelino, Silvio Silverio da Silva, and Antonio Ortiz Lopez.
© 2023 John Wiley & Sons Ltd. Published 2023 by John Wiley & Sons Ltd.

to traditional synthetic surfactants [5]. Biosurfactants (BS) or microbial surfactants (MS) constitute a diverse group of surface-active molecules. They occur in various chemical structures such a glycolipid, lipopeptides and lipoproteins, fatty acids, neutral lipids, phospholipids, and polymeric and particulate systems [6]. This chapter presents strategies for selecting surfactant-producing microorganisms, focusing on three aspects: i) traditional methods of detection, ii) a high throughput analysis method for the screening of potential biosurfactants producers, and iii) screening of microorganisms biosurfactants and lipases producers.

2.2 Traditional Methods of Detection

The search for new compounds must focus on molecules with high efficiency and capacity to withstand industrial conditions to be applied in the required areas. Establishing a bioprospecting strategy for microorganisms producing BS is crucial to ensure the quality of the final products. In this context, to search for microorganisms that produce these compounds, it is necessary to understand the metabolic processes that act in producing these compounds. In this sense, Felipe Da Silva et al. [7] precisely summarizes four theories that explain the evolution of the production of biosurfactants: 1) the ability to use hydrophobic substrates for the growth of microbial communities through the capacity of emulsification and solubilization, 2) the ability to separate from surfaces that are not suitable for the growth of microorganisms, in this case, the production of the surfactant suggests a regulation of the properties of the cell surface, increasing the permeability of the membrane, 3) the ability to produce an antimicrobial substance as protection from other organisms present in the evaluation ecosystem, and 4) the ability to use these compounds as energy reserves since they can enter the catabolic process as a possible source of energy during periods of nutrient limitation [7].

For this reason, understanding the physical-chemical properties of microbial surfactants is essential to understand the mechanisms of action of these compounds and establish a strategy for selecting microorganisms that produce these molecules. It should be borne in mind that one of the characteristics of biosurfactants is their tendency to adsorb at interfaces in an oriented fashion. The surface activity depends on the surfactant's structure, temperature, and solvent as the most relevant parameters [8]. Surface active tension (ST) affects the surface layer that causes that layer to behave as an elastic sheet. ST is the work required to expand (to break) the boundary between the liquid and the air. The units for surface tension are equivalent to a force per length ($mN\ m^{-1}$) or energy per area ($mJ\ m^{-2}$).

On the other hand, interfacial tension is required to expand the interface between two immiscible liquids. With this information, you can establish the critical micelle concentration (CMC): the concentration at which micelles begin to form. Beyond the CMC, no changes in interfacial properties are noticed. The desired strains produce biosurfactants with the lowest CMC values and the highest reduction of surface tension. Also, CMC values allow the evaluation of the performance of different biosurfactants in the interfacial phenomena, which is necessary to distinguish between the amount of surfactant required to produce a given amount of change in the sensation under investigation. This parameter is the efficiency of the surfactant [9]. Efficiency depends on the concentration of surfactant at the interface, which in turn affects the entropy of the molecule such that it can move from the bulk of the solution to the interface. An increase in the length of the C chain decreases the molecule's entropy, allowing it to enter the interface quickly.

Then again, the maximum change in the systems that the surfactant produce is related to the amount used, this parameter is the effectiveness of a surfactant. And it is used as a measure of the maximum effect the surfactant can produce in that interfacial process irrespective of concentration; effectiveness is measured by the minimum value to which it can lower the surface tension. The efficiency is determined by the ratio of surfactant concentration at the interface to that in bulk [10].

Based on the explanation given above, the search for new biosurfactants aims to find compounds with structures that have characteristics related to intense interfacial activity at low concentrations. Also, from an industrial point of view, the commerical viability of a biosurfactants depends on the compound's high emulsifying and solubilizing activity and overall stability at different temperature and pH conditions [11].

Currently, there is only a small number of commercially available biosurfactants: surfactin, sophorolipids, and rhamnolipids. Therefore, increased efforts to discover new BS must be applied by a broad range of different screening methods. In this aspect, various qualitative or quantitative methods can accurately assess the presence of biosurfactants in culture media.

2.2.1 Direct Measure of Surface/interfacial Activity

Historically, the measurement of surface tension is the primary measure to determine if a microorganism could synthesize compounds with surface activity. The direct action of the ST or IT of the culture supernatant is the the most straightforward screening method and is appropriate for preliminary screening; this gives a strong indication of the presence of surfactant. However, the ST does not measure the concentration of BS when it is above the CMC. This problem can be solved by serial diluting until a sharp ST is observed; the corresponding dilution of the supernatant is called the critical micelle dilution (CMD). The most widely used surface and interfacial tension measurement methods are du Nouy ring, Wilhelmy plate, spinning drop, and pendent drop. The du Noy ring method is the easiest and most frequently applied [12]. Generally, it is considered to be a culture promising a decrease to 40 mN m^{-1} or ≥ 20 mN m^{-1} in ST. The ST of distilled water is 72 mN m^{-1}, and the mineral culture media is about 55 mN m^{-1}. Indeed, a good BS producer is defined as reducing the surface tension of the growth medium by ≥ 20 mN m^{-1} compared with the initial medium [8, 11, 13, 14]. These strategies are force-based methods but other alternatives also exist, such as the pendant drop done with an optical tensiometer.

The pendant drop shape is an optical method for measuring IT. A drop is allowed to hang at the end of a capillary and it adopts an equilibrium profile that is a unique function of the tube radius, the IT, its density, and the gravitational field. The drop profile is extracted from the corrected drop image. Usually, the droplet contour is observed at the point of maximum slope of the light intensity. This contour is fitted to the Laplace–Young equation using modern computational methods; the input parameters of this algorithm are the reference framework, an angular correction for the vertical alignment, the radius of the droplet, and the interfacial tension [15]. Although surface tension is the most used method, the pendant drop method is gaining popularity due to its various benefits. The advantages of this techniques are: a) small sample volumes (100 ul) can be used, b) the material used (pipettes) in the measurement can be easily discarded and avoid contamination between samples, c) the probe or the container used during the measurement does not influence the results, since this technique focuses on the shape of the drop. In the force-based methods, the dimensions of the probe are critical for successful surface and interfacial tension measurements due to being fragile and need to be handled with care to ensure the correct measure [16].

According to the trend of rapid and reliable screening and selecting microbes capable of producing surfactants, the axisymmetric drop shape analysis is as strategy widely used in this bioprospecting process. The method was developed in colloid and surface science to simultaneously determine the contact angle and liquid surface tension from the profile of a droplet resting on a solid surface [17]. The underlying principle is that the shape of a liquid droplet depends significantly on the liquid surface tension. Droplets of liquids with a low ST are more apt to deviate from a perfectly spherical shape than droplets of liquids with a high ST. The efficacy of the method was demonstrated by selecting more than 30 strains of bacterium isolated from oil-contaminated land; the researcher reported that the assay was sensitive, rapid, and easy to perform [18].

2.2.2 Indirect Measure of Surface/interfacial Activity

The drop collapse assay provides information related to the modification of liquid droplets in the presence of a surfactant. Bioprospection of endophytic microorganisms that produce surface-active compounds was evaluated using this method [19]. The classical process consists of taking a sample of the supernatant of the microbial cultures and placing them on a solid surface covered by a thin layer of oil. According to published scientific information, if the liquid contains surfactants, the polar water molecules are repelled from the surface, and the drop remains stable [20]. This technique provides information related to the diameter of the drop and the amount of surfactant produced; therefore, it can be considered a semi-quantitative parameter; thus, the stability of drops depends on surfactant concentration and correlates with interfacial surface tension [21]. The advantage of this method is that the researcher does not need specific equipment or instrumentation.

On the other hand, the oil spreading test is based on decreasing the interfacial/superficial tension between the water and oil phases. This technique measures the diameter of clear zones caused when a drop of a surfactant-containing solution is placed on an oil–water surface; thus, the oil displacement area is directly proportional to the surface-active compound in the solution [22]. It is more accurate and reliable than the direct methods, and no specialized equipment or chemicals are used in this strategy. The oil spreading method is easier to perform and standardize in terms of duration, making it applicable to determine many samples [23]. These techniques are helpful to quickly establish a ranking about strains with higher surfactant producing capacity. The report of Rani et al. 2020 confirms that they used this technique for 34 isolates associated with *Bacillus, Streptomyces, Microbacterium, Micrococcus, Rhodococcus, Pseudomonas Arthrobacter,* and *Staphylococcus* genera, and found that *Bacillus methylotrophicus* OB9 had an elevated ability to produce biosurfactants [24]. An additional aspect that should be highlighted is the correlation between the screening techniques for biosurfactant producing microorganisms. Various reports indicated that the oil spreading assay showed corroboration with the drop collapse assay. The organisms found positive with drop collapse assay were positive for oil applying assay [25].

Another interesting technique for the selection of surfactant-producing microorganisms is to determine the emulsifying capacity. This technique involves contacting a supernatant solution of the cultures in studies with oils of different longer carbon chains. The ability of the surfactant to form macroscopic emulsions can be assessed by spectrophotometric analysis [26] or determined by the emulsion index. Emulsion index (E_{24}) is calculated as the ratio of the height of the emulsion layer and the total height of the liquid and correlates with surfactant concentration. This technique is beneficial for the first screening of biosurfactant-producing microbes. However, this technique should be considered independent of reducing surface tension; it is recommended to complement it with any previously described approach [7].

Another aspect to consider is that the emulsifier activity depends on the affinity of biosurfactant for oily substrates and oil polarity. Coronel et al. determined that a solution of biosurfactants in contact with hydrocarbon substrates bearing longer carbon chains, isopropyl palmitate (C_{16}) and isopropyl myristate (C_{17}), formed an O/W emulsion. In contrast, in oils with longer carbon chains (C_8–C_{12}), a W/O emulsion was formed. With this information, in addition, to selecting surfactant-producing strains, the application of these compounds in different industrial areas can be better focused on [11, 27].

Another indirect method often used in screening surfactant-producing microorganisms is that related to the cell surface hydrophobicity evaluation. This technique determines a direct correlation between the cell surface hydrophobicity and its adhesion in hydrocarbon samples favoring the emulsification process. This strategy is often used for environmental pollution with hydrophobic hydrocarbons where surfactants can play a role in bioremediation activities [28]. The effectiveness of biosurfactants in these processes is comparable to that of synthetic ones, making them a better choice for environmental applications concerning the indigenous ecosystem [29]. An important aspect is that the hydrophobicity of microorganisms depends on physiological elements. In this context, Al-Hawash et al. reported a decrease of 15% in cell surface hydrophobicity of strain *Aspergillus* sp. RFC-1 is dependent on the time of incubation (from 2 to 10 days) [28].

On the other hand, the amphipathic structure of BS and their structural similarity with the lipid bilayer of the erythrocyte membrane allows an interaction that triggers echinocyte-stomatocyte formation. Hemolysis implies several complex phenomena not fully clarified. However, according to the literature, the BS can affect the erythrocytes by three mechanisms: osmotic lysis, solubilization, and pore formation in the membrane [30]. In this context, the blood hemolysis test was developed to select bacteria producing biosurfactants but has recently been used for prospecting for filamentous fungi and yeasts producing biosurfactants [31, 32]. The results found with this technique are questioned in that they can give false negatives because low concentrations of BS can cause weak diffusion in the solid medium and the non-formation of halos [7]. Indeed, Eldin et al. [32] reported that this strategy had limitations. In this research, the isolated yeast; *Geotrichum candidum*, *Galactomyces pseudo candidum* and *Candida tropicalis* showed higher surfactant activity, emulsification index, and surface tension reduction but no exhibited hemolysis activity [32]. Therefore, due to the variability of the results, other screening tests must be associated with this method to corroborate biosurfactant production.

2.2.3 Effects of Culture Media Based on Agro-industrial By-products on Properties of BS

BS is a suitable alternative for application in several industrial processes that have been extensively studied due to remarkable advantages over a synthetic one. However, despite all the potential benefits, its industrial production is undeveloped, mainly because of the high cost of raw materials, low yields, and inefficiency in the recovery and purification methods. Thus, using agro-industrial waste or byproducts as feedstock is a potential approach for reducing production costs and would make BS economically viable and commercially competitive with synthetic surfactants [33–35]. In this context, the strategy to search for new BS also depends on inexpensive ways to produce them, using agro-industrial waste or byproducts to contribute to the circular economy.

Generally, the medium to produce BS contains carbon sources such as carbohydrates, oils and fats, and hydrocarbon groups. These compounds allow microorganisms to grow and produce the amphiphilic characteristic structure of surface-active compounds. For instance, microorganisms use water-soluble substrates like carbohydrates to form the hydrophilic moiety. Likewise, microorganisms consume nutrients like fats and oils to build up the hydrophobic portion of BS [36]. However, the use of carbon sources from agro-industrial byproducts/waste could alter the physicochemical properties of BS. Although these changes can be beneficial when specific properties are desired for a given application [37].

According to Kim et al. [38], the use of rapeseed oil as a carbon source influenced the surfactant properties of sophorolipids produced from *Starmerella bombicola*, obtaining a new esterified sophorolipid structure with lower CMC value and surface tension. These changes can occur due to inhibitory compounds in the rapeseed oil substrates. Similarly, modifications in the properties of BS are also influenced by changes in the nitrogen sources used. For example, Ishaq et al. [39] reported that substituting peptone for tryptone in the culture medium reduced the emulsification index of BS produced by *Aspergillus flavus*. On the other hand, the cultivation of microorganisms using alkanes as carbon source (C_{16} to C_{36}) for synthesizing glycolipids, phospholipids, and lipopeptides, increased cell surface hydrophobic activity, which improved the emulsification activity [40]. Also, in some instances, different raw materials used to produce BS contain an essential quantity of salt and according to Ikhwani et al. [41], the production of BS from *Pseudomonas aeruginosa* ATCC 15442 was not affected by salt concentrations up to 10% (weight/volume); however, a reduction in the CMC was reported.

The methods described as traditional are not unique; each strategy must be adapted to the conditions and equipment that researchers have available. Criteria for efficacy and efficiency must always be evaluated to select the strains with the most significant potential for producing surfactants. In addition, agro-industrial waste is a potential approach for reducing production costs; however, BS screening methods must be developed and established considering the presence of compounds that could change the properties of BS. The following section discusses an overview of high-throughput analysis methods for screening potential biosurfactants producers.

2.3 High-throughput Analysis Method for the Screening of Potential Biosurfactants Producers

Biosurfactants are a diverse group of biomolecules produced by many living organisms, having demonstrated their applications in the industrial, biomedical, and environmental fields (i.e. antivirals, antibiotics, antifungals, antitumor, emulsifiers, moisturizers, oil recovery, etc.) [42–48]. Consequently, finding high-yielding strains in nature or developing them through biotechnological approaches requires effective methods capable of high-throughput screening. Surfactants currently marketed are mostly of synthetic production; however, the green chemistry revolution is leading the industry and consumers' interest in environment-friendly biosurfactants, especially of microbial origin, because of the relatively low toxicity and unique biodegradability structural properties. The vast heterogeneity of biosurfactants makes it challenging to standardize screening methods to obtain potential biosurfactant-producing microorganisms and detection methods for the highly diverse biomolecules. Screening methods rely on the physical effects and molecular interaction of biosurfactants when exposed to water/oil-based surfaces or on the genetics of the producing microorganisms. These methods can be qualitative or quantitative with variable precisions and aim for screening of potent biosurfactant producers. Many existing methods have shown disadvantages, such as the difficulty of evaluating more than a couple of tens of microorganisms simultaneously, being time-consuming and requiring specific equipment and chemicals, having a specialized preparation for assaying, or a limited sensitivity for the biosurfactant activity. Detailed and extensive reviews on these methods have been previously discussed [49–51]. Molecular biology techniques are becoming more relevant with the rising volume of next-generation sequencing data obtained by metagenomic studies and have been extended for screening the genetic capacity of biosurfactant-producing microorganisms. Because bacteria are an essential source of bioactive compounds and only a tiny percentage of them can be cultivated and chemically studied in the lab, metagenomic screening methods are indispensable for discovering new and novel biosynthetic genes and biosurfactants from the uncultivated microbial diversity.

Due to the high diversity of possible biomolecules available in nature and biosurfactant-producing microorganisms, several research groups have focused on developing easy, reliable, fast, and economical high-throughput methods for screening potential active microorganisms or extensive culture collections for its biosurfactants. Various screening methods can be automated, making them ready for high-throughput screening. One of the first high-throughput screening accepted methods was developed by Chen and colleagues [18] by taking advantage of the optical distortion that a biosurfactant produces in the flat surface of pure water using a 96-microwell plate. Through this method, they were able to screen for biosurfactant-producing microbes by inspecting the wetting at the edges of the flat surface of water when the fluid surface becomes concave. The presence of biosurfactants affects the surface, thus making the fluid take the shape of a diverging lens and distorting the image of a black and white grid under the assaying microplate when viewed from above. The results were easy and fast to evaluate by visual inspection, giving qualitative detections in 96-well plates using small volumes of sample and being more reliable and less expensive than other methods of that time.

In 2010, a novel screening method was described to evaluate bacterial surfactant production with high-throughput potential [52]. The "atomized oil assay" was able to test multiple strains simultaneously within a few seconds, thus enabling thousands of strains to be screened for surfactant production in a reasonable time without requiring specialized equipment for its discovery. By misting oil droplets onto agar plates, the atomized oil assay can detect the presence of low levels of surfactants by altering the interaction of the oil with the agar surface. The application of the atomized oil assay to a wide variety of environmental bacterial strains and synthetic surfactants revealed it to be both more versatile and sensitive (it detects much lower concentrations of many surfactants) than the more commonly used drop collapse assay. It is a semi-quantitative technique and therefore has broad applicability for uses such as high-throughput mutagenesis screens of biosurfactant-producing bacterial strains [52,53].

Surfactin is a lipopeptide-type biosurfactant and is a powerful microbial surfactant not only for its superior surface activity but also because it combines interesting physicochemical properties and biological activities [54,55]. However, the high cost caused by its low productivity hampers the commercial application of surfactin [54, 56]. Hence, strategies for high-throughput screening of high-yielding and cost-effective strains are in urgent demand. Zhu and colleagues developed a colorimetric method capable of screening tens of thousands of surfactin producer organisms based on the blue-to-red color change of polydiacetylene (PDA) vesicles [57]. The presence or absence of biosurfactants, detected by the change in color of PDA vesicles, can be visualized within 10 minutes and by testing as little as 0.05 gr L^{-1} of surfactin. This method proved advantageous over the atomized oil assay due to its very high sensitivity in detecting biosurfactants and not depending on indirectly visible results [57]. Because of the versatility of the PDA polymer to be modified to detect cationic surfactants [58, 59] and ionic surfactants [60], it is possible to be adapted to detect not only surfactins but many other biosurfactants. However, this hasn't yet been tested. Even though the advantages of the PDA assay still require the time-consuming preparation of PDA vesicles precursors and fresh preparation of vesicles for each testing round.

In 2015, a similar chromatic screening method for surfactins was proposed by Yang et al. [61]. Instead of using PDA, they used bromothymol blue (BTB) as a color indicator for detecting the presence of surfactins. Since surfactins are negatively charged lipopeptides and do not directly interact with sulfonephthalein dyes [62], they described a novel use of BTB as a color indicator and cetylpyridinium chloride (CPC) as a mediator to form the (CPC–BTB) complex and competitively determine concentration by chromatic change based on the intense binding of the surfactin with the cationic CPC (Figure 2.1). In the presence of surfactants, the solution shifts from faint yellow–green to dark green or bright blue (Figure 2.1). Both colorimetric methods described before, PDA and CPC-BTB based methods, were tested under pH conditions similar to the fermentation broth and medium used for *in vitro* microorganism cultivation, demonstrating their capacity to test for biosurfactants without complex preparations.

Another method inspired by the CPC-BTB chromatic change was developed in 2019. Heuson and colleagues replaced BTB and incorporated the anionic fluorescent dye fluorescein (FL) as an indicator dye, resulting in a more stable CPC-FL complex, with increased

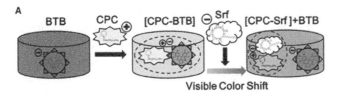

Figure 2.1 Modified figure from [61]. The mechanism for the chromatic response is described by the following equations [1]: BTB + CPC → [CPC–BTB] [2], [CPC–BTB] + Surfactin → [CPC-Srf] + BTB.

sensitivity for detecting surfactants [63]. The functional principle of the method remains the same as reported by Yang et al. [61], where the surfactin binding releases FL from the formed complex CPC-FL and fluorescence is detected. They also described a fully automated workflow for screening and quantifying surfactins from environmental isolates in a single run, enabling a high-throughput screening method.

At the same time, Ong et al. used only BTB [64], in contrast to other BTB complex-forming methods, in a high-throughput screening assay to detect the significant lipopeptide classes: surfactin, iturin, and fengycin. This new alternative method differentiated surfactin, iturin, and fengycin by different colors at concentrations of 1 g l^{-1} and below. Change from blue to yellow was the most intense color change, observed in the presence of iturin; from blue to light green indicated the action of fengycin; lastly, a change to a green color denoted the presence of surfactin. Ong et al. [64] went further and positively correlated the results of the chromatic test for predicting the existence of biosurfactants with the presence of biosurfactant-coding genes by PCR. This more straightforward method for screening lipopeptide-producing strains has the advantages of simple operation, no usages of toxic chemicals (CPC) and expensive equipment, instant and quantitative measurement, and applicability to all three classes of lipopeptides.

More recently, a colorimetric screening method based on the Victoria pure blue BO (VPBO) has been reported to quantify biosurfactants [65]. During their experiments, rhamnolipids produced by recombinant *Pseudomonas putida* KT2440 were used as a model. This method is based on the immobilization of VPBO on polystyrene microtiter plates; when surfactant molecules are present in sufficient concentration, they will form micelles, which results in the solubilization of the dye by weakening or even disturbing these interactions. One of the advantages shown by this method is the readily available and inexpensive VPBO reagent. Although the process requires at least 150 minutes for the preparation and proper coating of the microtiter plates with VPBO, once ready, the plates can be stored at 4 °C sealed with an aluminum foil plate seal. The VPBO assay can detect the surfactant properties of different compounds directly via their ability to solubilize the dye from a plastic surface but without the ability to discern the type of biosurfactant produced. Despite the various strategies used by all five color-based screening methods, they can test and detect the presence of biosurfactants directly from cell-free broth, thus eliminating the purification step and saving time.

Previously described methods for high throughput screening of biosurfactant producers depend on culturable microorganisms; therefore, only those susceptible to artificial cultivation can be isolated and studied. To give a broad overview of the yet-to-be-discovered capability of unculturable organisms, a mathematical model was applied to estimate the number of antibiotics produced only by the culturable members of the actinomycete genus Streptomyces is of the order of hundreds of thousands of compounds. We can think of many compounds if we extrapolate this number of natural products derived from this group and include the entire bacterial kingdom [66]. Due to the incredible diversity and lack of culturability of most microorganisms found in most natural microbiomes, metagenomics has become a powerful strategy to rescue the "functional" information and their ability to produce bioactive biosurfactants if these unculturable microorganisms. Metagenomics is a culture-independent genomic analysis technique for studying microbial communities. There are two general approaches to analyze and exploit metagenomic DNA for high-throughput screening of potential biosurfactant producers: sequence-based and functional activity screening of the microbiome.

Sequence-based metagenomics for the genetic identification of genes involved in the production of biosurfactants is an alternative method for screening microorganisms capable of producing these biomolecules in a faster and less laborious way than other screening methods. The advent of next-generation sequencing (NGS) platforms enabled massive metagenomic data for protein-coding sequences in curated sequence databases. Sequence-based metagenomics can be achieved by directly sequencing entire microbial communities in a PCR-free process (shotgun metagenomics) or a PCR-guided strategy. PCR-based techniques can be implemented to detect the presence of potential biosurfactant-producing microorganisms [66–69]. However, the existence of the genes in the microorganism genome is no guarantee they will be expressed, nor a prediction of the biosurfactant capacity or activity. The structural and compositional diversity of biosurfactants makes this technique incomplete, rendering it complementary to other screening methods. Therefore, this PCR-guided approach has mainly focused on the discovery of only a tiny group of biosurfactants like lipopeptides and rhamnolipids, mainly due to the high homology of NRPS operons in different species, the number of already designed primers in the literature, and the possibility of utilizing other databases and platforms for their identification [70]. According to Biniarz, Łukaszewicz, and Janek, this may be the most cost-effective and efficient way to analyze lipopeptides produced by microorganisms in different environments [70].

Nevertheless, the high diversity of structures, genes, and peptide sequences that can have a role as biosurfactants makes the design of a universal sequence-based screen challenging to achieve. This was shown by Tripathi et al. in 2019 when they discovered the biosynthesis of rhamnolipids by marine bacteria Marinobacter sp. MCTG107b and could not amplify any DNA sequence related to rhamnolipid synthesis gene orthologues [71]. They used degenerated primers designed from *P. aeruginosa* and *B. thailandensis* sequences, suggesting the rhamnolipid biosynthesis in *Marinobacter sp.* Enzymes catalyze MCTG107b with significantly different peptide sequences and even genetic sequences. The PCR sequence-based techniques allow the identification of genes based on homology with genes described in databases. However, the major problem for gene identification from metagenomic sequences is when the sequences or coding regions have no homology in biological databases. This makes it clear that a better understanding of the diversity and biosynthesis systems of existing biosurfactant molecules is required [72].

On the other hand, the PCR-free metagenomic strategy can be conducted relatively unbiasedly and does not require previous knowledge about the possible organisms present in the sample [66]. Because all DNA in the environmental sample is sequenced, the shotgun metagenomics strategy offers access to the functional gene composition of microbial communities to capture operons or gene encoding pathways that may direct the synthesis of novel biosurfactants [68, 73]. A significant advantage of PCR-free sequence-based metagenomics is that no previous knowledge is needed on biosurfactant synthetic pathways, and still, complete pathways can be reconstructed from the generated data [74–76]. Shotgun metagenomics can provide valuable insights into the potential function of the microbiome, representing an essential source for the discovery of novel biosynthetic pathways for biosurfactants and other similar biomolecules [70, 75]. We encourage readers interested in shotgun metagenomics to review best practices strategies that have been described extensively and are out of the scope of this review [77–79]. With the current developments in sequencing technologies (long-read and short-read sequencing) and the price dropping of technologies, researchers can now harness the power of hybrid sequencing strategies. Hybrid sequencing uses the high-throughput and high accuracy of short-read data to correct for long-read errors, reducing the amount of more costly long-read sequence data and resulting in more complete assemblies. This approach has been applied for the completion of complex single genomes, but more recently, it has been used for metagenomics to identify potential biosurfactants and biosurfactant-like active molecules. It facilitated the retrieval of metagenome-assembled genomes from activated sludges samples and the identification of several biosynthetic gene clusters [80]. The analysis of 23 Danish wastewater treatment plants for the recovery and assembly of over 1000 high-quality metagenome-assembled genomes to estimate the functional potential of the microbes involved in critical wastewater processes such as nitrification, denitrification, enhanced biological phosphate removal, floc formation, and solid–liquid separation [81]; both of these studies highlighted the potential to discover new natural products. Hybrid sequencing also enables complete genome sequencing from isolated microorganisms identified as biosurfactant producers, making it possible to identify genetic elements and biosynthetic gene clusters responsible for synthesizing bioactive compounds and providing valuable information on the genetic makeup of the biosynthetic pathways [82, 83].

Advances in computational genomics and database mining methods are enhancing retrieval of biosynthetic clusters and secondary metabolite prediction. A variety of software programs are available. PRISM (PRediction Informatics for Secondary Metabolomes), AntiSMASH (Antibiotics and Secondary Metabolites Analysis Shell), NaPDoS (Natural Product Domain Seeker), NP.searcher, and BioSurfDB [75, 84–88] are just a few of the tools available, but this is by no means an exhaustive list [89, 90]. New tools are continually being developed [91, 92]. Further improvements in these methods could lead to the potential genetic identification of biosurfactants producer strains.

Functional activity screening of the microbiome is another strategy in the search for biosurfactants. It depends on the biosurfactant-producing clone from a shotgun metagenomic clone library; however, it must be followed by a high-throughput screening method like the atomized oil assay, microwell optical distortion, and other practices previously described in this chapter. Despite this limitation, function-based searches hold the promise of uncovering novel enzymes or functionalities that are genetically and structurally distinct from existing enzymes/molecules. Screening metagenomic libraries for novel bioactive compounds is possible, but it needs the use of a suitable host organism, a functional expression system, and a robust screening methodology [93]. Thies and colleagues reported the expression of metagenomics genes encoding hydrolytic and biosurfactant biosynthetic enzymes, identifying the biosurfactant N-acyltyrosine [93]. The metagenomic library's phenotypic screening (atomized oil assay) constructed from environmental DNA from hydrocarbon-contaminated lake soil detected the biosurfactants ornithine lipid and lyso-ornithine lipid [94]. However, they screened the metagenomic library in multiple hosts (*Escherichia coli* and *Pseudomonas putida*), given the diversified collection of biochemical and regulatory conditions to access a more significant number of genes. This strategy allowed them to detect ornithine lipids in *P. putida*, confirming the lack of capability in *E. coli*. Another successful example of functional metagenomics was reported by Araújo et al. when they performed an available screening to detect the presence of genes with activity in oil degradation and biosurfactant production [95]. The soil samples were collected from the Jundiaí River (Natal, Brazil) because it showed intermittent drainage and salinity reaching four times seawater concentrations. The study screened 1240 clones, from which a single cloned showed biosurfactant and oil-degrading activity. By filtering the functional activity of this microbiome, they identified and characterized a hypothetical protein that showed surfactant properties. Despite several technical challenges such as heterologous expression, a lack of reliable high-throughput screening assays, and the extensive analytical work required to characterize the chemical structure of new biomolecules, these results highlight the opportunity of functional metagenomics to screen metagenomic libraries to access novel biomolecules without prior knowledge of their sequences; hence, envisioning the future discovery of entirely new biosurfactant types and producing organisms.

When combined high-throughput metagenomic screening methods (sequencing and function-based) with bioinformatics analysis for extensive sequence screening and subsequent validation by direct and indirect measurement of biosurfactant activity, there is enormous potential for discovering new bioactive molecules and microorganisms. The continuous development of novel high-throughput screening methods for bioactive molecules has a promising future.

2.4 Screening of Microorganisms Biosurfactants and Lipases Producers

The production of biosurfactants and lipases related to different beneficial microorganisms provides essential components in various industries. Microorganisms such as bacteria, fungi, and yeasts have been described as producing this type of compound with some biological activity. Among the different industrial applications of biosurfactants, the medicine and bioremediation industries stand out, as they are not toxic and precise in other conditions of use. As well, additional properties are detailed, such as modulator, flocculant, dispersion, etc. [96].

Moreover, the type of compound as a surfactant has good emulsifying attributes because it contains hydrophobic and hydrophilic compounds [97–99]. On the other hand, several microorganisms secrete enzymes such as lipases, which contribute to biocatalysis in biotransformation within different industrial processes [100]. One of the advantages of lipases and biosurfactants is the low toxicity, low production cost, basing its production on the use of waste from the agricultural industry [101]. Microorganism compound production generates new alternatives for its scaling at an industrial level and its subsequent use in different areas from agriculture, medicine, and cleaning [102].

Fungi produce this type of compound depending on the formulation or components of the substrate (malt extract, vegetable oils, molasses, etc.) and the production process parameters [103–105]. Biosurfactant-producing microorganisms belong to the phyla; firmicutes, actinobacteria, proteobacteria, ascomycota (yeasts), ascomycota (filamentous fungi), and basidiomycota [106]. Thus, filamentous fungi have been described as producers of both compounds on an industrial scale, of which it is mentioned; *Aspergillus fumigatus*, *Aspergillus niger*, *Candida bombicola*, *Candida rugosa*, *Pseudozyma* sp. *Penicillium* sp. and *Yarrowia lipolytica* [89, 112, 195–198].

The compound produces microorganisms of interest in various industries ranging from agriculture, bioremediation to potential use in medicine. In this sense, the genera that mainly belong to bacteria, which due to the conditions of the microorganisms belonging to this genus, adapting to different fermentation conditions, both solid and liquid [106], making them part of other investigations in the search for potential compounds and its different uses in various industries (Table 2.1, Table 2.2).

One of the primary uses of biosurfactants and lipases is related to bioremediation by oils or heavy metals [101], generating an alternative for the different contaminations generated in this type of industry. Another line of use for this type of compound is the food industry, where coating developed by biosurfactants is a viable alternative for storing and preserving packaged foods [107]. Likewise, several biosurfactant compounds have been described as potential antimicrobial compounds with immunomodulatory activities and antigens in vaccines [108].

In addition, fungal species related to the production of biosurfactants in different conditions of solid or liquid fermentation can produce these compounds. However, there have been only a few species described as capable of producing biosurfactants. Although the strains used are few, those producing biosurfactants are used in different industrial processes [109, 110].

On the other hand, the production of lipases by bacterial and fungal microorganisms which among the properties of this ecological or "green" catalyst, lipase has high catalytic efficiency, it can work under comparatively mild reaction conditions, it does not involve coenzymes or other cofactors or by-products; as such, it is widely used in various industrial fields, including the production of food, pharmaceuticals, paper, detergents, and biodiesel [111]. Again, bacterial genera are the most used in obtaining lipases, but this does not mean that there are no investigations into the use of fungi to get lipases within the fungal genus. The most common bacterial strains of lipases include *Bacillus* spp., *Pseudomonas* spp., *Staphylococcus* sp., and *Burkholderia* sp [112]. Bacterial lipase-catalyzed most hydrolytic reactions and the highest activity levels; bacterial enzymes are also among the most stable. Of this group, the lipase secreted by *Pseudomonas* spp. [113], is worth mentioning, which generates a better performance in general. At the same time, fungal lipases have been widely used in various biotechnology applications due to their stability, specificity, and ease of production. The lipases isolated from *Thermomyces lanuginosus*, *Rhizopus oryzae*, and *Aspergillus niger* have a high industrial value [111].

2 Selection of Biosurfactant-Producing Microorganisms

Table 2.1 Microbial strains used in studies on lipases.

Fungal species	Applications	References	Bacterial species	Applications	References
Fusarium solani	Biotechnological applications	[114]	Achromobacter sp	Multipurpose biological catalyst	[111]
Yarrowia lipolytica	Bioconservation and food industry	[115]	Virgibacillus pantothenticus	Application in the detergent industry	[116]
Aspergillus oryzae	Preparation of cheese and synthesized saturated fatty acids	[117]	Pseudomonas mendocina	Lipase in enzyme cocktails for deinking applications.	[118]
Rhizomucor javanicus	Non-hydrogenated solid fats	[119]	Acinetobacter radioresistens	Lipase production	[120]
Rhizomucor miehei	Cocoa-butter complement	[121]	Bacillus sp	Physicochemical parameters for optimized lipase production	[122]
Geotrichum candidum	Pharmaceutical compounds related to the elimination of bad cholesterol	[123]	Staphylococcus pasteuri	Biotechnological applications	[124]
C. Antarctica	Enriched oils, sizing lubricant removal, denim finishing	[125]	P. fluorescens	Production of biodiesel	[125]
Candida rugosa	Breast milk fat substitute	[127]	Staphylococcus warneri	Lipase industrial applications	[113]
Candida lipolytica	Cheese maturation, fatty acid production	[128]	S. xylosus	Biodiesel production	[129]
Penicillium camemberti					
Thermomyces lanuginosus	Produced a lipase containing detergent	[130]	Micrococcus sp	Lipase production	[131]
Penicillium roqueforti	Production of the characteristic flavor of blue cheese in dairy products.	[132]	Bacillus cereus	Lipase industrial applications	[113]
Aspergillus niger	Faster cheese ripening	[133]	Marinobacter lipolyticus	Lipase production and nanoparticles fabrication for biomedical application	[134]

Table 2.1 (Continued)

Fungal species	Applications	References	Bacterial species	Applications	References
Meyerozyma guilliermondii	Cheese whey is a by-product	[135]	Haloarcula sp	Biodiesel production	[136]
Aspergillus flavus	Biotechnological application	[137]	Bacillus subtilis	Biodetergent and for bioremediation of wastewater	[122]
Candida Antarctica	Biodiesel production	[121]	Geobacillus stearothermophilus	Detergent formulations	[138]
Rhizomucor miehei	Biodiesel production	[121]	Pseudomonas aeruginosa	Lipase industrial applications	[113]
Candida tropicalis	Biofilm production	[139]	Pseudomonas sp	Lipase industrial applications	[113]
Penicillium abeanum	Enzymes production	[140]			
Rhizopus nodosus	Enzymes in biotechnology and organic chemistry	[110]	P. alcaligenes	Lipase industrial applications	[113]
P. chrysogenum	Biocatalysts for application in organic synthesis	[141]	Pseudomonas plantarii	Lipase industrial applications	[113]
Bacterial species			Chromobacterium viscosum	Multipurpose biological catalyst	[111]
Pseudomonas sp.	Lipase industrial applications	[113]	Acinetobacter sp.	Lipase production	[142]
Pseudomonas cepacia	Lipase industrial applications	[113]	Bacillus thermocatenulatus	Enzymes production	[143]
Lactobacillus plantarum	Lipases for polymer degradation	[144]	Lactobacillus casei	Lipases for polymer degradation	[144]
Lactobacillus rhamnosus	Lipases for polymer degradation	[144]	Lactobacillus paracasei	Lipases for polymer degradation	[144]

Table 2.2 Microbial strains used in studies on biosurfactants.

Biosurfactants Microorganism					
Eubacteria			Lactobacillus spp Serratia	Biotechnology industry	[145]
Pedobacter sp.	Bioremediation	[146]	Serratia marcescens	Antifungal activity and antibacterial activity	[147] Admed

(Continued)

Table 2.2 (Continued)

Biosurfactants Microorganism					
Azotobacter sp.	Bioremediation	[148]	Variovorax paradoxus	Environmental applications	[149]
Paenibacillus macerans	Bioremediation	[150]	Pseudomonas aeruginosa	Bioremediation	[151]
Lactis. Helveticus	Food industry	[152]	Terricolous lichen	Bioremediation	[153]
Breadyrhizobium elkanii	Biotechnology applications	[154]	Peltigera membranacea	Bioremediation	[153]
Lactobacillus acidophilus	biosurfactant as an antimicrobial agent	[155]	Lactobacillus casei	Environmental and industrial application	[145]
Lactobacillus pentosus	Antimicrobial activity against several pathogens	[157]	Bacillus sphaericus	Environmental and industrial application	[145]
Methylobacterium sp.	Plant growth-promoting	[157]	Bacillus azotoformans	Environmental and industrial application	[145]
Pseudomonas aeruginosa	Production of rhamnolipids	[158]	Sphingobacterium spiritivorum	Bioremediation of soils	[159]
Pseudomonas & Rhodococcus	Oil-degrading microorganisms	[160]	Dunaliella salina	Bioremediation	[161]
Callyspongia diffusa	Bioremediation in extreme environments	[162]	Microalgae	Biotechnological applications	[164]
Brevibacterium and Vibrio	Polymer production	[164]	Fungi		
Rhodococcus ruber	Bacterial adhesion and biofilm formation	[165]	Aspergillus flavus	Antifungal activity	[39]
Virgibacillus salarius	Bioremediation	[166]	Trichosporon montevideense	Commercial applications, extreme environmental conditions	[167]
Rhodococcus ruber	Bacterial adhesion and biofilm formation	[165]	Geotrichum sp	Commercial applications, extreme environmental conditions	[167]
Pseudoalteromonas sp	Biotechnological industries	[168]	Pantoea ananatis	Biomedicine	[169]
Halomonas sp.	biotechnological industries	[168]	Trichosporon mycotoxinivorans	Biodiesel production	[170]
Antarctobacter sp.	Biotechnological industries	[168]	Fusarium sp	Environmental applications	[1]

Table 2.2 (Continued)

Biosurfactants Microorganism						
A. calcoaceticus	Bioremediation and enhanced oil recovery	[172]	Penicillium sp.	Biotechnology applications	[173]
Enterobacter sp	Bioremediation and production of antimicrobial products	[174]	Cunninghamella echinulata	Biotechnological processes	[175]
B. subtilis	Remediation of oil-polluted sites	[176]	Myroides odoratus	Medical applications	[177]
B. licheniformis	Bioremediation	[178]	M. odoramitimus	Medical applications	[177]
Solibacillus silvestris	Biotechnological applications	[179]	C. albicans	Medical applications	[180]
Lactobacillus plantarum	Food industry	[181]	Yarrowia lipolytica	Hydrocarbon remediation	[182]
Brevibacterium luteolum	Environmental purposes	[183]	Candida tropicalis BPU1	Dispersant for application in the oil industry	[4]
Bacillus licheniformis	Bioremediation	[185]	Yeasts		
Paenibacillus dendritiformis	Bioremediation	[186]	S. cerevisiae	Food industry	[107]
Bacillus sp	Environmental protection	[187]	Candida bombicola	Bioremediation	[188]
Pseudomonas sp.	Biotechnological applications	[189]	Vibrio alginolyticus	Hydrocarbon remediation	[182]
Xanthomonas campestris	Biotechnological applications	[189]	Trichosporon montevideense	Biotecnological industry	[190]
Pseudomonas syringae	Biotechnological applications	[189]	C. krusei	Biotecnological industry	[190]
Bacillus velezensis	Bioremediation	[191]	Aureobasidium pullulans	Various industrial applications	[192]
Bacillus licheniformis	Cosmetic, industrial processes	[193]	Actinobacteria		
Staphylococcus hominis	Aquaculture production	[108]	Streptomyces sp	Biotecnology industry	[194]

2.5 Conclusion and Future Perspectives

Efforts to find new biosurfactants have increased over the past decade since they can be utilized in different industrial sectors due to their unique properties. For this purpose, the environment is the primary source to isolate fungi, bacteria, and yeasts to produce surface-active compounds. Consequently, the need to discover and isolate new molecules has promoted the development of several methods for selecting microbial surfactant-producing strains discussed in this chapter. These methods range from the simplest to the most complex and sophisticated; the vast majority are based on the determination of surface tension or emulsification activity. However, based on the diverse structure and properties of BS, the application of various screening approaches is pivotal to obtaining accurate and reliable results. Likewise, the development of molecular biology techniques has driven those classical methods to be automated for high-throughput screening of biosurfactants and biosurfactant-producing microorganisms. Another aspect in the screening and production of BS is using agricultural wastes, or by-products focused on reducing the environmental impact and improving the production costs of these biocompounds, contributing to the circular economy. However, the composition of these complex substrates can modify the properties of BS, including affecting the growth of BS-producing strains.

A potential strategy for future biosurfactant screening could be the promotion of novel techniques that combine several detection strategies to provide a reliable diagnosis. In this context, the possibility of targeting the components of bioprospecting culture media can help the production of new BS with tailored properties. For this reason, the use of a carrier in the growth medium proved to be a novel approach for enhanced biosurfactant production [114]. In adition, strategies to promote the co-production of other compounds of interest can be beneficial from the economic point of view, because these alternative compounds can be exploited at an industrial level, for instance, the production of lipases discussed in this chapter. Similarly, the fact of using agro-industrial substrates as a low-cost source of production allows the possibility of applying current molecular biology techniques focused on the genetic modification of BS producing strains to broaden the use of complex substrates found in nature.

Finally, these beneficial microorganism are related to ecofriendly compounds can be commercialy used, increased the circular bioeconomy concept focus in sustainable industrial development.

References

1 Rebello S, Asok AK, Mundayoor S, Jisha MS. Surfactants: toxicity, remediation and green surfactants. *Environ Chem Lett.* 2014;12:275–287.
2 Ivankovic T, Hrenovic J, Gudelj I. Toxicity of commercial surfactants to phosphate-accumulating bacterium. *Acta Chim Slov.* 2009;56:1003–1009.
3 Palmer M, Hatley H. The role of surfactants in wastewater treatment: impact, removal and future techniques: a critical review. *Water Res.* 2018;147:60–72.
4 Ivanković T, Hrenović J. Surfactants in the environment. *Arh Hig Rada Toksikol.* 2010;61:95–110.
5 Marchant R, Banat IM. Microbial biosurfactants: challenges and opportunities for future exploitation. *Trends Biotechnol.* 2012;30(11):558–565.
6 Fenibo EO, Douglas SI, Stanley HO. A review on microbial surfactants: production, classifications, properties and characterization. *J Adv Microbiol.* 2019;18(3):1–22.

7. da Silva AF, Banat IM, Giachini AJ, Robl D. Fungal biosurfactants, from nature to biotechnological product: bioprospection, production and potential applications. *Bioprocess Biosyst Eng*. Springer Science and Business Media Deutschland GmbH; 2021;44:2003–2034.
8. Jahan R, Bodratti AM, Tsianou M, Alexandridis P. Biosurfactants, natural alternatives to synthetic surfactants: physicochemical properties and applications. *Adv Colloid Interface Sci*. Elsevier B.V.; 2020:275.
9. Pessôa MG, Vespermann KAC, Paulino BN, Barcelos MCS, Pastore GM, Molina G. Newly isolated microorganisms with potential application in biotechnology. *Biotechnol Adv*. Elsevier Inc.; 2019;37:319–339.
10. Rosen MJ Surfactants and interfacial phenomena surfactants and interfacial phenomena. 2004.
11. Coronel-León J, de Grau G, Grau-Campistany A, Farfan M, Rabanal F, Manresa A, et al. Biosurfactant production by AL 1.1, a Bacillus licheniformis strain isolated from Antarctica: production, chemical characterization and properties. *Ann Microbiol*. 2015;65(4)2065–2078.
12. Du Noüy L. An Interfacial Tensiometer For Universal Use [Internet]. http://rupress.org/jgp/article-pdf/7/5/625/1247172/625.pdf.
13. Burgos-Díaz C, Pons R, Teruel JA, Aranda FJ, Ortiz A, Manresa A, et al. The production and physicochemical properties of a biosurfactant mixture obtained from Sphingobacterium detergens. *J Colloid Interface Sci*. 2013;394:368–379.
14. Burgos-Díaz C, Pons R, Espuny MJ, Aranda FJ, Teruel JA, Manresa A, et al. Isolation and partial characterization of a biosurfactant mixture produced by Sphingobacterium sp. isolated from soil. *J Colloid Interface Sci*. 2011;361(1):195–204.
15. Pinazo A, Angelet M, Pons R, Lozano M, Infante MR, Pérez L. Lysine–bisglycidol conjugates as novel lysine cationic surfactants. *Langmuir*. 2009;25(14):7803–7814.
16. Chittepu OR. Isolation and characterization of biosurfactant producing bacteria from groundnut oil cake dumping site for the control of foodborne pathogens. *Grain Oil Sci Technol*. 2019 Mar;2(1):15–20.
17. van der Vegt W, van der Mei HC, Noordmans J, Busscher HJ. Assessment of bacterial biosurfactant production through axisymmetric drop shape analysis by profile. *Appl Microbiol Biotechnol*. 1991;35(6):766–770.
18. Chen C-Y, Baker SC, Darton RC. The application of a high throughput analysis method for the screening of potential biosurfactants from natural sources. *J Microbiol Methods*. 2007;70(3):503–510.
19. Serrano L, Moreno AS, Sosa D, Castillo D, Bonilla J, Romero CA, et al. Biosurfactants synthesized by endophytic Bacillus strains as control of Moniliophthora perniciosa and Moniliophthora roreri Lizette. 2021; (78): e20200165.
20. Thavasi R, Sharma S, Jayalakshmi S. Evaluation of screening methods for the isolation of biosurfactant producing marine bacteria. *J Pet Environ Biotechnol*. 2013;04(02).
21. Jain DK, Collins-Thompson DL, Lee H, Trevors JT. A drop-collapsing test for screening surfactant-producing microorganisms. *J Microbiol Methods*. 1991;13(4):271–279.
22. Fracchia L, Cavallo M, Martinotti M, Banat I. Biosurfactants and bioemulsifiers biomedical and related applications; present status and future potentials. In: Dhanjoo N. Ghista, editor. *Biomedical Science and Engineering Technology*. IntechOpen;2012;325–370

23 Zhao F, Liang X, Ban Y, Han S, Zhang J, Zhang Y, et al. Comparison of methods to quantify rhamnolipid and optimization of oil spreading method. *Tenside Surfactants Deterg*. 2016;53(3):243–248.

24 Rani M, Weadge JT, Jabaji S. Isolation and characterization of biosurfactant-producing bacteria from oil well batteries with antimicrobial activities against food-borne and plant pathogens. *Front Microbiol*. 2020;11:64.

25 Nayarisseri A, Singh P, Singh SK. Screening, isolation and characterization of biosurfactant producing Bacillus subtilis strain ANSKLAB03. *Bioinformation*. 2018;14(06):304–314.

26 Mohanram R, Jagtap C, Kumar P. Isolation, screening, and characterization of surface-active agent-producing, oil-degrading marine bacteria of Mumbai Harbor. *Mar Pollut Bull*. 2016;105(1):131–138.

27 Rodrigues LR. Microbial surfactants: fundamentals and applicability in the formulation of nano-sized drug delivery vectors. *J Colloid Interface Sci*. 2015;449:304–316.

28 Al-Hawash AB, Zhang J, Li S, Liu J, Ghalib HB, Zhang X, et al. Biodegradation of n-hexadecane by Aspergillus sp. RFC-1 and its mechanism. *Ecotoxicol Environ Saf*. 2018;164:398–408.

29 Kaczorek E, Pacholak A, Zdarta A, Smułek W. The impact of biosurfactants on microbial cell properties leading to hydrocarbon bioavailability increase. *Colloids Interfaces*. 2018;2(3):35.

30 Manaargadoo-Catin M, Ali-Cherif A, Pougnas JL, Perrin C. Hemolysis by surfactants – a review. *Adv Colloid Interface Sci*. Elsevier; 2016;228:1–16.

31 Marcelino PRF, Peres GFD, Terán-Hilares R, Pagnocca FC, Rosa CA, Lacerda TM, et al. Biosurfactants production by yeasts using sugarcane bagasse hemicellulosic hydrolysate as new sustainable alternative for lignocellulosic biorefineries. *Ind Crops Prod*. 2019;129:212–223.

32 Eldin AM, Kamel Z, Hossam N. Isolation and genetic identification of yeast producing biosurfactants, evaluated by different screening methods. *Microchem J*. 2019;146:309–314.

33 Mohanty SS, Koul Y, Varjani S, Pandey A, Ngo HH, Chang JS, et al. A critical review on various feedstocks as sustainable substrates for biosurfactants production: a way towards cleaner production. *Microb Cell Fact*. 2021;20:120.

34 Gaur VK, Sharma P, Sirohi R, Varjani S, Taherzadeh MJ, Chang JS, et al. Production of biosurfactants from agro-industrial waste and waste cooking oil in a circular bioeconomy: an overview. *Bioresour Technol*. 2022;343.

35 Domínguez Rivera Á, Martínez Urbina MÁ, López Y López VE. Advances on research in the use of agro-industrial waste in biosurfactant production. *World J Microbiol Biotechnol*. Springer Netherlands; 2019;35:155.

36 Nurfarahin AH, Mohamed MS, Phang LY. Culture medium development for microbial-derived surfactants production—an overview. *Molecules*. 2018;23:1049.

37 Farias CBB, Almeida FCG, Silva IA, Souza TC, Meira HM, de Soares da Silva RCF, et al. Production of green surfactants: market prospects. *Electron J Biotechnol*. Pontificia Universidad Catolica de Valparaiso; 2021;51:28–39.

38 Kim JH, Oh YR, Hwang J, Jang YA, Lee SS, Hong SH, et al. Value-added conversion of biodiesel into the versatile biosurfactant sophorolipid using Starmerella bombicola. *Clean Eng Technol*. 2020;1.

39 Ishaq U, Akram MS, Iqbal Z, Rafiq M, Akrem A, Nadeem M, et al. Production and characterization of novel self-assembling biosurfactants from Aspergillus flavus. *J Appl Microbiol*. 2015;119(4):1035–1045.

40 Wang XB, Nie Y, Tang YQ, Wu G, Wu XL. N-Alkane Chain Length Alters Dietzia sp. strain DQ12-45-1b biosurfactant production and cell surface activity. *Appl Environ Microbiol*. 2013;79(1):400–402.

41 Ikhwani AZN, Nurlaila HS, Ferdinand FDK, Fachria R, Hasan AEZ, Yani M, et al. Preliminary study: optimization of pH and salinity for biosurfactant production from Pseudomonas aeruginosa in diesel fuel and crude oil medium. In: IOP Conference Series: Earth and Environmental Science. Institute of Physics Publishing; 2017.

42 McClements DJ, Gumus CE. Natural emulsifiers — biosurfactants, phospholipids, biopolymers, and colloidal particles: molecular and physicochemical basis of functional performance. *Adv Colloid Interface Sci*. 2016;234:3–26.

43 Çelik A, Manga EB, Çabuk A, Banat IM. Biosurfactants' potential role in combating COVID-19 and similar future microbial threats. *Appl Sci*. 2021;11:334.

44 Rodrigues L, Banat IM, Teixeira J, Oliveira R. Biosurfactants: potential applications in medicine. *J Antimicrob Chemother*. 2006;57(4):609–618.

45 Subramaniam MD, Venkatesan D, Iyer M, Subbarayan S, Govindasami V, Roy A, et al. Biosurfactants and anti-inflammatory activity: a potential new approach towards COVID-19. *Curr Opin Environ Sci Health*. 2020;17:72.

46 Adu SA, Naughton PJ, Marchant R, Banat IM. Microbial biosurfactants in cosmetic and personal skincare pharmaceutical formulations. *Pharmaceutics*. 2020;12(11):1099.

47 De Almeida DG, de Soares Da Silva RCF, Luna JM, Rufino RD, Santos VA, Banat IM, et al. Biosurfactants: promising molecules for petroleum biotechnology advances. *Front Microbiol*. 2016;7:1718.

48 Li G, McInerney MJ. Use of biosurfactants in oil recovery. In: Lee S. editor. *Consequences of Microbial Interactions with Hydrocarbons, Oils, and Lipids: Production of Fuels and Chemicals*. Springer, Cham; 2016. pp. 1–16.

49 Walter V, Syldatk C, Hausmann R. *Screening Concepts for the Isolation of Biosurfactant Producing Microorganisms.* In: Sen, R.editor. Biosurfactants. Advances in Experimental Medicine and Biology. New York: Springer;. 2010, pp. 1–13.

50 Varjani SJ, Upasani VN. Critical review on biosurfactant analysis, purification and characterization using rhamnolipid as a model biosurfactant. *Bioresour Technol*. 2017;232:389–397.

51 Satpute SK, Banpurkar AG, Dhakephalkar PK, Banat IM, Chopade BA. Methods for investigating biosurfactants and bioemulsifiers: a review. *Crit Rev Biotechnol*. 2010;30:127–144.

52 Burch AY, Shimada BK, Browne PJ, Lindow SE. Novel high-throughput detection method to assess bacterial surfactant production. *Appl Environ Microbiol*. 2010;76(16):5363–5372.

53 Burch AY, Browne PJ, Dunlap CA, Price NP, Lindow SE. Comparison of biosurfactant detection methods reveals hydrophobic surfactants and contact-regulated production. *Environ Microbiol*. 2011;13(10):2681–2691.

54 Banat IM, Makkar RS, Cameotra SS. Potential commercial applications of microbial surfactants. *Appl Microbiol Biotechnol*. 2000;53(5):495–508.

55 Rodrigues L, Banat IM, Teixeira J, Oliveira R. Biosurfactants: potential applications in medicine. *J Antimicrob Chemother*. 2006;57:609–618.

56 Hu F, Liu Y, Li S. Rational strain improvement for surfactin production: enhancing the yield and generating novel structures. *Microb Cell Fact.* 2019;18(1):42.

57 Zhu L, Xu Q, Jiang L, Huang H, Li S. Polydiacetylene-based high-throughput screen for surfactin producing strains of Bacillus subtilis. *PLoS ONE.* 2014;9(2).

58 Chen X, Lee J, Jou MJ, Kim J-M, Yoon J. Colorimetric and fluorometric detection of cationic surfactants based on conjugated polydiacetylene supramolecules. *Chem Commun.* 2009;(23):3434–3436.

59 Lee S, Lee KM., Lee M, Yoon J. Polydiacetylenes bearing boronic acid groups as colorimetric and fluorescence sensors for cationic surfactants. *ACS Appl Mater Interfaces.* 2013;5(11):4521–4526.

60 Thongmalai W, Eaidkong T, Ampornpun S, Mungkarndee R, Tumcharern G, Sukwattanasinitt M, et al. Polydiacetylenes carrying amino groups for colorimetric detection and identification of anionic surfactants. *J Mater Chem.* 2011;21(41):16391.

61 Yang H, Yu H, Shen Z. A novel high-throughput and quantitative method based on visible color shifts for screening Bacillus subtilis THY-15 for surfactin production. *J Ind Microbiol Biotechnol.* 2015;42(8):1139–1147.

62 Changyin Wylgl LC. Study on the chromogenic reaction of anionic surfactants with quaternary ammonium salts and sulphonphthalein dyes and its applications. *Environ Chem.* 1997;5.

63 Heuson E, Etchegaray A, Filipe SL, Beretta D, Chevalier M, Phalip V, et al. Screening of lipopeptide-producing strains of bacillus sp. using a new automated and sensitive fluorescence detection method. *Biotechnol J.* 2019 April 1;14(4).e1800314

64 Ong SA, Wu JC. A simple method for rapid screening of biosurfactant-producing strains using bromothymol blue alone. *Biocatal Agric Biotechnol.* 2018;16:121–125.

65 Kubicki S, Bator I, Jankowski S, Schipper K, Tiso T, Feldbrügge M, et al. A straightforward assay for screening and quantification of biosurfactants in microbial culture supernatants. *Front Bioeng Biotechnol.* 2020;8:958.

66 Piel J. Approaches to capturing and designing biologically active small molecules produced by uncultured microbes. *Annu Rev Microbiol.* 2011;65(1):431–453.

67 Tapi A, Chollet-Imbert M, Scherens B, Jacques P. New approach for the detection of non-ribosomal peptide synthetase genes in Bacillus strains by polymerase chain reaction. *Appl Microbiol Biotechnol.* 2010;85(5):1521–1531.

68 Sachdev DP, Cameotra SS. Biosurfactants in agriculture. *Appl Microbiol Biotechnol.* 2013;97(3):1005–1016.

69 Williams W, Trindade M. Metagenomics for the discovery of novel biosurfactants. In: Charlies, T. *Functional Metagenomics: Tools and Applications.* Springer, Cham; 2017. pp. 95–117.

70 Biniarz P, Łukaszewicz M, Janek T. Screening concepts, characterization and structural analysis of microbial-derived bioactive lipopeptides: a review. *Crit Rev Biotechnol.* 2017;37:393–410.

71 Tripathi L, Twigg MS, Zompra A, Salek K, Irorere VU, Gutierrez T, et al. Biosynthesis of rhamnolipid by a Marinobacter species expands the paradigm of biosurfactant synthesis to a new genus of the marine microflora. *Microb Cell Fact.* 2019;18(1):164.

72 Tripathi L, Irorere VU, Marchant R, Banat IM. Marine derived biosurfactants: a vast potential future resource. *Biotechnol Lett.* 2018;40(11–12):1441–1457.

73 Owen JG, Reddy BVB, Ternei MA, Charlop-Powers Z, Calle PY, Kim JH, et al. Mapping gene clusters within arrayed metagenomic libraries to expand the structural diversity of biomedically relevant natural products. *Proc Natl Acad Sci*. 2013;110(29):11797–11802.

74 Abraham BS, Caglayan D, Carrillo NV, Chapman MC, Hagan CT, Hansen ST, et al. Shotgun metagenomic analysis of microbial communities from the Loxahatchee nature preserve in the Florida Everglades. *Environ. Microbiome*. 2020;15(1):2.

75 Da Silva GF, Gautam A, Silveira Duarte IC, Delforno TP, de Oliveira VM, Huson DH. *Interactive Analysis of Biosurfactants in Fruit-Waste Fermentation Samples Using BioSurfDB and MEGAN*. Sci Rep2021;(12):7769.

76 Montiel D, Kang H-S, Chang F-Y, Charlop-Powers Z, Brady SF. Yeast homologous recombination-based promoter engineering for the activation of silent natural product biosynthetic gene clusters. *Proc Natl Acad Sci*. 2015;112(29):8953–8958.

77 Quince C, Walker AW, Simpson JT, Loman NJ, Segata N. Shotgun metagenomics, from sampling to analysis. *Nat Biotechnol*. 2017;35(9):833–844.

78 Thomas T, Gilbert J, Meyer F. Metagenomics – a guide from sampling to data analysis. *Microb Inform Exp*. 2012;2(1):3.

79 Bharti R, Grimm DG. Current challenges and best-practice protocols for microbiome analysis. *Brief Bioinform*. 2021;22(1):178–193.

80 Liu L, Wang Y, Yang Y, Wang D, Cheng SH, Zheng C, et al. Charting the complexity of the activated sludge microbiome through a hybrid sequencing strategy. *Microbiome*. 2021;9(1):205.

81 Singleton CM, Petriglieri F, Kristensen JM, Kirkegaard RH, Michaelsen TY, Andersen MH, et al. Connecting structure to function with the recovery of over 1000 high-quality metagenome-assembled genomes from activated sludge using long-read sequencing. *Nat Commun*. 2021;12(1):2009.

82 Benaud N, Edwards RJ, Amos TG, D'Agostino PM, Gutiérrez-Chávez C, Montgomery K, et al. Antarctic desert soil bacteria exhibit high novel natural product potential, evaluated through long-read genome sequencing and comparative genomics. *Environ Microbiol*. 2021;23(7):3646–3664.

83 Bernat P, Nesme J, Paraszkiewicz K, Schloter M, Płaza G. Characterization of extracellular biosurfactants expressed by a pseudomonas putida strain isolated from the interior of healthy roots from Sida hermaphrodita grown in a heavy metal contaminated soil. *Curr Microbiol*. 2019 November 20;76(11):1320–1329.

84 Arulprakasam KR, Dharumadurai D. Genome mining of biosynthetic gene clusters intended for secondary metabolites conservation in actinobacteria. *Microb Pathog*. 2021 December;161:105252.

85 Oliveira JS, Araújo W, Lopes Sales AI, de Brito Guerra A, da Silva Araújo SC, de Vasconcelos ATR, et al. BioSurfDB: knowledge and algorithms to support biosurfactants and biodegradation studies. *Database*. 2015;2015:bav033.

86 Skinnider MA, Johnston CW, Gunabalasingam M, Merwin NJ, Kieliszek AM, MacLellan RJ, et al. Comprehensive prediction of secondary metabolite structure and biological activity from microbial genome sequences. *Nat Commun*. 2020;11(1):6058.

87 Skinnider MA, Dejong CA, Rees PN, Johnston CW, Li H, Webster ALH, et al. Genomes to natural products prediction informatics for secondary metabolomes (PRISM). *Nucleic Acids Res*. 2015;439645–9662.

88 Weber T, Blin K, Duddela S, Krug D, Kim HU, Bruccoleri R, et al. AntiSMASH 3.0-a comprehensive resource for the genome mining of biosynthetic gene clusters. *Nucleic Acids Res*. 2015;43:237–243.
89 Machado H, Tuttle RN, Jensen PR. Omics-based natural product discovery and the lexicon of genome mining. *Curr Opin Microbiol*. 2017;39:136–142.
90 Ziemert N, Alanjary M, Weber T. The evolution of genome mining in microbes – a review. *Nat Prod Rep*. 2016;33(8):988–1005.
91 Twigg MS, Baccile N, Banat IM, Déziel E, Marchant R, Roelants S, et al. Microbial biosurfactant research: time to improve the rigour in the reporting of synthesis, functional characterization and process development. *Microb Biotechnol*. 2021;14(1):147–170.
92 Santana-Pereira ALR, Sandoval-Powers M, Monsma S, Zhou J, Santos SR, Mead DA, et al. Discovery of novel biosynthetic gene cluster diversity from a soil metagenomic library. *Front Microbiol*. 2020;11:585398.
93 Thies S, Rausch SC, Kovacic F, Schmidt-Thaler A, Wilhelm S, Rosenau F, et al. Metagenomic discovery of novel enzymes and biosurfactants in a slaughterhouse biofilm microbial community. *Sci Rep*. 2016;627035.
94 Williams W, Kunorozva L, Klaiber I, Henkel M, Pfannstiel J, van Zyl LJ, et al. Novel metagenome-derived ornithine lipids identified by functional screening for biosurfactants. *Appl Microbiol Biotechnol*. 2019;103(11):4429–4441.
95 da Araújo SCS, Silva-Portela RCB, de Lima DC, da Fonsêca MMB, Araújo WJ, da Silva UB, et al. MBSP1: a biosurfactant protein derived from a metagenomic library with activity in oil degradation. *Sci Rep*. 2020;10(1):1340.
96 Pacwa-Płociniczak M, Płaza GA, Piotrowska-Seget Z, Cameotra SS. Environmental applications of biosurfactants: recent advances. *Int J Mol Sci*. 2011;12(1):633–654.
97 Chiewpattanakul P, Phonnok S, Durand A, Marie E, Thanomsub BW. Bioproduction and anticancer activity of biosurfactant produced by the dematiaceous fungus Exophiala dermatitidis SK80. *J Microbiol Biotechnol*. 2010;20(12):1664–1671.
98 Morita T, Fukuoka T, Imura T, Kitamoto D. Formation of the two novel glycolipid biosurfactants, mannosylribitol lipid and mannosylarabitol lipid, by Pseudozyma parantarctica JCM 11752T. *Appl Microbiol Biotechnol*. 2012;96(4):931–938.
99 Sarafin Y, Donio MBS, Velmurugan S, Michaelbabu M, Citarasu T. Kocuria marina BS-15 a biosurfactant producing halophilic bacteria isolated from solar salt works in India. *Saudi J Biol Sci*. 2014;21(6):511–519.
100 Dalmaso GZL, Ferreira D, Vermelho AB. Marine extremophiles a source of hydrolases for biotechnological applications. *Mar Drugs*. 2015;13(4):1925–1965.
101 Bhange K, Chaturvedi V, Bhatt R. Simultaneous production of detergent stable keratinolytic protease, amylase and biosurfactant by Bacillus subtilis PF1 using agro industrial waste. *Biotechnol Rep*. 2016;10:94–104.
102 Aktar L, Khan FI, Islam T, Mitra S, Saha ML. Isolation and characterization of indigenous lipase producing bacteria from lipid-rich environment. *Plant Tissue Cult Biotechnol*. 2016;26(2):243–253.
103 de Carvalho-gonçalves LCT, Gorlach-Lira K. Lipases and biosurfactants production by the newly isolated Burkholderia sp. *Braz J Biol Sci*. 2018;5(9):57–68.

104. Kumar S, Mathur A, Singh V, Nandy S, Khare SK, Negi S. Bioremediation of waste cooking oil using a novel lipase produced by Penicillium chrysogenum SNP5 grown in solid medium containing waste grease. *Bioresour Technol*. 2012;120:300–304.
105. Colla LM, Rizzardi J, Pinto MH, Reinehr CO, Bertolin TE, Costa JAV. Simultaneous production of lipases and biosurfactants by submerged and solid-state bioprocesses. *Bioresour Technol*. 2010;101(21):8308–8314.
106. Banat IM, Carboué Q, Saucedo-Castañeda G, de Jesús Cázares-marinero J. Biosurfactants: the green generation of speciality chemicals and potential production using Solid-State fermentation (SSF) technology. *Bioresour Technol*. 2021;320:124222.
107. Ribeiro BG, Guerra JMC, Sarubbo LA. Potential food application of a biosurfactant produced by saccharomyces cerevisiae URM 6670. *Front Bioeng Biotechnol*. 2020;8:1–13.
108. Rajeswari V, Kalaivani Priyadarshini S, Saranya V, Suguna P, Shenbagarathai R. Immunostimulation by phospholipopeptide biosurfactant from Staphylococcus hominis in Oreochromis mossambicus. *Fish Shellfish Immunol*. 2016;48:244–253.
109. Fakruddin M. Biosurfactant: production and application. *J Pet Environ Biotechnol*. 2012;03(04).1000124
110. Szymczak T, Cybulska J, Podleśny M, Frąc M. Various perspectives on microbial lipase production using agri-food waste and renewable products. *Agriculture (Switzerland)*. 2021;11(6)540.
111. Chandra P, Enespa SR, Arora PK. Microbial lipases and their industrial applications: a comprehensive review. *Microb Cell Fact*. BioMed Central; 2020;19:1–42.
112. Santos DKF, Rufino RD, Luna JM, Santos VA, Salgueiro AA, Sarubbo LA. Synthesis and evaluation of biosurfactant produced by Candida lipolytica using animal fat and corn steep liquor. *J Pet Sci Eng*. 2013;105:43–50.
113. Javed S, Azeem F, Hussain S, Rasul I, Siddique MH, Riaz M, et al. Bacterial lipases: a review on purification and characterization. *Prog Biophys Mol Biol*. 2018;132:23–34.
114. Geoffry K, Achur RN. Optimization of novel halophilic lipase production by Fusarium solani strain NFCCL 4084 using palm oil mill effluent. *J Genet Eng Biotechnol*. 2018;16(2):327–334.
115. Fickers P, Benetti PH, Waché Y, Marty A, Mauersberger S, Smit MS, et al. Hydrophobic substrate utilisation by the yeast Yarrowia lipolytica, and its potential applications. *FEMS Yeast Res*. 2005;5(6–7):527–543.
116. Al-ghanayem AA. Purification and characterization of thermo-alkaline stable lipase from Bacillus coagulans and its compatibility with commercially available detergents. 2021;26(5):2994–3001.
117. Celligoi MAPC, Baldo C, De Melo MR, Gasparin FGM, Marques TA, De Barros M. Lipase properties, functions and food applications. In: Ray RC, Rosell CM. editors. *Microbial Enzyme Technology in Food Applications*.Boca Raton: CRC Press; 2017; pp. 214–240.
118. Nathan VK, Rani ME. A cleaner process of deinking waste paper pulp using Pseudomonas mendocina ED9 lipase supplemented enzyme cocktail. *Environ Sci Pollut Res*. 2020;27(29):36498–36509.
119. Badgujar VC, Badgujar KC, Yeole PM, Bhanage BM. Immobilization of Rhizomucor miehei lipase on a polymeric film for synthesis of important fatty acid esters: kinetics and application studies. *Bioprocess Biosyst Eng*. 2017;40(10):1463–1478.

120 Gupta KK, Nigam A, Jagtap S, Krishna R. Scale-up and inhibitory studies on productivity of lipase from Acinetobacter radioresistens PR8. *J Biosci Bioeng*. 2017;124(2):150–155.

121 Shahedi M, Yousefi M, Habibi Z, Mohammadi M, As'habi MA. Co-immobilization of Rhizomucor miehei lipase and Candida antarctica lipase B and optimization of biocatalytic biodiesel production from palm oil using response surface methodology. *Renew Energy*. 2019;141:847–857.

122 Haniya M, Naaz A, Sakhawat A, Amir S, Zahid H, Syed SA. Optimized production of lipase from Bacillus subtilis PCSIRNL-39. *Afr J Biotechnol*. 2017;16(19):1106–1115.

123 Andualema B, Gessesse A. Microbial lipases and their industrial applications: review. *Biotechnology*. 2012;11:100–118.

124 Alkabee HJJ, Alsalami AK, Alansari BM. Determination the lipase activity of staphylococcus sp. strain isolated from clinical specimens. *J Pure Appl Microbiol*. 2020;14(1):437–446.

125 Kundys A, Białecka-Florjańczyk E, Fabiszewska A, Małajowicz J. Candida antarctica lipase B as catalyst for cyclic esters synthesis, their polymerization and degradation of aliphatic polyesters. *J Polymers Environ*. 2018;26(1):396–407.

126 Irimie F, Paizs C, Tos MI Molecules efficient biodiesel production catalyzed by nanobioconjugate of lipase from pseudomonas fluorescens 2020;25:651.

127 Binhayeeding N, Yunu T, Pichid N, Klomklao S, Sangkharak K. Immobilisation of Candida rugosa lipase on polyhydroxybutyrate via a combination of adsorption and cross-linking agents to enhance acylglycerol production. *Process Biochem*. 2020;95:174–185.

128 Zhou A, Jing N, Wang G, Xu Q. Construction of a "battlements" structure by zirconium glyphosate on silica nanospheres and the catalytic stability promotion of candida lipolytica lipase on the hybrid materials. *Ind Eng Chem Res*. 2018;57(44):15031–15038.

129 Marques RV, Guidoni LLC, Araujo TR, Santos MAZ, de Pereira CMP, Duval EH, et al. Modification of bovine fatty waste with strains of Staphylococcus xylosus: feedstock for biodiesel. *Environ Challenges*. 2021;4:100180.

130 Šibalić D, Šalić A, Tušek AJ, Sokač T, Brekalo K, Zelić B, et al. Sustainable production of lipase from thermomyces lanuginosus: process optimization and enzyme characterization. *Ind Eng Chem Res*. 2020;59(48):21144–21154.

131 Sumarsih S, Hadi S, Andini DGT, Nafsihana FK. Carbon and nitrogen sources for lipase production of micrococcus sp. isolated from palm oil mill effluent-contaminated soil. *IOP Conf Ser: Earth Environ Sci*. 2019;217(1) 012029.

132 Araujo SC, Ramos MRMF, Do Espírito Santo EL, de Menezes LHS, de Carvalho MS, de Tavares IMC, et al. Optimization of lipase production by Penicillium roqueforti ATCC 10110 through solid-state fermentation using agro-industrial residue based on a univariate analysis. *Prep Biochem Biotechnol*. 2021;0(0):1–6.

133 Mandari V, Nema A, Devarai SK. Sequential optimization and large scale production of lipase using tri-substrate mixture from Aspergillus niger MTCC 872 by solid state fermentation. *Process Biochem*. 2020;89:46–54.

134 Oves M, Qari HA, Felemban NM, Khan MZ, Rehan ZA, Ismail IMI. Marinobacter lipolyticus from Red Sea for lipase production and modulation of silver nanomaterials for anti-candidal activities. *IET Nanobiotechnol*. 2017;11(4):403–410.

135 Knob A, Izidoro SC, Lacerda LT, Rodrigues A, de Lima VA. A novel lipolytic yeast Meyerozyma guilliermondii: efficient and low-cost production of acid and promising feed lipase using cheese whey. *Biocatal Agric Biotechnol*. 2020;24.

136 Li X, Yu HY. Characterization of an organic solvent-tolerant lipase from Haloarcula sp. G41 and its application for biodiesel production. *Folia Microbiol (Praha)*. 2014;59(6):455–463.

137 Kareem SO, Adebayo OS, Balogun SA, Adeogun AI, Akinde SB. Purification and characterization of Lipase From *Aspergillus flavus* PW2961 using Magnetic Nanoparticles. *Niger J Biotechnol*. 2017;32(1):77.

138 Abol-Fotouh D, AlHagar OEA, Hassan MA. Optimization, purification, and biochemical characterization of thermoalkaliphilic lipase from a novel Geobacillus stearothermophilus FMR12 for detergent formulations. *Int J Biol Macromol*. 2021;181:125–135.

139 Prasath KG, Tharani H, Kumar MS, Pandian SK. Palmitic acid inhibits the virulence factors of candida tropicalis: biofilms, cell surface hydrophobicity, ergosterol biosynthesis, and enzymatic activity. *Front Microbiol*. 2020;11:1–21.

140 Paluzar H, Tuncay D, Aydogdu H. Production and characterization of lipase from Penicillium aurantiogriseum under solid-state fermentation using sunflower pulp. *Biocatal Biotransform*. 2021;39(4):333–342.

141 Sadaf A, Grewal J, Jain I, Kumari A, Khare SK. Stability and structure of Penicillium chrysogenum lipase in the presence of organic solvents. *Prep Biochem Biotechnol*. 2018;48(10):977–983.

142 Patel R, Prajapati V, Trivedi U, Patel K. Optimization of organic solvent-tolerant lipase production by Acinetobacter sp. UBT1 using deoiled castor seed cake. *3 Biotech*. 2020;10(12):1–13.

143 Kajiwara S, Yamada R, Matsumoto T, Ogino H. N-linked glycosylation of thermostable lipase from Bacillus thermocatenulatus to improve organic solvent stability. *Enzyme Microb Technol*. 2020;132(May 2019):109416.

144 Khan I, Ray Dutta J, Ganesan R. Lactobacillus sps. lipase mediated poly (ε-caprolactone) degradation. *Int J Biol Macromol*. 2017;95:126–131.

145 Satpute SK, Kulkarni GR, Banpurkar AG, Banat IM, Mone NS, Patil RH, et al. Biosurfactant/s from Lactobacilli species: properties, challenges and potential biomedical applications. *J Basic Microbiol*. 2016;56(11):1140–1158.

146 Beltrani T, Chiavarini S, Cicero DO, Grimaldi M, Ruggeri C, Tamburini E, et al. Chemical characterization and surface properties of a new bioemulsifier produced by Pedobacter sp. strain MCC-Z. *Int J Biol Macromol*. 2015;72:1090–1096.

147 Ahmed EF, Hassan SS. Antimicrobial activity of a bioemulsifier produced by Serratia marcescens S10. *J Al-Nahrain Univ Sci*. 2013;16(1):147–155.

148 Sianipar M, Kardena E, Hidayat S. The application of biosurfactant produced by Azotobacter sp. for oil recovery and reducing the hydrocarbon loading in bioremediation process. *Int J Environ Sci Dev*. 2016;7(7):494–498.

149 Franzetti A, Gandolfi I, Raimondi C, Bestetti G, Banat IM, Smyth TJ, et al. Environmental fate, toxicity, characteristics and potential applications of novel bioemulsifiers produced by Variovorax paradoxus 7bCT5. *Bioresour Technol*. 2012;108:245–251.

150 Liang TW, Wu CC, Cheng WT, Chen YC, Wang CL, Wang IL, et al. Exopolysaccharides and antimicrobial biosurfactants produced by paenibacillus macerans TKU029. *Appl Biochem Biotechnol*. 2014;172(2):933–950.

151 Thavasi R, Nambaru VRMS, Jayalakshmi S, Balasubramanian T, Banat IM. Biosurfactant production by pseudomonas aeruginosa from renewable resources. *Indian J Microbiol*. 2011;51(1):30–36.

152 Shah N, Prajapati JB. Effect of carbon dioxide on sensory attributes, physico-chemical parameters and viability of Probiotic L. helveticus MTCC 5463 in fermented milk. *J Food Sci Technol*. 2014;51(12):3886–3893.

153 Sigurbjörnsdóttir MA, Vilhelmsson O. Selective isolation of potentially phosphate-mobilizing, biosurfactant-producing and biodegradative bacteria associated with a sub-Arctic, terricolous lichen, Peltigera membranacea. *FEMS Microbiol Ecol*. 2016;92(6):1–7.

154 Moretto C, Castellane TCL, Lopes EM, Omori WP, Sacco LP, de Lemos EGM. Chemical and rheological properties of exopolysaccharides produced by four isolates of rhizobia. *Int J Biol Macromol*. 2015;81:291–298.

155 Shokouhfard M, Kasra Kermanshahi R, Vahedi Shahandashti R, Feizabadi MM, Teimourian S. The inhibitory effect of a Lactobacillus acidophilus derived biosurfactant on Serratia marcescens biofilm formation. *Iran J Basic Med Sci*. 2015;18(10):1001–1007.

156 Adamu A, Ijah UJJ, Riskuwa ML, Ismail HY, Ibrahim UB. Study on biosurfactant production by two bacillus species. *Int J Sci Knowl*. 2015;3(1):13–20.

157 Joe MM, Saravanan VS, Sa T. Aggregation of selected plant growth promoting methylobacterium strains: role of cell surface components and hydrophobicity. *Arch Microbiol*. 2013;195(3):219–225.

158 Dos Santos AS, Pereira N, Freire DMG. Strategies for improved rhamnolipid production by Pseudomonas aeruginosa PA1. *PeerJ*. 2016;2016(5):1–16.

159 Noparat P, Maneerat S, Saimmai A. Application of biosurfactant from Sphingobacterium spiritivorum AS43 in the biodegradation of used lubricating oil. *Appl Biochem Biotechnol*. 2014;172(8):3949–3963.

160 Petrikov K, Delegan Y, Surin A, Ponamoreva O, Puntus I, Filonov A, et al. Glycolipids of pseudomonas and rhodococcus oil-degrading bacteria used in bioremediation preparations: formation and structure. *Process Biochem*. 2013;48(5–6):931–935.

161 Mishra B, Varjani S, Agrawal DC, Mandal SK, Ngo HH, Taherzadeh MJ, et al. Engineering biocatalytic material for the remediation of pollutants: a comprehensive review. *Environ Technol Innov*. 2020;20:101063.

162 Dhasayan A, Kiran GS, Selvin J. Production and characterisation of glycolipid biosurfactant by halomonas sp. MB-30 for potential application in enhanced oil recovery. *Appl Biochem Biotechnol*. 2014;174(7):2571–2584.

163 Ivanova N, Gugleva V, Dobreva M, Pehlivanov I, Stefanov S, Andonova V We are intechopen, the world's leading publisher of open access books built by scientists, for scientists TOP 1 %. 2016;3.

164 Kiran GS, Lipton AN, Priyadharshini S, Anitha K, Suárez LEC, Arasu MV, et al. Antiadhesive activity of poly-hydroxy butyrate biopolymer from a marine Brevibacterium casei MSI04 against shrimp pathogenic vibrios. *Microb Cell Fact*. 2014;13(1):1–12.

165 Kuyukina MS, Ivshina IB, Korshunova IO, Stukova GI, Krivoruchko AV. Diverse effects of a biosurfactant from Rhodococcus ruber IEGM 231 on the adhesion of resting and growing bacteria to polystyrene. *AMB Express*. 2016;6(1):1–12.

166 Elazzazy AM, Abdelmoneim TS, Almaghrabi OA. Isolation and characterization of biosurfactant production under extreme environmental conditions by alkali-halo-thermophilic bacteria from Saudi Arabia. *Saudi J Biol Sci*. 2015;22(4):466–475.

167 Monteiro AS, Bonfim MRQ, Domingues VS, Corrêa A, Siqueira EP, Zani CL, et al. Identification and characterization of bioemulsifier-producing yeasts isolated from effluents of a dairy industry. *Bioresour Technol*. 2010;101(14):5186–5193.

168 Gutierrez T, Shimmield T, Haidon C, Black K, Green DH. Emulsifying and metal ion binding activity of a glycoprotein exopolymer produced by Pseudoalteromonas sp. strain TG12. *Appl Environ Microbiol*. 2008;74(15):4867–4876.

169 Smith DDN, Nickzad A, Stavrinides J, Déziel E. Amoeba Dictyostelium discoideum. 2016;1(1):3–6.

170 De Souza Monteiro A, Souza Domingues V, Souza MVD, Lula I, Bonoto Goṉalves D, Pessoa De Siqueira E, et al. Bioconversion of biodiesel refinery waste in the bioemulsifier by Trichosporon mycotoxinivorans CLA2. *Biotechnol Biofuels*. 2012;5(1):1.

171 Qazi MA, Subhan M, Fatima N, Ali MI, Qazi AS. International journal of bioscience, biochemistry and bioinformatics. 2013;3(6).

172 Ohadi M, Dehghannoudeh G, Shakibaie M, Banat IM, Pournamdari M, Forootanfar H. Isolation, characterization, and optimization of biosurfactant production by an oil-degrading Acinetobacter junii B6 isolated from an Iranian oil excavation site. *Biocatal Agric Biotechnol* 2017;12:1–9.

173 Luna-Velasco MA, Esparza-García F, Cañizares-Villanueva RO, Rodríguez-Vázquez R. Production and properties of a bioemulsifier synthesized by phenanthrene-degrading Penicillium sp. *Process Biochem*. 2007;42(3):310–314.

174 Ekprasert J, Kanakai S, Yosprasong S. Improved biosurfactant production by enterobacter cloacae B14, stability studies, and its antimicrobial activity. *Polish J Microbiol*. 2020;69(3):273–282.

175 Silva NRA, Luna MAC, Santiago ALCMA, Franco LO, Silva GKB, de Souza PM, et al. Biosurfactant-and-bioemulsifier produced by a promising Cunninghamella echinulata isolated from caatinga soil in the Northeast of Brazil. *Int J Mol Sci*. 2014;15(9):15377–15395.

176 Adetunji AI, Olaniran AO. Production and characterization of bioemulsifiers from Acinetobacter strains isolated from lipid-rich wastewater. *3 Biotech*. 2019;9(4):1–11.

177 Maneerat S, Bamba T, Harada K, Kobayashi A, Yamada H, Kawai F. A novel crude oil emulsifier excreted in the culture supernatant of a marine bacterium, Myroides sp. strain SM1. *Appl Microbiol Biotechnol*. 2006;70(2):254–259.

178 Suthar H, Hingurao K, Desai A, Nerurkar A. Selective plugging strategy based microbial enhanced oil recovery using Bacillus licheniformis TT33. *J Microbiol Biotechnol*. 2009;19(10):1230–1237.

179 Markande AR, Acharya SR, Nerurkar AS. Physicochemical characterization of a thermostable glycoprotein bioemulsifier from Solibacillus silvestris AM1. *Process Biochem*. 2013;48(11):1800–1808.

180 Solaiman DKY, Ashby RD, Birbir M, Caglayan P. Antibacterial activity of sophorolipids produced by candida bombicola on gram-positive and gram-negative bacteria isolated from salted hides. *J Am Leather Chem Assoc.* 2016;111(10):358–363.

181 Madhu AN, Prapulla SG. Evaluation and functional characterization of a biosurfactant produced by Lactobacillus plantarum CFR 2194. *Appl Biochem Biotechnol.* 2014;172(4):1777–1789.

182 Hu X, Wang C, Wang P. Optimization and characterization of biosurfactant production from marine Vibrio sp. strain 3B-2. *Front Microbiol.* 2015;6:1–13.

183 Vilela WFD, Fonseca SG, Fantinatti-Garboggini F, Oliveira VM, Nitschke M. Production and properties of a surface-active lipopeptide produced by a new marine brevibacterium luteolum strain. *Appl Biochem Biotechnol.* 2014;174(6):2245–2256.

184 Almeida DG, de da Silva RCFS, Luna JM, Rufino RD, Santos VA, Sarubbo LA. Response surface methodology for optimizing the production of biosurfactant by Candida tropicalis on industrial waste substrates. *Front Microbiol.* 2017;8:1–13.

185 Joshi SJ, Geetha SJ, Desai AJ. Characterization and application of biosurfactant produced by Bacillus licheniformis R2. *Appl Biochem Biotechnol.* 2015;177(2):346–361.

186 Bezza FA, Nkhalambayausi Chirwa EM. Biosurfactant from Paenibacillus dendritiformis and its application in assisting polycyclic aromatic hydrocarbon (PAH) and motor oil sludge removal from contaminated soil and sand media. *Process Saf Environ Prot.* 2015;98:354–364.

187 Shaligram S, Kumbhare SV, Dhotre DP, Muddeshwar MG, Kapley A, Joseph N, et al. Genomic and functional features of the biosurfactant producing Bacillus sp. AM13. *Funct Integr Genomics.* 2016;16(5):557–566.

188 Elshafie AE, Joshi SJ, Al-Wahaibi YM, Al-Bemani AS, Al-Bahry SN, Al-Maqbali D, et al. Sophorolipids production by Candida bombicola ATCC 22214 and its potential application in microbial enhanced oil recovery. *Front Microbiol.* 2015;6:1–11.

189 Renard P, Canet I, Sancelme M, Matulova M, Uhliarikova I, Eyheraguibel B, et al. Cloud microorganisms, an interesting source of biosurfactants. IN: Dutta A. editor. *Surfactants Deterg.* 2019;73–89.

190 Monteiro AS, Miranda TT, Lula I, Denadai ÂML, Sinisterra RD, Santoro MM, et al. Inhibition of Candida albicans CC biofilms formation in polystyrene plate surfaces by biosurfactant produced by Trichosporon montevideense CLOA72. *Colloids Surf B Biointerfaces.* 2011;84(2):467–476.

191 Meena KR, Dhiman R, Singh K, Kumar S, Sharma A, Kanwar SS, et al. Purification and identification of a surfactin biosurfactant and engine oil degradation by Bacillus velezensis KLP2016. *Microb Cell Fact.* 2021;20(1):1–12.

192 Kim JS, Lee IK, Yun BS. A novel biosurfactant produced by Aureobasidium pullulans L3-GPY from a tiger lily wild flower, Lilium lancifolium thunb. *Plos One.* 2015;10(4):1–12.

193 Coronel-León J, de Grau G, Grau-Campistany A, Farfan M, Rabanal F, Manresa A, et al. Biosurfactant production by AL 1.1, a Bacillus licheniformis strain isolated from Antarctica: production, chemical characterization and properties. *Ann Microbiol.* 2015;65(4):2065–2078.

194 Maniyar JP, Doshi DV, Bhuyan SS, Mujumdar SS. Bioemulsifier production by Streptomyces sp. S22 isolated from garden soil. *Indian J Exp Biol.* 2011;49(4):293–297.

195 Castiglioni GL, Stanescu G, Rocha LAO, Costa JAV. Modelagem analítica e otimização numérica da produção de biossurfactantes por Aspergillus fumigatus em fermentação sólida. *Acta Sci Technol*. 2014;36(1):61–67.
196 Kannahi M, Sherley M. Biosurfactant production by pseudomonas putida and aspergillus niger from oil contamined site. *Int J Res Dev Pharm Life Sci*. 2012;3(4):37–42.
197 Reinehr CO, Rizzardi J, Silva MF, de Oliveira D, Treichel H, Colla LM. Production of lipases with aspergillus niger and aspergillus fumigatus through solid state fermentation: evaluation of substrate specificity and use in esterification and alcoholysis reactions. *Quimica Nova*. 2014;37(3):454–460.
198 Bhardwaj G. Biosurfactants from Fungi: a review. *J Pet Environ Biotechnol*. 2013;04(06):10000160.

3

Metabolic Engineering as a Tool for Biosurfactant Production by Microorganisms

Roberta Barros Lovaglio[1],*, Vinícius Luiz da Silva[2], and Jonas Contiero[3]

[1] Universidade Federal de São Carlos – Centro de Ciências da Natureza – Campus Lagoa do Sino
[2] MicroGreen – Soluções Biotecnológicas
[3] Universidade Estadual Paulista "Julio de Mesquita Filho" – Instituto de Biociências – Campus Rio Claro
* Corresponding author

3.1 Metabolic Engineering and Biosurfactants

Biosurfactants have numerous advantages when compared with synthetic ones, despite this they are not widely used because of the high cost of production. According to Tiso et al. [1] the reasons that hinder the market success of biosurfactant, like rhamnolipids, are associated with factors such as complex regulation in wild-type strains, costly substrates, and low production rates. In addition, it must be considered that the main producers could be a pathogenic strain, as with *Pseudomonas aeruginosa*. The metabolic engineering tools are able to overcome these challenges, an efficient approach to developing industrial fermentation processes is using directed genetic changes in industrially useful bacterial strains for metabolite overproduction [2–6].

To allow the success of metabolic engineering the following steps need to be carried out sequentially: 1. careful analysis of cellular function, 2. based on the results of step 1, an improved strain is designed, and 3. construction of a genetically modified strain [6].

The analysis of cellular function may be obtained through metabolic flux analysis (MFA), which can be determined with measurements of the specific rates of production and consumption of intra- and extracellular metabolites [7]. This understanding of cellular biochemistry can be achieved using stoichiometric models representing the biochemistry in a system of linear equations [8]. Metabolic flux analysis can also be performed with tracker substrates using the stable isotope ^{13}C. The results of these experiments allow for the development of a metabolic map with a diagram of the main biochemical reactions involved and the estimated speed at each stage in the stationary state [9]. In addition, the use of metabolic flux analysis can facilitate the detection of the metabolic pathway's hard branch points and contribute to overcoming the same, generating significant improvements for metabolite production [10].

Biosurfactants and Sustainability: From Biorefineries Production to Versatile Applications, First Edition.
Edited by Paulo Ricardo Franco Marcelino, Silvio Silverio da Silva, and Antonio Ortiz Lopez.
© 2023 John Wiley & Sons Ltd. Published 2023 by John Wiley & Sons Ltd.

Microorganisms have evolved to survive under natural conditions and are not able to exhibit high production and accumulation of any biomolecule [11] Therefore, implementation of a new synthetic route for any product, including biosurfactants, which is both economical and environmentally friendly, requires the development of engineering strategies to meet productivity demands [12]. The limited understanding of microbial physiology regulation and the lack of predictive models for cellular biocatalysis as a whole are the main reasons for the low productivity of microbial compounds in general [13]. According to Müller et al. [11] to aid in the development of rhamnolipid synthetic processes, it is necessary to understand the regulation of rhamnolipid production during bioreactor cultivation and thus, develop strategies based on metabolic engineering aiming to increase the production.

The main challenges to achieve a broad industrial production of biosurfactants can be overcome by combining tools provided by metabolic engineering, synthetic and systems biology. These approaches allow the design and construction of strains with 1. optimized growth on a broader range of carbon sources, 2. optimized metabolic pathways for biosurfactant synthesis, 3. elimination or reduced by-product formation, and 4. uncoupling synthesis from cellular regulation, moreover, the construction of model heterologous hosts may bypass the downside of pathogenic strains [14, 15].

In this chapter we present the metabolic engineering applied to rhamnolipid synthesis using this glycolipid as a study model. The following section gives an overview on how to improve rhamnolipids production. The first point focusses on regulation and heterologous production and the second on strain optimization to broaden the range of carbon sources. The improvement of overall cellular physiology and elimination or reduction of by-product formation are described on third and fourth topics, respectively (Figure 3.1). Finally, some future perspectives will be addressed.

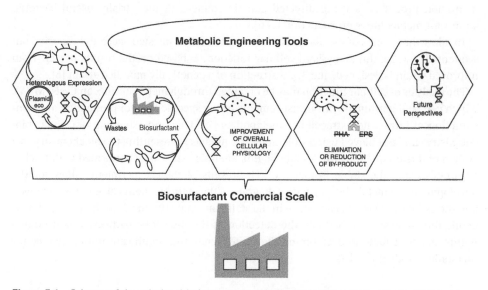

Figure 3.1 Schema of the relationship between metabolic engineering strategies for increased biosurfactant production.

3.2 Regulation and Heterologous Production of Biosurfactants

The mainchallenges that can be overcome by heterologous expression are:

- Strain pathogenicity [16] is what makes the bioprocess dangerous and expensive due to the need for a production with higher safety levels; furthermore, the bioproduct is not accepted for applications in the food and cosmetic sectors [11, 17].
- The wild type strain has a low production rate [1].
- Another important factor to take into account is the complex regulation network involved in rhamnolipids synthesis [18].

With these problems in mind some authors have been using molecular bioengineering to design new strains for biosurfactant production [19–21].

The biosurfactant structure is comprised of a hydrophobic and a hydrophilic portion, the synthesis of these moieties involve pathways where some enzymes participate in the synthesis and the regulation of this metabolite formation (Figure 3.2) [22].

Burger et al. [23], proposed the steps involved in rhamnolipid biosynthesis for the di-rhamnolipids (Rha-Rha C10C10):

1) 2β-hydroxydecanoyl-CoA → β-hydroxydecanoyl- β-hydroxydecanoate + 2 CoA-SH.
2) TDP-1-rhamnose + β-hydroxydecanoyl-β-hydroxydecanoate → TDP + 1-rhamnoyl-β-hydroxydecanoyl- β-hydroxydecanoate.
3) TDP-1-rhamnose + 1-rhamnosyl-β-hydroxydecanoyl- β-hydroxydecanoate → TDP + 1-rhamnosyl-1-rhamnosyl-β-hydroxydecanoyl-β-hydroxydecanoate.

The di-rhamnolipids formations are caried out by dimerization of two β-hydroxydecanoic acid moieties and two sequential rhamnosylation reactions, catalyzed by a specific rhamnosyl transferase in each step [23].

The identification of genes related to rhamnolipid biosynthesis started with Ochsner et al. [24]. These authors identified an open read frame (ORF), named rhlR, due to homology with other species proteins, they suggested that RhlR protein is a transcriptional activator. In subsequent studies, Ochsner et al. [25] identified two genes involved in rhamnolipid production, rhlA and rhlB, which are organized in an operon located upstream to the gene rhlR. The regulatory protein encoded by rh1R is necessary for the transcriptional activation of the operon rhlAB. The authors concluded that RhlB is a catalytic protein that promotes the binding of dTDP-L-rhamnose to the rhamnolipid acyl chain and that RhlA must be present for the synthesis of biosurfactants to occur; however, the function of RhlA has not been clearly determined. Later in 1995 Ochsner et al. [26] established the heterologous expression of *rhlAB* in different organisms *Escherichia coli, Pseudomonas fluorescens, Pseudomonas oleovoran*, and *Pseudomonas putida* (Figure 3.2).

Déziel et al. [27] presented evidence that the rhlA gene is involved in the synthesis of β-3-(3-hydroxyalkanoyloxy) alkanoic acid (Haa), which is a rhamnolipid precursor and rhamnosyltransferase 1 (RhlB) substrate, later this was confirmed by Zhu and Rock. [28].

Concerning to hydrophilic moiety synthesis, Olvera et al. [29] concluded that the conversion of glucose to rhamnose occurs through the action of phosphoglucomutase AlgC since mutants disrupted in this gene, did not produce rhamnolipids. This enzyme converts

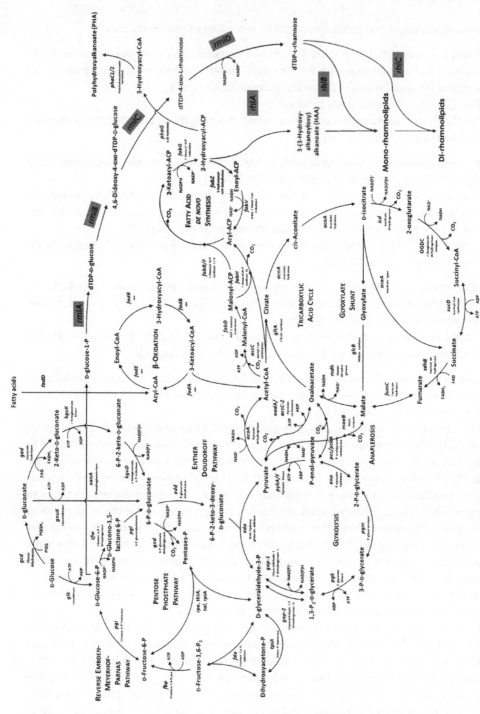

Figure 3.2 Metabolic pathways and genes involved in rhamnolipids biosynthesis.

D-glucose-6-phosphate to D-glucose-1-phosphate, which is used by RmlA, RmlB, RmlC, and RmlD to produce dTDP-L-rhamnose (Figure 3.2) [29, 30].

The production of many virulence factors and secondary metabolites, such as rhamnolipids, are regulated by cell density through a system called *quorum sensing* (QS) [31]. QS is characterized by the production of diffusible compounds called autoinducers, which accumulate with increasing population density, promoting cellular interaction and activation of transcriptional regulators [32–34]. *P. aeruginosa* strains have two QS systems, Las and Rhl. The RhlR/RhlI and LasR/LasI (regulatory/autoinducer protein) complexes modulate the transcription of 5–10% of the entire *P. aeruginosa* genome [18]. A third signal molecule, 2-heptyl-3-hydroxy-4-quinolone or *Pseudomonas* quinolone signal (PQS), acts as a bridge between the las and rhl QS systems. The regulatory protein LasR is required for PQS biosynthesis, while RhlR promotes the activation of this molecule [35]. According to Juhas et al. [36] there are additional regulators that control the fine tuning of the responses of Las and Rhl systems.

Dekimpe and Déziel. [18] reviewed the proposal that LasR would be the master regulator in the classical *quorum sensing* hierarchy. The results presented by the authors showed that the activity of the Rhl system is not shut down, but expressed late in LasR mutants, thus allowing the expression of virulence factors such as pyocyanin, rhamnolipids and homoserine lactone during the stationary growth phase. Furthermore, RhlR activates specific functions controlled by LasR, such as 3-oxo-C12-homoserine lactone and PQS, being able to bypass the absence of the Las system. These results demonstrate that the hierarchy model of QS system is more complex than the one that proposes the superposition of the action of Las to the Rhl system.

The complex regulatory network and the limited precursors synthesis led to the search for alternatives that would enable the expansion of the scale of rhamnolipid production, the heterologous expression has shown promising results over time, as described below.

Wang et al. [37] inserted the *rhlAB* operon with a native promoter into the *Pseudomonas aeruginosa* PAO1-rhlA⁻ and *Escherichia coli* BL21 chromosomes. The T7 promoter was used in this work. The engineered *E. coli* TnERAB produced, respectively, 65–80 mg L^{-1} and 167.5 mg L^{-1} of rhamnolipids in MS and LB media, both of them supplemented with glucose [37].

Han et al. [38] also constructed an engineered *E. coli* strain capable of producing rhamnolipids, they used the T7 promoter, the mutant with successful rhamnolipid production was submitted to a directed evolution and subsequent site directed mutagenesis in a single amino acid, from RhlB protein, at the position 168. This strategy resulted in rhamnolipids with the increased performance in reduce interfacial tension as well as enhanced oil recovery [38].

Two others works reported the successful heterologous production of this biosurfactant in *E. coli* using the T7 promoter [39, 40]. Jafari et al. [39] showed that the constructed *E. coli* was able to produce more rhamnolipids than the positive control *P. aeruginosa* PG201, using the oil displacement method. Du et al. [40] co-overexpressed *rfb*D, gene from rhamnoses biosynthesis, and *rhl*AB–*rhl*C, the authors reported a production of 0. 64 g L^{-1} of rhamnolipids, however without the *rfb*D gene, the modified strain produced 0.446 g L^{-1}.

In another work, the operon *rhlAB*, from *P. aeruginosa* PA14, was inserted into pNOT19 plasmid to produce a vector pF1bR4 that was used for *E. coli* TG2 transformation, giving rise the recombinant strain *E. coli* pF1bR4. A MALDI-TOF analysis was not able to detecting rhamnolipids production by *E. coli* pF1bR4, however a minor component of the commercial rhamnolipid (*m/z* 361.96 – 361.99) was found, in addition, it was able to emulsify hexadecane, on the other hand, the culture broth from wild type *E. coli* TG2 did not emulsify. The authors conclude that the *rhlAB* operon is not sufficient for effective production of rhamnolipids by *E. coli* pF1bR4 recombinant strain [41].

Cabrera-Valladares et al. [19] hypothesized that dTDP-L-rhamnose synthesis is the bottleneck for the low productivity of rhamnolipids in the heterologous expression. In order to check this, the authors co-expressed operons rhlAB and rmlBDAC in an *E. coli* strain. The approach applied overcame the problem of restricted availability of rhamnose and an increased rhamnolipid production reaching 120.6 mg L^{-1} was observed.

In summary, several microorganisms, with different genetic strategies being applied in each case, were employed as hosts for *P. aeruginosa* genes, such as *E. coli* [19, 25], *Cellvibrio japonicus* [42], *P. chloraphis* [43], *P. fluorescenes* [26], *P. oleovorans* [26], *P. stutzeri* [44], *P. putida* [20, 21, 27, 45] and *P. taiwanensis* [46].

According to the compiled works, *Pseudomonas putida* KT2440, wild type or engineered strains, was the most promising host strain for rhamnolipid biosynthesis. Here are some reported advantages for *P. putida* KT2440:

- This is a safety certified strain, which facilitates industrial-scale production procedures and is less expensive [1, 47].
- *P. putida* KT2440 is resistant to high rhamnolipid concentrations (up to 90 g L^{-1}) [20].
- This strain has the pathways necessary for rhamnolipid precursor biosynthesis [48].

Different promotors were evaluated, trying to optimize the rhamnolipids production in *P. putida*. There were some works that aimed to understand the relationship between the host regulation network and rhamnolipids production in this strain, and based on this information, they tried to direct the carbon flux to rhamnolipids production deleting genes involved in other metabolites [20, 21, 47, 49–54]. Table 3.1 compiles the highest productions of rhamnolipids using *Pseudomonas putida* as host for heterologous expression.

The maximum rhamnolipid heterologous production was 14.9 g L^{-1} [50], even though it was not as high as *P. aeruginosa* production, 39 g L^{-1} [11], this is a promising result that may provide a basis for other studies to reach higher biosurfactant concentrations using heterologous expression in *P. putida*.

Table 3.1 Compilation of strains and high amounts of rhamnolipids produced using heterologous expression on *P. putida* KT2440 in the selected works.

Paper	Strain	Rhamnolipids (g L^{-1})
Wittgens et al. [20]	*P. putida* KT42C1 pVLT31_rhlAB	1.5
Beuker et al. [50]	*P. putida* KT2440 pSynPro8oT_rhlAB	14.9
Tiso et al. [21]	*P. putida* KT2440 pSynPro8	3.2
Tiso et al. [47]	*P. putida* KT2440 with the rhlC containing vector	3.3
Tiso et al. [51]	*P. putida* KT2440 Δflag SK4	1.5
Anic et al. [52]	*P. putida* EM383	6.0
Noll et al. [54]	*P. putida* KT2440 rhlAB with native *P. aeruginosa* 5′-UTR region	1.2

3.3 Extension of Substrate Range for Biosurfactant Production

The accelerated industrial growth of sectors such as agriculture, livestock, and the production and processing of food has occurred to keep up with the rapid world population growth [55]. With the increase in industrialization, there has been a growing generation of agro-industrial waste. The use of these by-products as a carbon source for microorganisms helps to reduce environmental pollution and the biobased product costs, such as biosurfactants.

Lignocellulosic biomasses are considered to be an abundant resource of low-cost, renewable and organic carbon-rich raw materials. The main sources of vegetal biomass are agricultural waste, forestry, and urban residues. This substrate contains significant amounts of polysaccharides and lignin, which can be converted into products with high added value [56, 57]. Lignocellulose chemical constitution corresponds to 35–50% cellulose, 20–35% hemicellulose, and 15–20% lignin [58]. D-glucose molecules linked by β – (1,4) glycosidic and intramolecular hydrogen bonds form the rigid cellulose chain [59]. Hemicellulose heterogeneous biopolymer consists of different monosaccharides such as β-D-xylose, α-L-arabinose, β-D-mannose, β-D-glucose, α-D-galactose, with the greatest amount of xylose [60, 61]. Lignin is already a polyphenolic macromolecule that acts by hindering the degradation process of plant biomass [62]. For the full use of sugars derived from lignocellulose, the microorganism has to be able to metabolize both C6 sugars (cellulose) and C5 sugars (hemicellulose) as carbon sources [15].

Although the ability to degrade cellulose and hemicellulose is not present in many groups of microorganisms, engineered strains developed for this purpose, could overcome this hurdle. Through the tools of synthetic biology, gene encoding for transporters and/or enzymes can be introduced in the wild bacteria, making it able to convert different carbon sources, increasing the extension of substrate for biosurfactant production [15]. Henkel et al. and Meijnen et al. [63, 64] insert in *P. putida* S12 the genes xylA (xylose isomerase) and xylB (xylulose kinase) from *E. coli*. The engineered strains were able to efficiently consume xylose and L-arabinose as a substrate for growth. The authors came to the conclusion that xylA and xylB may show non-specific activity for L-arabinose.

The annual solid and/or effluent waste from dairy corresponds to around 4–11 million tonnes globally. The dairy industry has a significant impact on the environment, due to the use of large amounts of water and the emission of effluents into the environment [65]. The dairy effluent main component is whey, which contains around 4 – 5% of lactose [66]. This carbohydrate could be another alternative carbon source for biosurfactant production. With this purpose in mind, Koch et al. [66] introduced in *Pseudomonas aeruginosa* PAO1 and PG-201 the genes lacY (beta-galactoside permease) and lacZ (beta-galactosidase) from *E. coli*. The engineered strains were able to grow on lactose as a sole carbon source, as well as whey waste.

Bahia et al. [16] produced mono rhamnolipids, for the first time, from sucrose by recombinant GRAS *Saccharomyces cerevisiae* strains. These authors transferred to the yeast, six enzymes from *P. aeruginosa* (RmlA, RmlB, RmlC, RmlD, RhlA and RhlB genes) involved in mono-rhamnolipid biosynthesis. The sucrose phosphorylase gene from *Pelomonas saccharophila*, was inserted and the invertase gene *suc2* was disrupted to reduce the pathway's overall energy requirement. The strains displayed potential for rhamnose and mono rhamnolipid production. The approach presented by this work, may contribute to overcoming some common challenges of biosurfactant production, such as pathogenicity, complex regulation network, and high-cost substrates. Furthermore, the engineered yeast could be further developed to utilize C5 sugars for RL production from cellulosic residues [16].

Other approaches consist of the use of the most successful host for heterologous rhamnolipid production, *P. putida* KT2440, to explore some non-conventional carbon sources e.g. ethanol, pyrolysis oil, or alternative sugars like xylose and arabinose [42, 67–70].

3.4 Improvement of Overall Cellular Physiology

Rhamnolipids produced by *P. aeruginosa* are well studied biosurfactants, they have already been investigated for more than 70 years [71].

A phenotypical or genotypical alteration in the wild type strain may lead to an improvement in the rhamnolipid production, different strategies can be applied to improve the overall cellular physiology. First, a random mutation may be applied using transposons, exposure to mutagenic agent, such as UV, gamma ray, high concentrations of antibiotics, and others [66, 72–76].

It was demonstrated that the random mutation demands a lot of laboratory work for mutant screening, however, it can be very successfull. Lovaglio [74] related an improvement of 70% in the rhamnolipid production using the random mutant strain *P. aeruginosa* LBI 2A1.

The use of chemical mutagenic was described by Tahzibi et al. [73] it was reported that the N-methyl-N'-nitro-N-nitrosoguanidine generated a mutant strain able to produce 10 times more rhamnolipids than the wild type strain. Corroborating these results, mutants developed by gamma ray exposure, also demonstrated an improvement of 1.5 to 3 times the ratios of this biosurfactant synthesis [75,76].

Another strategy is the adaptive laboratory evolution, it is an approach that allows evolutionary phenomena to be analyzed in a controlled laboratory setting [77].

The changes in the cellular metabolism through directed evolution leads to an innovative approach to understanding metabolic pathways of biotechnological products [78]. The natural adaptation leads to mutations that are maintained in future generations through natural selection The mutations that are able to improve the organism performance and reproductive success of the organism in a given environment will be maintained through the subsequent generations [79].

Adaptive evolution has become a useful tool for biotechnological applications in industrial processes, due to the ability of the microorganism to survive in adverse conditions and quickly adapt to new environmental conditions [80]. The use of metabolic engineering strategies, such as adaptive evolution, allows for the optimization of yield and substrate bioconversion to the product of interest.

In their recent paper Bator et al. [69] performed an adaptive laboratory evolution to improve the growth ratio of *Pseudomonas putida* KT2440 in ethanol, for heterologous rhamnolipid production, the approach allowed a growth improve and also produced 5 g L^{-1} of biosurfactants using this carbon source.

Salazar-Bryam. [81] also used adaptive laboratory evolution, to adapt *P. aeruginosa* LBI 2A1, to use glycerol more efficiently as a carbon source, after 65 generations *P. aeruginosa* LBI 2A1-T16A produced 70% more than the wild type strain.

Both, random mutations and adaptive laboratory evolution approaches if used together with a genomic study could bring knowledge to base a rational direct mutation.

A third approach consists of direct mutations, Dobler et al. [82] overexpressed the *estA* gene, encoding gene for a membrane-bound esterase, into *Pseudomonas aeruginosa*. The mutant rhamnolipid production was up to 3.9 times higher than the wild type strain, the maximum concentration reached by modified bacteria was 14.62 g L^{-1}.

He et al. [83] showed that increasing the *rhlAB* genes copy number, efficiently enhanced rhamnolipid production by the recombinant *P. aeruginosa* DAB. Huang et al. [84] introduced *rhlAB* and *rhlC* genes into *Pseudomonas aeruginosa* wild-type strain, giving rise DNAB and DNC strains, respectively.

The engineered bacteria DNAB and DNC showed higher rhamnolipid yield than the DN1, proving that the overexpression of *rhlAB* and *rhlC* genes had a positive impact on rhamnolipid production. When compared to wild type, the yields obtained by the modified DNAB and DNAC strains were, respectively, 1.28 and 1.43 times higher [84].

3.5 Elimination or Reduction of By-product

Another strategy to overcome the challenges of rhamnolipid production, is related to the elimination or blocking of the competitive pathways, this approach drives the carbon flux to the biosurfactant synthesis.

The hydroxyalkanoyloxy-alkanoic acid (HAA) and dTDP-rhamnose are the precursors required for rhamnolipids biosynthesis. Some metabolic pathways, compete for the glycosyl and lipid precursors in *Pseudomonas aeruginosa*, such as exopolysaccharide (EPS) and polyhydroxyalkanoates (PHA) [85, 86].

With limitation in mind Lei et al. [87] constructed a strain disrupted for the synthesis of exopolysaccharide Psl and PHA. The constructed mutants, *P. aeruginosa* SG ΔpslAB and *P. aeruginosa* SG ΔphaC1DC2, were able to grow and presented an increase in rhamnolipid production of 21% and 25.3%, respectively. A double disrupted mutant, *P. aeruginosa* SG ΔpslAB ΔphaC1DC2, produced 69.7% more rhamnolipids than the wild type [88].

3.6 Future Perspectives

In this section more strategies are presented that can be applied to contribute to the effective establishment of biosurfactants in the market. The cost-effective production of biosurfactants mainly depends on the producing strain. As displayed, metabolic engineering, metabolic flux analysis, and synthetic biology approaches are tools employed to obtain non-pathogenic strains with high rhamnolipid productivity.

According to Chong et al. [88] besides the possibility of building hyperproducing strains, metabolic engineering can be applied for reducing strain pathogenicity, which are critical factors for economical and safe production of biosurfactants.

The reduction of the most important rhamnolipid producer pathogenicity can be achieved inhibiting pyocyanin production, this redox-active secondary metabolite is critical for *P. aeruginosa* infection [89]. First, pyocyanin synthesis can be prevented by supply restriction of precursors from shikimic acid biosynthetic pathway [88, 89]. A second strategy is to make pyocyanin harmless to human cells, through the methyl group removal by demethylase enzyme (PodA) [90]. Furthermore, this structural modification hinders biofilm formation, which is an important step in the infectious process of this bacterium. The demethylase PodA gene introduction into *P. aeruginosa* might increase the potential of this strain in industrial applications [88].

Another contribution that arises from the union of genetic and engineering approaches is the development of tailor-made biosurfactants. Changes in the molecular composition of the hydrophilic and hydrophobic moieties increase the bioproduct specificity, furthermore, it expands the areas where they can be applied [91].

Regarding rhamnolipids, heterologous expression in *E. coli* of RhlA variants, with specificity for 3-hydroxyfatty acids with differents chain lengths, resulted in a HAA composition alteration, when compared to the typical homologues produced by *Pseudomonas* (C_{10}-C_{10}) and *Burkholeria* (C_{14}-C_{14}) [91]. Manipulation of enzymes involved in the 3-hydroxyfatty acids incorporation into HAA and rhamnolipids, by single amino acid or whole domain exchanges, is yet another alternative for the hydrophobic moiety development of this molecule [92].

Increasing the biosurfactant's structural diversity could bring greater specificity and efficiency to the product. In addition, it can favor the expansion of activities where it is employed. The pharmaceutical industry, where many products are classified as being of low volume and high cost, characteristics similar to those encountered in the biosurfactant production, the high final cost of the product is justified, this is another aspect to be considered for these natural products.

Another point of view is to look at the cell as a biorefinery, from where different kinds of green products can be obtained. Instead of deleting genes related to by-products, a strategy for multiple biotechnological productions could be developed, even including foreign metabolite pathways through synthetic biology [93].

3.7 Conclusions

As described, the application of a combination of techniques such as metagenome, DNA sequencing, bioinformatics, genome editing, proteomics, metabolomics, and fluxomics will support the industrial production of biosurfactants, improving the strains already known or through screening of new pathways with greater production and structural diversification of known compounds, in addition to the discovery of new biosurfactants. The market establishment of this bioproduct is in line with the sustainable development goals (SDG), environmental preservation policies, and combating climate change, since they can be produced from renewable resources, including agro-industrial waste, are biodegradable and ecofriendly. In this way, this microbial production process contributes to the establishment of the circular economy, and has a positive socioeconomic impact.

References

1. Tiso T, Germer A, Küpper B, Wichmann R, Blank LM. Methods for recombinant rhamnolipid production. In: McGenity TJ, Timmis KN, Fernández BN, editors. *Hydrocarbon and Lipid Microbiology Protocols*. Springer-Verlag Berlin Heidelberg; 2015. pp. 65–94.
2. Liu Y, Shin H, Li J, Liu L. Toward metabolic engineering in the context of system biology and synthetic biology: advances and prospects. *Appl Microbiol Biotechnol*. 2015;99:1109–1118.
3. Chen Y, Nielsen J. Advances in metabolic pathway and strain engineering paving the way for sustainable production of chemical building blocks. *Curr Opin Biotechnol*. 2013;24:965–972.
4. Paddon CJ, Keasling JD. Semi-synthetic artemisinin: a model for the use of synthetic biology in pharmaceutical development. *Nat Rev Microbiol*. 2014;12:355–367.
5. Stephanopoulos G. Synthetic biology and metabolic engineering. *ACS Synth Biol*. 2012;1(11):514–525.
6. Nielsen J. Metabolic engineering. *Appl Microbiol Biotechnol*. 2001;55:263–283.
7. Stephanopoulos GN, Aristidou AA, Nielsen J. *Metabolic Engineering Principles and Methodologies*. San Diego: Academic Press, 1998.
8. Kuepfer L. Stoichiometric modelling of microbial metabolism. In: Krömer JO, Nielsen LK, Blank LM, editors. *Metbolic Flux Analysis*. Humana Press, New York; 2014. pp. 3–18.
9. Gombert AK, Nielsen J. Mathematical modeling of metabolism. *Curr Opin Biotechnol*. 2000;11:180–186.
10. Stephanopoulos GN, Vallino JJ. Network rigidity and metabolic engineering in metabolite overproduction. *Science*. 1991;252:1675–1681.
11. Müller MM, Hausmann R. Regulatory and metabolic network of rhamnolipids biosynthesis: traditional and advanced engineering towards biotechnological production. *Appl Microbiol Biotechnol*. 2011;91:251–264.
12. Kuhn D, Blank LM, Schmid A, Buhler B. Systems biotechnology—rational whole-cell biocatalyst and bioprocess design. *Eng Life Sci*. 2010;10:384–397.
13. Woodley JM. New opportunities for biocatalysis: making pharmaceutical processes greener. *Trends Biotechnol*. 2008;26:321–327.
14. Alam K, Hao J, Zhang Y, Li A. Synthetic biology-inspired strategies and tools for engineering of microbial natural product biosynthetic pathways. *Biotechnol Adv*. 2021;49:1–20.
15. Henkel M, Müller MM, Kügler JH, Lovaglio RB, Contiero J, Syldatk C, Hausmann R. Rhamnolipids as biosurfactants from renewable resources: concepts for next-generation rhamnolipid production. *Process Biochem*. 2012;47:1207–1219.
16. Bahia FM, Almeida GC, Andrade LP1, Campos CG, Queiroz LR, da Silva RLV, Abdelnur PV, Corrêa R, Bettiga M, Parachin NS. Rhamnolipids production from sucrose by engineered *Saccharomyces cerevisiae*. *Sci Rep*. 2018;8:2905–2915.
17. Toribio J, Escalante AE, Soberón-Chávez G. Rhamnolipids: production in bacteria other than *Pseudomonas aeruginosa*. *Eur J Lipid Sci Technol*. 2010;112:1082–1087.

18 Dekimpe V, Deziel E. Revisiting the quorum-sensing hierarchy in *Pseudomonas aeruginosa*: the transcriptional regulator RhlR regulates LasR-specific factors. *Microbiol.* 2009;155:712–723.
19 Cabrera-Valladares N, Richardson AP, Olvera C, Treviño LG, Déziel E, Lépine F, et al. Monorhamnolipids and 3-(3- hydroxyalkanoyloxy)alkanoic acids (HAAs) production using *Escherichia coli* as a heterologous host. *Appl Microbiol Biotechnol.* 2006;73:187–194.
20 Wittgens A, Tiso T, Arndt TT, Wenk P, Hemmerich J, Müller C, et al. Growth independent rhamnolipid production from glucose using the non-pathogenic *Pseudomonas putida* KT2440. *Microb Cell Fact.* 2011;10:80.
21 Tiso T, Sabelhaus A, Behrens B, Wittgens A, Rosenau F, Hayen H, et al. Creating metabolic demand as an engineering strategy in *Pseudomonas putida* – rhamnolipid synthesis as an example. *Metab Eng Commun.* 2016;3:234–244.
22 Desai JD, Banat IM. Microbial production of surfactants and their commercial potential. *Microbiol Mol Biol.* 1997;61:47–64.
23 Burger MM, Glaser L, Burton RM. The enzymatic synthesis of a rhamnose containing glycolipid by extracts of Pseudomonas aeruginosa. *J Biol Chem.* 1963;238:2595–2602.
24 Ochsner UA, Koch AK, Fiechter A, Reiser JT. Isolation and characterization of aregulatory gene affecting rhamnolipid biosurfactant synthesis in *Pseudomonas aeruginosa*. *J Bacteriol.* 1994a;176:2044–2054.
25 Ochsner UA, Fiechter A, Reiser J. Isolation, characterization, and expression in Escherichia coli of the *Pseudomonas aeruginosa* rhlAB genes encoding a rhamnosyltransferase involved in rhamnolipid biosurfactant synthesis. *J Biol Chem.* 1994b;269(31):19787–19795.
26 Ochsner UA, Reiser J, Fiechter A, Witholt B. Production of *Pseudomonas aeruginosa* rhamnolipid biosurfactants in heterologous hosts. *Appl Environ Microbiol.* 1995;61:3503–3506.
27 Déziel E, Lépine F, Milot S, Villemur R. rhlA is required for the production of a novel biosurfactant promoting swarming motility in *Pseudomonas aeruginosa*: 3-(3-hydroxyalkanoyloxy)alkanoic acids (HAAs), the precursors of rhamnolipids. *Microbiology.* 2003;149:2005–2013.
28 Zhu K, Rock CO. RhlA converts β-hydroxyacyl-acyl carrier protein intermediates in fatty acid synthesis to the β-hydroxydecanoyl-β-hydroxydecanoate component of rhamnolipids in *Pseudomonas aeruginosa*. *J Bacteriol.* 2008;190(9):3147–3154.
29 Olvera C, Goldberg JB, Sanchez R, Soberón-Chávez G. The *Pseudomonas aeruginosa* algC gene product participates in rhamnolipid biosynthesis. *FEMS Microbiol Lett.* 1999;179:85–90.
30 Robertson BD, Frosch M, van Putten JP. The identification of cryptic rhamnose biosynthesis genes in *Neisseria gonorrhoeae* and their relationship to lipopolysaccharide biosynthesis. *J Bacteriol.* 1994;176:6915–6920.
31 Latifi A, Foglino M, Tanaka K, Williams P, Lazdunski A. A hierarchical quorum-sensing cascade in *Pseudomonas aeruginosa* links the transcriptional activators LasR and RhlR (VsmR) to expression of the stationary-phase sigma factor RpoS. *Mol Microbiol.* 1996;21:1137–1146.
32 Fuqua WC, Greenberg EP. Self perception in bacteria: molecular mechanisms of stimulus-response coupling. *Curr Opin Microbiol.* 1998;1:183–189.

33 Salmond GPC, Bycroft BW, Stewart GSAB, Willams P. The bacterial enigma: cracking the code of cell-cell communication. *Mol Microbiol Salem.* 1995;16:615–624.
34 Williams P. Quorum sensing, communication and cross-kingdom signaling in the bacterial world. *Microbiol.* 2007;153:3923–3938.
35 Wagner VE, Filiatrault MJ, Picardo KF, Iglewski BH. *Pseudomonas aeruginosa* virulence and pathogenesis issues. In: Cornelis P, editors. *Pseudomonas Genomics and Molecular Biology.* caister academic press, norfolk; 2008. pp. 129–158.
36 Juhas M, Eberl L, Tümmler B. Quorum-sensing: the power of cooperation in the world of pseudomonas. *Environ Microbiol.* 2005;7:459–471.
37 Wang Q, Fang X, Bai B, Liang X, Shuler PJ, Goddard WA III, et al. Engineering bacteria for production of rhamnolipid as an agent for enhanced oil recovery. *Biotechnol Bioeng.* 2007;98:842–853.
38 Han L, Liu P, Peng Y, Lin J, Wang Q, Ma Y. Engineering the biosynthesis of novel rhamnolipids in *Escherichia coli* for enhanced oil recovery. *J Appl Microbiol.* 2014;117:139–150.
39 Jafari A, Raheb J, Bardania H, Rasekh B. Isolation, cloning and expression of rhamnolipid operon from *Pseudomonas aeroginosa* ATCC 9027 in logarithmic phase in *E. coli* BL21. *Am J Life Sci.* 2014;2:22–30.
40 Du J, Zhang A, Hao J, Wang J. Biosynthesis of di-rhamnolipids and variations of congeners composition in genetically-engineered *Escherichia coli. Biotechnol Lett.* 2017;39:1041–1048.
41 Kryachko Y, Nathoo S, Lai P, Voordouw J, Prenner EJ, Voordouw G. Prospects for using native and recombinant rhamnolipid producers for microbially enhanced oil recovery. *Int Biodeter Biodegr.* 2013;81:133–140.
42 Horlamus F, Wang Y, Steinbach D, Vahidinasab M, Wittgens A, Rosenau F, et al. Potential of biotechnological conversion of lignocellulose hydrolyzates by *Pseudomonas putida* KT2440 as a model organism for a bio-based economy. *Glob Change Biol Bioenergy.* 2019;11:12.
43 Solaiman DKY, Ashby RD, Gunther NW IV, Zerkowski JA. Dirhamnose-lipid production by recombinant nonpathogenic bacterium *Pseudomonas chlororaphis. Appl Microbiol Biotechnol.* 2015;99:4333–4342.
44 Zhao F, Shi R, Zhao J, Li G, Bai X, Han S, et al. Heterologous production of *Pseudomonas aeruginosa* rhamnolipid under anaerobic conditions for microbial enhanced oil recovery. *J Appl Microbiol.* 2015;118:379–389.
45 Cha M, Lee N, Kim M, Kim M, Lee S. Heterologous production of *Pseudomonas aeruginosa* EMS1 biosurfactant in Pseudomonas putida. *Bioresour Technol.* 2008;99:2192–2199.
46 Tiso T, Zauter R, Tulke H, Leuchtle B, Li WJ, Behrens B, et al. Designer rhamnolipids by reduction of congener diversity: production and characterization. *Microb Cell Fact.* 2017;16:225.
47 Timmis KN. *Pseudomonas putida*: a cosmopolitan opportunist par excellence. *Environ Microbiol.* 2002;4:779–781.
48 Abdel-Mawgoud AM, Lépine F, Déziel E. A stereospecific pathway diverts b-oxidation intermediates to the biosynthesis of rhamnolipid biosurfactants. *Chem Biol.* 2014;21:156–164.
49 Behrens B, Engelen J, Tiso T, Blank LM, Hayen H. Characterization of rhamnolipids by liquid chromatography/mass spectrometry after solid-phase extraction. *Anal Bioanal Chem* 2016;408:2505–2514.

50 Beuker J, Barth T, Steier A, Wittgens A, Rosenau F, Henkel M, et al. High titer heterologous rhamnolipid production. *AMB Express.* 2016a;6:124.

51 Tiso T, Ihling N, Kubicki S, Biselli A, Schonhoff A, Bator I, et al. Integration of genetic and process engineering for optimized rhamnolipid production using Pseudomonas putida. *Front Bioeng Biotechnol.* 2020;8:976.

52 Anic I, Nath A, Franco P, Wichmann R. Foam adsorption as an ex situ capture step for surfactants produced by fermentation. *J Biotechnol.* 2017;258:181–189.

53 Anic I, Apolonia I, Franco P, Wichmann R. Production of rhamnolipids by integrated foam adsorption in a bioreactor system. *AMB Express.* 2018;8:122.

54 Noll P, Treinen C, Müller S, Senkalla S, Lilge L, Hausmann R, et al. Evaluating temperature-induced regulation of a ROSE-like RNA-thermometer for heterologous rhamnolipid production in *Pseudomonas putida* KT2440. *AMB Express.* 2019;9:154.

55 Usmani Z, Sharma M, Gaffey J, Sharma M, Dewhurst RJ, Moreau B, Newbold J, Clark W, Thakur VK, Gupta VK. Valorization of dairy waste and by-products through microbial bioprocesses. *Biores Technol.* 2022;346:1–10.

56 Kawaguchi H. Bioprocessing of bio-based chemicals produced from lignocellulosic feedstocks. *Curr Opin Biotechnol.* 2016;42:30–39.

57 Vu HP, Nguyen LN, Vu MT, Johir MAH, McLaughlan R, Nghiem LD. A comprehensive review on the framework to valorise lignocellulosic biomass as biorefinery feedstocks. *Sci Total Environ.* 2020;743:1–16.

58 Mood SH, Golfeshan AH, Tabatabaei M, Jouzani GS, Najafi GH, Gholami M, Ardjmand M. Lignocellulosic biomass to bioethanol, a comprehensive review with a focus on pretreatment. *Renew Sust Energ Rev.* 2013;27:77–93.

59 Ten E, Vermerris W. Functionalized polymers from lignocellulosic biomass: state of the art. *Polymers.* 2013;5(2):600–642.

60 Gírio FM, Fonseca C, Carvalheiro F, Duarte LC, Marques S, Bogel-Lukasik R. Hemicelluloses for fuel ethanol: a review. *Biores Technol.* 2010;101(13):4775–4800.

61 Khaire KC, Moholkar VS, Goyal A. Bioconversion of sugarcane tops to bioethanol and other value added products: an overview. *Mat Sci Energy Technol.* 2021;4:54–68.

62 Agbor VB, Cicek N, Sparling R, Berlin A, LevinDB. Biomass pretreatment: fundamentals toward application. *Biotechnol Adv.* 2011;29:675–685.

63 Meijnen JP, de Winde JH, Ruijssenaars HJ. Engineering *Pseudomonas putida* S12 for efficient utilization of D-xylose and L-arabinose. *Appl Environ Microbiol.* 2008;74:5031–5037.

64 Meijnen JP, de Winde JH, Ruijssenaars HJ. Establishment of oxidative D-xylose metabolism in *Pseudomonas putida* S12. *Appl Environ Microbiol.* 2009;75:2784–2789.

65 Slavov KA. General characteristics and treatment possibilities of dairy wastewater–A review. *Food Technol Biotechnol.* 2017;55:14–28.

66 Koch AK, Reiser J, Kappeli O, Fiechter A. Genetic construction of lactoseutilizing strains of Pseudomonas aeruginosa and their application in biosurfactant production. *Bio-Technol.* 1988;6:1335–1339.

67 Arnold S, Henkel M, Wanger J, Wittgens A, Rosenau F, Hausmann R. Heterologous rhamnolipid biosynthesis by P. putida KT2440 on biooil derived small organic acids and fractions. *AMB Express.* 2019;9:80.

68 Wang Y, Horlamus F, Henkel M, Kovacic F, Schläfle S, Hausmann R, et al. Growth of engineered Pseudomonas putida KT2440 on glucose, xylose, and arabinose: hemicellulose

hydrolysates and their major sugars as sustainable carbon sources. *Glob Change Biol Bioenergy.* 2019;11:249–259.
69. Bator I, Karmainski T, Tiso T, Blank LM. Killing two birds with one stone – strain engineering facilitates the development of a unique rhamnolipid production process. *Front Bioeng Biotechnol.* 2020a;8:899.
70. Bator I, Wittgens A, Rosenau F, Tiso T, Blank LM. Comparison of three xylose pathways in Pseudomonas putida KT2440 for the synthesis of valuable products. *Front Bioeng Biotechnol.* 2020b;7:480.
71. Jarvis FG, Johson MJ. A glycolipid produced by Pseudomonas aeruginosa. *J Am Chem Soc.* 1949;71:4124–4126.
72. Mulligan CN, Gibbs BF. Correlation of nitrogen metabolism with biosurfactant production by *Pseudomonas aeruginosa*. *Appl Environ Microbiol*, Washington. 1989;55:3016–3019.
73. Tahzibi A, KamaL F, Assadi MM. Improved production o rhamnolipids by a *Pseudomonas aeruginosa* mutant. *Iran Biomed J.* 2004;8:23–31.
74. Lovaglio RB. Produção de ramnolipídios por mutantes de *Pseudomonas aeruginosa* LBI Tese de Doutorado, depositada na biblioteca virtual da UNESP. 2011.
75. Iqbal S, Khalid Z, Malik K. Enhanced biodegradation and emulsification of crude oil and hyperproduction of biosurfactants by a gamma ray-induced mutant of *Pseudomonas aeruginosa*. *Lett Appl Microbiol.* 1995;21:176–179.
76. Housseiny EL, Ghadir S, et al. Isolation, screening and improvement of rhamnolipid production by *Pseudomonas aeruginosa*. *Indian J Biotechnol.* 2017;16:611–619.
77. Dragosits M, Mattanovich D. Adaptive laboratory evolution–principles and applications for biotechnology. *Microb Cell Fact.* 2013;12:64.
78. Chatterjee R, Yuan L. Directed evolution of metabolic pathways. *Trends Biotechnol.* 2006;24:28–38.
79. Elena SF, Lenski RE. Evolution experiments with microorganisms: the dynamics and genetic bases of adaptation. *Nat Rev Genet.* 2003;4:457–469.
80. Portnoy VA, Bezdan D, Zengler K. Adaptive laboratory evolutionharnessing the power of biology for metabolic engineering. *Curr Opin Biotechnol.* 2011;22:590–594.
81. Salazar-Bryam AM. Estrategias up-stream para a otimização da produção de ramnolipídios a partir de fontes alternativas de carbono: um enfoque na linhagem. Tese de Doutorado, depositada na biblioteca virtual da UNESP, 2021.
82. Dobler L, de Carvalho BR, Alves WDS, Neves BC, Freire DMG. Almeida RV Enhanced rhamnolipid production by *Pseudomonas aeruginosa* overexpressing estA in a simple medium. *PLoS ONE.* 2017;12(8).
83. He C, Dong W, Li J, Li Y, Huang C, Ma Y. Characterization of rhamnolipid biosurfactants produced by recombinant Pseudomonas aeruginosa strain DAB with removal of crude oil. *Biotechnol Lett.* 2017;39(9):1381–1388.
84. Huang C, Li Y, Tian Y, Hao Z, Chen F, Ma Y. Enhanced rhamnolipid production of Pseudomonas aeruginosa DN1 by metabolic engineering under diverse nutritional factors. *J Pet Environ Biotechnol.* 2018;9:5.
85. Soberón-Chavez G, Aguirre-Ramirez M, Sanchez R. The *Pseudomonas aeruginosa* Rhla enzyme is involved in rhamnolipid and polyhydroxyalkanoate production. *J Ind Microbiol Biotechnol*, Hampshire. 2005;32:675–677.
86. Abdel-mawgoud MA, hausmann R, lépine F, müller MM, déziel E. Rhamnolipids: detection, analysis, biosynthesis, genetic regulation, and bioengineering of production.

In: Soberónchávez G, editors. *Biosurfactans*, Microbiology Monographs, Springer, Berlin; 2011. pp. 13–44.

87 Lei L, Zhao F, Han S, Zhang Y. Enhanced rhamnolipids production in *Pseudomonas aeruginosa* SG by selectively blocking metabolic bypasses of glycosyl and fatty acid precursors. *Biotechnol Lett*. 2020;42:997–1002.

88 Chong H, Li Q. Microbial production of rhamnolipids: opportunities, challenges and strategies. *Microb Cell Fact*. 2017;16:137–149.

89 Lau GW, Hassett DJ, Ran H, Kong F. The role of pyocyanin in Pseudomonas aeruginosa infection. *Trends Mol Med*. 2004;10:599–606.

90 Costa KC, Glasser NR, Conway SJ, Newman DK. Pyocyanin degradation by a tautomerizing demethylase inhibits *Pseudomonas aeruginosa* bioflms. *Science*. 2017;355:170–173.

91 Wittgens A, Rosenau F. Heterologous rhamnolipid biosynthesis: advantages, challenges, and the opportunity to produce tailor-made rhamnolipids. *Front Bioeng Biotechnol*. 2020;8:1–11.

92 Germer A, Tiso T, Müller C, Behrens B, Vosse C, Scholz K, Froning M, Hayen H, Blank LM. Exploiting the natural diversity of RhlA acyltransferases for the synthesis of the rhamnolipid precursor 3-(3-Hydroxyalkanoyloxy)alkanoic acid. *Appl Environ Microbiol*. 2020;86(6):e02317-19.

93 Lovaglio RB, Silva VL, Ferreira H, Hausmann R, Contiero J. Rhamnolipids know-how: looking for strategies for its industrial dissemination. *Biotechnol Adv*. 2015;33:1715–1726.

4

Biosurfactant Production in the Context of Biorefineries

Paulo Ricardo Franco Marcelino[1],, Carlos Augusto Ramos[1], Maria Teresa Ramos[1], Renan Murbach Pereira[1], Rafael Rodrigues Philippini[2], Emily Emy Matsumura[3], and Silvio Silvério da Silva[1]*

[1] Laboratory of Bioprocesses and Sustainable Products, Department of Biotechnology, Engineering School of Lorena, University of São Paulo (EEL/USP), Lorena 12602–810, SP, Brazil
[2] Bosque Foods GmbH, 10435 Berlin, Germany
[3] Laboratory of Virology, Department of Plant Sciences, Wageningen University and Research, Droevendaalsesteeg 1, Building 107, 6708 PB Wageningen, Netherlands
* Corresponding author

4.1 Biorefineries in Contemporary Society

In contemporary society, still dependent on the so-called petroeconomics, the concept of an oil refinery is of fundamental importance. A refinery can be defined as the full use of abundant and low-cost raw material, in this case oil, using physical and/or chemical treatments to produce commercial products with higher added value. It can also be added that a refinery operates through a succession of different stages, starting with fractionation operations, aiming at the separation of different substances. Subsequent to the fractionation, the conversion steps of the substances obtaining products for immediate use or to be made available as raw material for future transformations by the chemical industry will take place. Among the main products of oil refineries are gasoline, diesel oil, liquefied gas, kerosene for aviation, common kerosene, solvents, lubricants, petroleum coke, paraffins, resins, and polymers. Generally these refinery products are of low biodegradability, toxic, and have low compatibility [1].

In recent years, due to the various environmental, economic, and social problems that the oil-dependent society has been facing, the intensification of the call for sustainable development has been highlighted. Several organizations and world powers have sought the concept of bioeconomy and circular economy, based on sustainable processes and products. With this, the advent of the concept of biorefineries was also observed, based on principles of sustainability and green chemistry, whose main raw material in these green refineries is biomass.

Biorefineries can be defined as industrial facilities that integrate processes for the sustainable conversion of biomass (organic materials of starchy, lignocellulosic, oleaginous origin, and others) into value-added products [2–4]. These sustainable industries integrate several conversion routes – biochemical, microbial, chemical, and thermochemical – in the search for the best use of biomass and the energy it contains. The objective of biorefineries is to optimize the use of resources, minimize effluents, and obtain ecofriendly products,

Biosurfactants and Sustainability: From Biorefineries Production to Versatile Applications, First Edition.
Edited by Paulo Ricardo Franco Marcelino, Silvio Silverio da Silva, and Antonio Ortiz Lopez.
© 2023 John Wiley & Sons Ltd. Published 2023 by John Wiley & Sons Ltd.

maximizing benefits and profit [4]. In the literature, there are several classifications for biorefineries based on: (i) type of raw material (starch, lignocellulosic, oleaginous, algal, and others), (ii) type of technology used in the conversion of biomass (mechanical/physical, biochemical, or thermochemical), (iii) platform technology status (conventional, advanced or first and second generation biorefineries), (iv) main product (bioethanol, biodiesel, etc.) and (v) intermediate production (synthesis gas, sugar, or lignin) [5]. Among the classifications presented, the most common, in the technical literature and in the industrial environment, are related to the type of raw material used and the type of technology used in the conversion of biomass.

4.2 Biomass and Biorefineries: Industrial By-products as Raw Materials for Biorefineries

In recent decades, with the advancement of sustainable research focused on bioprocesses (fermentations and biocatalysis), several industrial sectors, mainly the food, chemical, and pharmaceutical sectors, prioritized the use of materials previously considered industrial by-products, as raw materials for obtaining products with high added value and ecofriendly characteristics, the so-called bio-based products or sustainable products. The adoption of this type of raw material stands out not only for its environmental appeal, but also for its very low cost, enabling a good cost/benefit ratio when compared to those obtained by traditional chemical syntheses, which use expensive raw materials, many of them derivatives of petrochemicals [6].

In addition to adding value to by-products, the following are highlighted as advantages of bioprocesses in relation to traditional chemical syntheses: (i) the use of mild process conditions, as the reactions are carried out by microorganisms and enzymes at a temperature close to room temperature, which entails lower energy expenditure; (ii) pH of the culture or reaction medium close to neutrality; (iii) the low load of waste generated during the process, in addition to having low or zero toxicity, causing no impact on the environment; (iv) regio and stereoselectivity of reactions, mainly in biocatalytic processes, resulting in the production of enantiomerically pure compounds, not requiring the use of expensive resolution (separation) methods, and (v) the possibility of carrying out biological reactions in the solid medium (solid state fermentations or biocatalysis) [7].

The use of by-products in biotechnological processes is not new to society. Although bioprocesses using agro-industrial by-products have only now received greater emphasis, they have already been used, mainly in solid-state fermentation (SSF), since the seventeenth century. It is known that in France, producers of *Agaricus bisporus (Champignon)* already cultivated these mushrooms on the "beds" used for storing fruits, made up of different types of straw (rice, wheat, oats) mixed with animal manure (poultry, cattle, horses) [8]. In 1896, Takamine produced a digestive enzyme, takadiastase, in solid-state fermentation by *Aspergillus oryzae* on wheat bran. Between 1900 and 1920, the use of by-products as substrates in the production of microbial enzymes and kojic acid by SSF was observed. In the second half of the twentieth century and the beginning of the twentyfirst century, the production of other microbial metabolites using agricultural and agro-industrial residues as raw material was improved and several other products were obtained, such as: protein-enriched foods for animal and human consumption, antibiotics, alkaloids, organic acids, polyols, biopesticides, and others [9].

In addition to by-products of lignocellulosic and starchy origin, oilseeds, such as residual oils and fats, also stand out as sources of nutrients in fermentation processes. They usually come from the food industry and are widely used in the production of lipases by microorganisms, mainly by yeasts and filamentous fungi [10].

Residual glycerol from biodiesel synthesis has also come to the fore in recent years as a potential raw material for industry. According to Mota et al. [11], glycerol from biodiesel production can be used in the production of acetals, ethers, esters, acrolein, acrylic acid, glyceraldehyde, glyceric acid, and glycerin carbonate. On the other hand, application of residual glycerol in biotechnological processes, according to the works of Rivaldi et al. [12] and Mattam et al. [13], can be observed in the production of 1,3 – propanediol, organic acids, ethanol, polyhydroxyalkanoates (PHA), omega 3 polyunsaturated acid, enzymes, antibiotics, microbial biomass, and pigments.

It should also be noted that the by-products rich in nitrogenous compounds are extremely important for bioprocesses. The low-cost nitrogen sources commonly used in fermentation processes are: cornstarch, corn gluten meal, peanut meal, soybean meal, and dairy residues [14]. Soybean and rice bran have been studied and applied in several bioprocesses, such as enzyme production, microbial biomass, bioflavors, second-generation ethanol, and xylitol [15, 16]. The mineral composition of these grains, in addition to supplementing crops with organic nitrogen, makes these extracts promising sources of micronutrients and growth factors. Both in soybean and rice grains, as well as in their extracts, vitamins (niacin, riboflavin, pantothenic acid, and thiamine), and amino acids (arginine, cysteine, glycine, histidine, isoleucine, leucine, lysine, methionine, phenylalanine, threonine, tryptophan, tyrosine, and valine) [17]. Thus, these soybean and rice bran extracts can serve as low-cost alternatives to replace enriched yeast extracts in various bioprocesses. Making a comparison, the price of yeast extract can reach U$ 216.00 per kg, while soy bran costs around U$ 0.47 per kg and rice R$ 0.65 per kg [18, 19].

In addition to the bioproducts mentioned above, biosurfactants are also viable product alternatives to be produced in biorefineries. As previously reported in the introductory chapter of this book, biosurfactants are metabolites of animal, vegetable, or microbial origin, with an amphipathic structure, with outstanding physical-chemical and biological properties, and can be used in various industrial sectors. In addition to the versatility of applications, biosurfactants are considered ecofriendly products and may be possible substitutes for synthetic surfactants [20].

The ecofriendly characteristics of biosurfactants are due to their high biodegradability, low toxicity, biocompatibility and the possibility of production from agro-industrial by-products, making these bioproducts alternatives for biorefineries [20–22].

As can be seen in the topics below, there are studies in the literature that report the production of biosurfactants using oleaginous biomass (rich in lipids), starchy and lignocellulosic (rich in sugars), and residual glycerol from biodiesel as raw materials.

4.3 Biosurfactant Production in the Context of Lignocellulosic Biorefineries

The valorization of residual vegetable biomass in bioprocesses, mainly those of lignocellulosic origin, is due to the fact that they are materials rich in carbohydrates. In the composition of these materials 20–50% of cellulose was found, a linear homopolymer composed of glucose and 15–35% of hemicellulose, a heteropolymer composed mainly of xylose and arabinose. In addition to the glycidic portion, lignin is also found (10–30%), a polyphenolic macromolecule [23–25]. These compounds are commonly found in plant cell walls, arranged so that the cellulose is "cemented" by hemicellulose and lignin (Figure 4.1).

The cellulose, hemicellulose and lignin contents vary according to the parts, ages, and plant species (Table 4.1). The rest of the biomass consists, in a minority, of substances such as proteins, pectin, oils, minerals, terpenes, alkaloids, and various pigments. Due to the rich organic composition, lignocellulosic materials are used in bioprocesses mainly as sources of carbon, nitrogen, sulfur, micronutrients and growth factors of fundamental importance for microbial nutrition [26]. It should be noted that among the lignocellulosic biomasses, the most frequently used in studies to obtain bioproducts has been sugarcane bagasse and straw, mainly with a view to obtaining second-generation ethanol.

The production of biosurfactants using lignocellulosic by-products is already a reality. Various studies in the last decade were carried out using both cellulosic and hemicellulosic fractions, as can be seen in the following sections.

The use of lignocellulosic biomass as a raw material to obtain biosurfactants, or any other bioproducts, requires a pre-treatment step. During the pre-treatment, the recalcitrant fraction, in this case lignin, will be removed, and then the hydrolysis of the polysaccharides into fermentable sugars (glucose and xylose) will be carried out. (Figure 4.2).

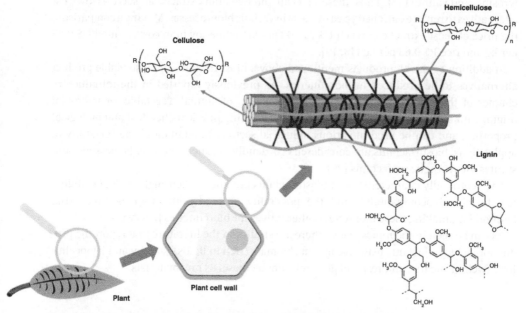

Figure 4.1 Structure and chemical composition of lignocellulosic biomass.

4.3 Biosurfactant Production in the Context of Lignocellulosic Biorefineries

Table 4.1 Chemical composition of some lignocellulosic biomasses.

Lignocellulosic biomass	Cellulose (%)	Hemicellulose (%)	Lignin (%)
Sugarcane straw	40–44	30–32	22–25
Sugarcane bagasse	32–48	19–24	23–32
Hardwood	43–47	25–35	16–24
Softwood	40–44	25–29	25–31
Corn stalk	35	25	35
Corn cob	45	35	15
Cotton	95	2	0,3
Wheat straw	30	50	15
Sisal	73.1	14.2	11
Rice straw	43.3	26.4	16.3
Corn fodder	38–40	28	7–21
Coconut fiber	36–43	0.15–0.25	41–45
Banana fiber	60–65	6–8	5–0
Barley straw	31–45	27–8	14–9

Source: [27].

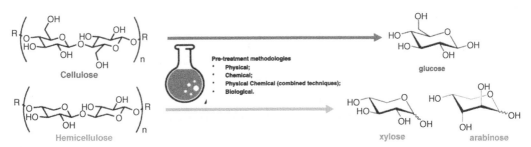

Figure 4.2 Fermentable sugars obtained in the pre-treatment step of lignocellulosic biomass.

In the literature, several pre-treatment methods used to obtain cellulosic or hemicellulosic hydrolysates can be found. However, it should be noted that the adoption of a particular technique will depend on the physical and chemical properties of the lignocellulosic material used as raw materials [28]. In addition, the chosen technique must be capable of producing high yields of pentose sugars, not releasing significant amounts of compounds that act as inhibitors of microbial metabolism (furans and phenols) (Figure 4.3), not producing high levels of residues, being simple and cost-effective [29, 30].

The pre-treatment methods currently used are divided into physical, chemical, physical–chemical and biological processes, as can be seen in Table 4.2.

In studies on the production of biosurfactants using lignocellulosic biomass, one can observe the use of techniques for obtaining the hemicellulosic and cellulosic fraction already used in obtaining other bioproducts, such as second-generation ethanol and xylitol, for example.

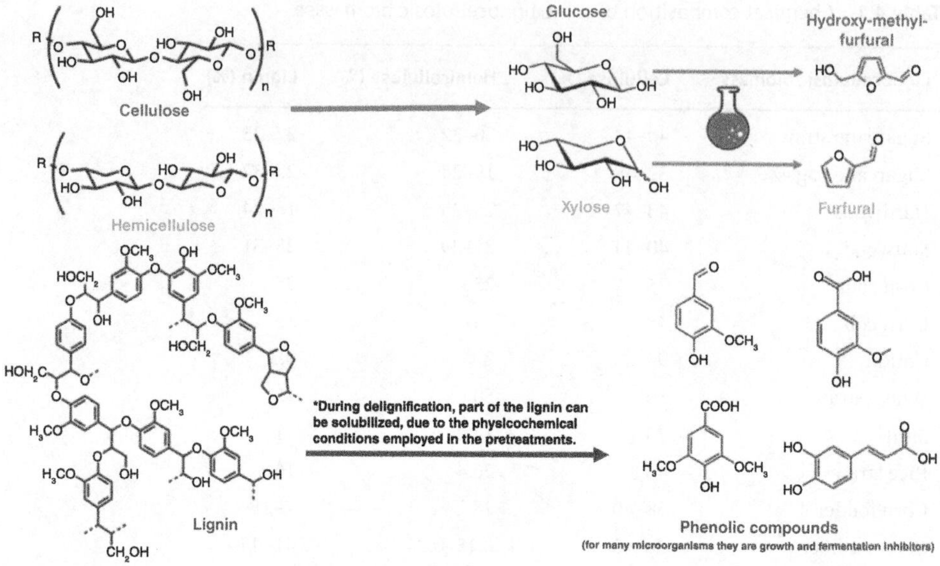

Figure 4.3 Main inhibitor compounds formed during pre-treatment of lignocellulosic biomass.

Table 4.2 Main types of pre-treatments and techniques used in the lignocellulosic biomasses.

Type of pre-treatment	Technique
Physical	Grinding and crushing
	Ultrasonication
	Centrifugal grinding
	Extrusion
Chemical	Pre-treatment with diluted mineral acid
	Pre-treatment with acid-acetone mixture
	Pre-treatment with ionic liquid
	Alkaline pre-treatment with potassium permanganate
	Organosolv
	Pre-treatment with metal chlorides
Physical–chemical	Steam explosion
	Hot water pre-treatment
	Wet oxidation
	Explosion with super critical carbon dioxide
	IHRW pre-treatment
	Plasma
	AFEX
Biological	Microbial consortium
	Pre-treatment with fungal species
	Enzymatic pre-treatment

Source: [30, 31].

Portilla-Rivera et al. [32] and Portilla-Rivera et al. [33] reported the production of biosurfactants by a *Lactobacillus pentosus* strain, using the hemicellulose hydrolysate of grape pomace as the raw material. The characterization of grape pomace showed that this lignocellulosic material has a low sugar content (10.8% cellulose and 11.2% hemicellulose), while the lignin and extractive content are respectively 50.9% and 14.7%, considered high. The pre-treatment used for delignification of the material and the hydrolysis of polysaccharides to obtain fermentable sugars, in this case xylose, was of the acid type. In this study, dilute sulfuric acid concentrations (1–5% mV^{-1}) were evaluated at time intervals of 0–180 minutes, as little was known about this biomass. The hydrolytic process was optimized using an acid concentration of 3.3%, at a temperature of 130 °C for 125 minutes. The use of this hemicellulosic hydrolysate as a carbon source in the culture medium showed that after 18 hours of fermentation, 4.8 mg L^{-1} of intracellular surfactin was obtained. This data represents a yield of 0.60 mg of intracellular biosurfactant per g of consumed sugars. It is noteworthy that despite a low yield, the surfactin produced by *L. pentosus* under the studied conditions presented a critical micellar concentration of 2.9 g L^{-1}.

One of the probable problems for the low yield in the process reported by Portilla-Rivera et al. [32] is that the lignocellulosic material used had a marked chemical composition of compounds that can generate inhibitors for the fermentation process (lignin and extractives). Furthermore, the conditions adopted for fermentation also contributed to the production of lactic acid, competing with the production of the biosurfactant. Furthermore, the surfactin obtained was intracellular, and the process of extracting an intracellular product may result in lower yields due to several factors intrinsic to the process. It was observed that the severity of the parameters adopted in this pre-treatment also resulted in obtaining significant concentrations of inhibitor compounds (acetic acid, phenolics, and furans).

Studies by Cortés-Camargo [34] showed the production of biosurfactants by a halotolerant *Bacillus tequilensis* ZSB10 strain in cellulosic and hemicellulose hydrolysates from vine-trimming shoots. The pre-treatment of the lignocellulosic material, occurred in three stages, in this case: (i) acid hydrolysis, (ii) an alkaline reaction, and (iii) an enzymatic hydrolysis of cellulignin using cellulases. At the end of the pre-treatment, a cellulosic hydrolysate with 21.57 g L^{-1} of glucose was obtained. The authors reported that after the fermentation process in a 2 L stirred tank reactor, 1.52 g L^{-1} of extracellular biosurfactant was obtained with a critical micellar concentration (CMC) of 177.14 mg L^{-1}, lowering the surface tension up to 38.6 mN m^{-1}. The emulsifying potential of the biosurfactant obtained in kerosene, measured by the emulsification index (IE), was below 50%. It should be noted that in this work, the authors used a strategy of combining pre-treatments in order to obtain higher concentrations of sugars, since the lignocellulosic material used was not trivial, with pre-treatment conditions previously studied in the literature.

Marcelino and collaborators [35] showed the possibility of the production of glycolipid biosurfactants by yeast in hemicellulosic hydrolysate of sugarcane bagasse. In this study, at first, a screening of several yeasts isolated from Brazilian biomes was carried out. Most of the microorganisms used in the selection were collected from decomposing lignocellulosic materials. Based on this strategy, the strains that produce biosurfactants in detoxified and non-detoxified sugarcane bagasse hemicellulosic hydrolysate can be found. It was observed that the studied yeasts produced biosurfactants in culture medium based on detoxified and non-detoxified hydrolysate. A smaller number of yeasts produced biosurfactant in a medium based on hemicellulosic hydrolysate. This difference can be explained by the ionic strength and the concentration of inhibitor compounds in both situations.

In culture media based on non-detoxified hemicellulosic hydrolysate, the ionic strength is high due to the particles present from the preparation of the hydrolysate. Thus, the osmolarity of the culture medium is altered when compared to synthetic media or media submitted to the detoxification process, in which successive filtrations and centrifugations occur. Furthermore, in culture media based on non-detoxified hemicellulose hydrolysate, there is a significant concentration of inhibitory compounds, such as phenolics, furans, and acetic acid. One of the ways to mitigate the effect of inhibitors in the non-detoxified hydrolysate is the use of some yeasts with the capacity to degrade xenobiotic compounds, which will reduce the concentration of inhibitors in the culture medium. Due to the various disadvantages of working with non-detoxified hydrolysate, the authors continued with studies of biosurfactant production in culture medium based on detoxified hydrolysate, obtaining, after 48 h of fermentation, volumetric productivities of biosurfactants of $0.006-0.167 \, g \, L^{-1} \, h^{-1}$ [35].

Chaves et al. [36] reported for the first time the production of biosurfactants in hemicellulosic hydrolysate of sugarcane straw detoxified and non-detoxified by yeasts isolated from the Antarctic continent. The biosurfactant produced by *Naganishia adellienses* L95 showed outstanding emulsifying properties at low temperatures. The results of this study are interesting for biorefineries, as they show the production of biosurfactants in lignocellulosic biomass; the use of extremophile microorganisms; and biosurfactants with different properties that can be applied in industrial processes at low temperatures in the future.

In the study by Barbosa et al. [37], the biosurfactant production by the yeast *Scheffersomyces shehatae* was observed in a culture medium based on hemicellulose hydrolysate of detoxified sugarcane bagasse and residual vegetable oil. The biosurfactant obtained showed outstanding emulsifying characteristics and a volumetric productivity of $0.076 \, g \, L^{-1} \, h^{-1}$. It is interesting to emphasize the strategy used in this study, which mixed substrates of different polarities. According to Fontes et al. [38], although the production of biosurfactants occurs in the presence of polar carbon sources, such as sugars, several studies have shown that the highest biosurfactant productions are obtained when non-polar substrates are added. Many works describe the importance of the combination of a substrate insoluble in water and a carbohydrate, as constituents of the culture medium.

In addition to the hemicellulose, cellulosic hydrolysate can also be used as a substrate in biorefineries. The cellulosic hydrolysate can be obtained by enzymatic processes, in which a pool of commercial cellulase enzymes are used, and have a majority composition of glucose. Furthermore, these hydrolysates do not contain inhibitory compounds, since the enzymatic process for obtaining them is specific and uses mild physicochemical conditions. The main disadvantage of cellulosic hydrolysates is that compared to hemicelluloses they can be expensive due to the enzymes used in the hydrolysis process [39].

Faria et al. [40] reported the production of biosurfactants of the type mannosylerythritol lipids (MELs) by a yeast of the genus *Pseudozyma* in semi-synthetic media using purified xylan as a carbon source. This work is important because it shows the possibility of directly converting a hemicellulose into a biosurfactant, without the need for a pre-treatment with drastic physicochemical conditions. Microorganisms that produce cellulases and xylanases can be used in this type of process.

Da Mata et al. [41] reported the production of biosurfactants by *Aureobasidium pullulans* LB83 in SSF. In this work, the authors showed the direct conversion of cellulose from pretreated sugarcane bagasse into biosurfactants, due to the yeast's potential to produce cellulase enzymes.

In the review article by Tan and Li [42] several works were cited that reported the production of rhamnolipids by bacteria such as *Pseudomonas aeruginosa*, *Acinetobacter calcoaceticus* and some of the genus *Lactobacillus* using the cellulosic hydrolysate of wheat straw and waste from fruit products.

As can be seen, studies on the production of biosurfactants in the context of lignocellulosic biorefineries have increased, mainly due to the urgency of implementing sustainable processes and products in the scenario of transition from the petroeconomy to the bioeconomy. However, there is still a need for advance in works that focus on the selection of microbial strains that overproduce biosurfactants; in production in bioreactors, from bench to industrial scale, in adjustments of operating parameters such as agitation and aeration; and also solve problems related to foam formation during fermentation.

4.4 Biosurfactant Production in the Context of Oleaginous Biorefineries

In this section, the production of biosurfactants using oleaginous industrial by-products in the context of biorefineries will be emphasized. In the literature, the main oily raw materials used in the production of biosurfactants are animal fat waste, the food processing industry (frying edible oils and fats, olive oil, rape seed oil, sunflower, and vegetable oils) and oil processing mills (coconut cake, canola meal, olive oil mill waste water, palm oil mill, peanut cake, effluent, soybean cake, soapstock, and waste from lubricating oil) [43, 44].

In the production of biosurfactants using oleaginous by-products, mainly waste oils from the food industry, published studies have shown the use of bacteria such as *Pseudomonas*, *Bacillus* and *Serratia*, but focus mainly on yeasts such as *Candida*, *Starmerella* and *Pseudozyma*. However, in the literature there are also reports of the production of biosurfactants by the yeast genera *Rhodotorula*, *Pichia*, *Debaryomyces*, *Aureobasidium*, *Kluyveromyces*, *Issatchenkia*, *Cryptococcus*, *Rhizopus*, *Yarrowia*, and *Trichosporon* [45, 46]. It should be noted that yeasts recently described as positive lipases and oleaginous (accumulators of intracellular lipids) have also stood out as better producers of biosurfactants. In the literature there are also some reports of filamentous fungi producing biosurfactants using oily substrates.

As can be seen yeasts have highlighted in the production of biosurfactants due to the GRAS *status* (generally recognized as safe) of several species used, not presenting risks of pathogenicity and toxicity, allowing the application of biosurfactants without restrictions [38, 47]. In addition, when compared to bacteria, yeasts are more resistant to the biosurfactant secreted and accumulated in the medium during fermentation, suffering fewer morphological changes, mainly in the plasmatic membrane and remaining viable for a longer period of time [48].

Oleaginous raw materials are rich in hydrophobic substrates, which are considered inducers of the production of biosurfactants, and because of this they are known as hydrophobic inducers. Biosurfactant inducers are used in the culture medium in order to promote the induction of the production of these compounds and increase the final yield. They are defined as supplementary sources of carbon, which may be present in low or medium concentrations in the culture medium. Studies show that hydrophobic inducers can cause changes in the chemical structure, and consequently modify the physical-chemical and biological properties of biosurfactants [49–52].

The main substrates considered to be hydrophobic inducers used in the production of biosurfactants are hydrocarbons and vegetable oils. These compounds are considered secondary sources of carbon and can act in microbial growth and also in the synthesis of the non-polar portion of biosurfactants [52–54]. However, one of the main problems in the production of biosurfactants using oleaginous by-products is the low transfer of oxygen, due to the higher viscosity of the oils. The high concentration of these compounds in the fermentation process can lead to low oxygen transfer, in such a way as to affect cell growth and also the production of biosurfactants/bioemulsifiers, so when applying these inducers it is necessary to know the process that is being carried out [52].

Table 4.3 presents a summary of the main microorganisms producing biosurfactants, the oleaginous raw materials used and the type of biosurfactant produced.

The production of biosurfactants using waste oils as raw material is used more routinely than agricultural by-products rich in sugars. Generally, the oils used in these processes must

Table 4.3 Some studies on the production of biosurfactants by yeast using oleaginous industrial by-products.

Microorganism	Raw material/substrate	Biosurfactant produced	References
Candida bombicola	Animal fat and glucose	Glycolipid	[55]
Candida antarctica	Oil refinery residual hydrocarbons	Glycolipid	[56]
Candida bombicola ATCC 22214	Corn oil and honey	Glycolipid	[57]
Candida lipolytica	Waste from vegetable oil industries	Protein-carbohydrate-lipid complex	[58]
Candida bombicola	Waste oil from restaurants	Glycolipid	[59]
Candida lipolytica	Waste from soybean oil industries	Not identified	[60]
Candida sphaerica UCP0995	Residual peanut oil	Not identified	[61]
Scheffersomyces anomala PY1	Soybean oil	Glycolipid	[62]
Candida sp.	Residual peanut oil	Glycolipid	[63]
Candida bombicola	Cane molasses and soybean oil	Glycolipid	[64]
Trichosporon montevideense CLOA 72	Sunflower oil	Glycolipid	[48]
Candida glabrata UCP1002	Waste from vegetable oil industries	Not identified	[65]

also be pre-treated, aiming mainly at the removal of solids that can interfere with fermentation. To remove these solids, unitary operations such as centrifugation or filtration are usually used. In addition, the peroxide/hydroperoxide index of the residual oils must also be analyzed, as high concentrations of these compounds can interfere with the fermentation process [66].

4.5 Biosurfactant Production in the Context of Starchy and Biodiesel Biorefineries

In addition to lignocellulosic and oily by-products, the production of biosurfactants can also be carried out using starchy raw materials. Starch is defined as a renewable, natural, and biodegradable polymer prepared by many plants as a source of stored energy. Two different polysaccharides are present in starch structure: (1) linear (1,4)-linked α-d-glucan amylose, and (2) highly (1,6) branched α-d-glucan amylopectin [67, 68].

Among the starchy agro-industrial by-products that can be used as raw materials in the production of biosurfactants and other bioproducts, we can mention wastes from: soybean, potato, sweet potato, and sweet sorghum. The starch found in these raw materials must be depolymerized, releasing glucose as the main fermentable sugar [43]. Because of this, starchy raw materials are also subjected to pre-treatments aimed at starch hydrolysis [69]. The pre-treatments used for starch hydrolysis can be carried out chemically or enzymatically. Before pre-treatment, the starch gelatinization stage is necessary, in which it will be hydrated at a given temperature in the range 55–80 °C, depending on the starch source.

In the food industries, acid hydrolysis of starch occurs when a high concentration of starch (30–40 g per 100 g of solids) is treated with diluted inorganic acids at a temperature lower than that of gelatinization (30–60 °C) during one or more hours of reaction [70]. Temperature control in acid hydrolysis is a critical point in the process, since glucose in an acid medium and high temperature can suffer dehydration and form hydroxymethylfurfural.

The enzymatic hydrolysis of starch consists of the action of a pool of enzymes called amylase or amylolytic complex. The starchy raw materials, after the gelatinization process, become a suitable substrate for the action of the amylases that break the existing bonds in the biopolymers, amylose and amylopectin, and releasing glucose molecules and small biopolymers called dextrins into solution, which in turn will also be decomposed. Amylolytic enzymes are mostly produced by fungi or bacteria that modern technologies in bioprocesses have made very efficient and available at affordable costs for industrial use. For an efficient conversion of already gelatinized starch macromolecules to glucose and other low molecular mass derivatives, the coordinated action of the enzymes of the amylolytic complex is necessary: α-amylases, β-amylases, α–D-glucosidase, exo-1, 4-α-D-glucanases, glucosidases, pullulanases, and isoamylases.

The enzymatic hydrolysis of starch is also used in the food industry to produce glucose syrup. In addition, it is also used in the process of producing alcohol from starchy sources (cassava, corn, sweet potato, and others). It should be noted that, when compared to chemical hydrolysis, the enzymatic method still has a high cost and lower yield, requiring studies aimed at optimizing the process to reduce costs [71, 72].

In the literature, works can be found on the use of starchy by-products in the biosurfactant production. Potato processing effluents have been used in the production of surfactin by bacteria of *Bacillus genus* [73–77].

In addition to starchy by-products, another raw material that can be used in the production of biosurfactants is glycerol. In the production of biodiesel, glycerol is the main by-product obtained after the transesterification of vegetable oils and animal fats. Consequently, glycerol production has become increasingly abundant as worldwide biodiesel production increases [78]. Among the microorganisms that produce biosurfactants that can use glycerol as a nutrient, *Bacillus subtilis* and *Pseudomonas aeruginosa* strains have been highlighted [79, 80]. Howerer, There are also studies showing the production of biosurfactants in glycerol-based culture media using *Yarrowia lipolytica*, a type of lipolytic yeast, as a fermentation agent [81].

The main problem with using residual glycerol from biodiesel as a substrate for the production of biosurfactants and other by-products in fermentation processes are impurities such as methanol, fatty acid methyl esters, and salts left over from the transesterification reaction [78]. Thus, a pre-treatment step is required, in which unit operations such as the use of exchange resins acidification and extraction processes can be used [82].

4.6 Conclusion

As can be seen in this chapter, biosurfactants are promising value-added products to be produced in the broader context of biorefineries using different types of raw materials. Through clean processes, these bioproducts can be obtained and contribute to the sustainable development of society, making the transition from the petroeconomy to the bioeconomy smooth. However, there are still several technical and economic obstacles to be studied and overcome, requiring academic and industrial efforts to overcome these barriers and make biosurfactants popular products.

References

1 *Antonio Aprigio da Silva Curvelo – revista Opiniões* https://florestal.revistaopinioes.com.br/revista/detalhes/24-biorrefinaria-materia-prima-definindo-o-process accessed January 26th, 2023.
2 National Renewable Energy Laboratory (NREL). https://www.nrel.gov/research/re-biomass.html Accessed January 26th, 2023.
3 The IEA Bioenergy Technology Collaboration Programme (TCP) - Task 42: biorefining in a Circular Economy https://task42.ieabioenergy.com/#:~:text=Task%2042%20provides%20an%20international,development%2C%20demonstration%20and%20policy%20analysis Accessed January 26th, 2023.
4 Dwi Prasetyo W, Putra ZA, Bilad MR, Mahlia TMI, Wibisono Y, Nordin NAH, et al. Insight into the sustainable integration of bio-and petroleum refineries for the production of fuels and chemicals. *Polymers.* 2020;12(5):1091.
5 Cherubini F, Jungmeier G, Wellisch M, Willke T, Skiadas I, Van Ree R, de Jong E. Toward a common classification approach for biorefinery systems. *Biofuel, Bioprod Bioref.* 2009;3(5):534–546.
6 Pelizer LH, Pontieri MH, de Oliveira Moraes I. Utilização de resíduos agro-industriais em processos biotecnológicos como perspectiva de redução do impacto ambiental. *J Technol Management Innov.* 2007;2(1):118–127.

7 Comasseto J. A reunião da química e da biologia. In: Marsaioli AJ, Porto ALM editors. *Biocatálise e biotransformação: Fundamentos e aplicações*. São Paulo: Schoba, 2010; 11–22.
8 Jesus JPFD Desenvolvimento de cinco linhagens de *Agaricus bisporus lange* (Imbach) ("*champignon* de Paris") em diferentes formulações de composto e meios de cultura. Agronomia (Energia na Agricultura) [dissertação]. Botucatu (SP): UNESP; 2011.
9 Pandey A, Soccol CR, Nigam P, Soccol VT. Biotechnological potential of agro-industrial residues. I: sugarcane bagasse. *Biores Technol*. 2000;74(1):69–80.
10 Castanha RF Utilização de soro de queijo para a produção de lipídeos por leveduras oleaginosas. Microbiologia Agrícola [dissertação]. Piracicaba (SP): Universidade de São Paulo; 2012.
11 Mota CJ, da Silva CX, Gonçalves VL. Gliceroquímica: novos produtos e processos a partir da glicerina de produção de biodiesel. *Química Nova*, 2009;32:639–648.
12 Rivaldi JD, Sarrouh BF, Silva SS, Fiorilo R. Biotechnological strategies for glycerol utilization derived from biodiesel production; Glicerol de biodiesel: estrategias biotecnologicas para o aproveitamento do glicerol gerado da producao de biodiesel. *Biotecnologia Ciencia & Desenvolvimento*, 2008;37:44–51.
13 Mattam AJ, Clomburg JM, Gonzalez R, Yazdani SS. Fermentation of glycerol and production of valuable chemical and biofuel molecules. *Biotechnol Lett*. 2013;35:831–842.
14 Vandamme EJ. Agro-industrial residue utilization for industrial biotechnology products. In: Nigam P, Pandey A, editors. *Biotechnology for Agro-industrial Residues Utilisation: Utilisation of Agro-residues*. Springer, 2009. 3–11.
15 Chaud LCS, Arruda PV, de Almeida Felipe MDG. Potencial do farelo de arroz para utilização em bioprocessos. *Nucleus*. 2009;6(2):1–14.
16 Rossi SC Produção de aromas frutais por *Ceratocystis fimbriata* cultivado em polpa cítrica, farelo de soja e melaço de cana por fermentação no estado sólido – determinação da atividade de pectinase (poligalacturonase), esterases e lipases. Processos Biotecnológicos [dissertação]. Curitiba (PR): Universidade Federal do Paraná; 2011.
17 Miller TL, Churchill BW. *Substrates for Large-scale Fermentations*. American Society for Microbiology; 1986.
18 Surfactina – Sigma Aldrich/Merck) https://www.sigmaaldrich.com/BR/pt/search/24730-31-2?focus=products&gclid=EAIaIQobChMItsTgsteX_QIV0UFIAB3b4QGbEAAYAiAAEgJ-c_D_BwE&page=1&perpage=30&sort=relevance&term=24730-31-2&type=cas_number Accessed February 15th, 2023.
19 http://www.clicmercado.com.br/novo/cotacoes/buscacot.asp Accessed February 15th, 2023.
20 Barbosa FG, Ribeaux DR, Rocha T, Costa RA, Guzmán RR, Marcelino PR. Biosurfactants: sustainable and Versatile Molecules. *Journal of the Brazilian Chemical Society*, 2022;33:870–893.
21 Gil CV, Rebocho AT, Esmail A, Sevrin C, Grandfils C, Torres CA et al. Characterization of the thermostable biosurfactant produced by *Burkholderia thailandensis* DSM 13276. *Polymers*. 2022;14(10):2088.
22 Bjerk TR, Severino P, Jain S, Marques C, Silva AM, Pashirova T et al. Biosurfactants: properties and applications in drug delivery, biotechnology and ecotoxicology. *Bioengineering*. 2021;8(8):115.
23 Santos FA, Queiróz JHD, Colodette JL, Fernandes SA, Guimarães VM, Rezende ST. Potencial da palha de cana-de-açúcar para produção de etanol. *Química nova*, 2012;35:1004–1010.

24 Pellegrini VDOA, Sepulchro AGV, Polikarpov I. Enzymes for lignocellulosic biomass polysaccharide valorization and production of nanomaterials. *Curr Opin Green Sustainable Chem*. 2020;26:100397.

25 Hu L, Fang X, Du M, Luo F, Guo S. Hemicellulose-based polymers processing and application. *Am J Plant Sci*. 2020;11(12):2066–2079.

26 Singh nee' Nigam P, Gupta N, Anthwal A. Pre-treatment of agro-industrial residues. biotechnology for agro-industrial residues utilisation: utilisation of agro-residues. 2009:13–33.

27 Cortez LAB. *Bioetanol de Cana-de-Açúcar: P&D para Produtividade Sustentabilidade*. São Paulo: Edgard Blücher Ltda; 2010.

28 Hamelinck CN, Van Hooijdonk G, Faaij AP. Ethanol from lignocellulosic biomass: techno-economic performance in short-, middle-and long-term. *Biomass and Bioenergy*. 2005;28(4):384–410.

29 Lynd LR. Overview and evaluation of fuel ethanol from cellulosic biomass: technology, economics, the environment, and policy. *Ann Rev Energy Environm*. 1996;21(1):403–465.

30 Chen H, Wang L. Sugar strategies for biomass biochemical conversion. In: Chen H, Wang L editors. *Technologies for Biochemical Conversion of Biomass*. London: Elsevier, 2017; 137–164.

31 Ravindran R, Jaiswal AK. A comprehensive review on pre-treatment strategy for lignocellulosic food industry waste: challenges and opportunities. *Biores Technol*. 2016;199:92–102.

32 Rivera OMP, Moldes AB, Torrado AM, Domínguez JM. Lactic acid and biosurfactants production from hydrolyzed distilled grape marc. *Process Biochem*. 2007;42(6):1010–1020.

33 Rivera OMP, Menduiña ABM, Agrasar AMT, González JMD. Biosurfactants from grape marc: stability study. *J Biotechnol*. 2007;2(131):S136.

34 Cortés-Camargo S, Pérez-Rodríguez N, de Souza Oliveira RP, Huerta BEB, Domínguez JM. Production of biosurfactants from vine-trimming shoots using the halotolerant strain *Bacillus tequilensis* ZSB10. *Ind Crops Prod*, 2016;79:258–266.

35 Marcelino PRF, Peres GFD, Terán-Hilares R, Pagnocca FC, Rosa CA, Lacerda TM, et al. Biosurfactants production by yeasts using sugarcane bagasse hemicellulosic hydrolysate as new sustainable alternative for lignocellulosic biorefineries. *Ind Crops Prod*. 2019;129:212–223.

36 Chaves FDS, Brumano LP, Franco Marcelino PR, da Silva SS, Sette LD, Felipe MDGDA. Biosurfactant production by Antarctic-derived yeasts in sugarcane straw hemicellulosic hydrolysate. In: Kaltschmitt M, Hofbauer H, editors. *Biomass Conversion and Biorefinery*. Springer, 2021. 1–11.

37 Barbosa FG, Marcelino PRF, Lacerda TM, Philippini RR, Giancaterino ET, Mancebo MC. Production, physicochemical and structural characterization of a bioemulsifier produced in a culture medium composed of sugarcane bagasse hemicellulosic hydrolysate and soybean oil in the context of biorefineries. *Fermentation*. 2022;8(11):618.

38 Fontes GC, Amaral PFF, Coelho MAZ. Produção de biossurfactante por levedura. *Quím Nova*, 2008;31:2091–2099.

39 Maitan-Alfenas GP, Visser EM, Guimarães VM. Enzymatic hydrolysis of lignocellulosic biomass: converting food waste in valuable products. *Curr Opin Food Sci*. 2015;1:44–49.

40 Faria NT, Marques S, Fonseca C, Ferreira FC. Direct xylan conversion into glycolipid biosurfactants, mannosylerythritol lipids, by *Pseudozyma antarctica* PYCC 5048T. *Enzyme and Microbial Technology*, 2015;71:58–65.

41 RAM DC, Rubio-Ribeaux D, Carneiro BC, Franco PM, de Azevedo Mendes G, da Silva IL et al. Sugarcane bagasse pretreated by different technologies used as support and carbon source in solid-state fermentation by Aureobasidium pullulans LB83 to produce bioemulsifier. *Biomass Conversion and Biorefinery*. 2023:1–14.

42 Tan YN, Li Q. Microbial production of rhamnolipids using sugars as carbon sources. *Microbial Cell Fact*. 2018;17(1):1–13.

43 Banat IM, Satpute SK, Cameotra SS, Patil R, Nyayanit NV. Cost effective technologies and renewable substrates for biosurfactants' production. *Front Microbiol*, 2014;5:697.

44 Gaur VK, Sharma P, Sirohi R, Varjani S, Taherzadeh MJ, Chang JS et al. Production of biosurfactants from agro-industrial waste and waste cooking oil in a circular bioeconomy: an overview. *Biores Technol*. 2022;343:126059.

45 Christofi N, Ivshina IB. Microbial surfactants and their use in field studies of soil remediation. *J Appl Microbiol*. 2002;93(6):915–929.

46 Beopoulos A, Chardot T, Nicaud JM. Yarrowia lipolytica: a model and a tool to understand the mechanisms implicated in lipid accumulation. *Biochimie*. 2009;91(6):692–696.

47 Barth G, Gaillardin C. Physiology and genetics of the dimorphic fungus *Yarrowia lipolytica*. *FEMS Microbiol Rev*. 1997;19(4):219–237.

48 Monteiro AS, Coutinho JO, Júnior AC, Rosa CA, Siqueira EP, Santos VL. Characterization of new biosurfactant produced by *Trichosporon montevideense* CLOA 72 isolated from dairy industry effluents. *J Basic Microbiol*. 2009;49(6):553–563.

49 Ehrhardt DD, Secato JFF, Tambourgi EB. Produção de biossurfactante por Bacillus subtilis utilizando resíduo do processamento do abacaxi como substrato. In: *Anais do XX Congresso Brasileiro de Engenharia Química* São Paulo: Edgard Blücher. 2015; 1960–1965.

50 Nurfarahin AH, Mohamed MS, Phang LY. Culture medium development for microbial-derived surfactants production—an overview. *Molecules*. 2018;23(5):1049.

51 Gudiña EJ, Rodrigues AI, de Freitas V, Azevedo Z, Teixeira JA, Rodrigues LR. Valorization of agro-industrial wastes towards the production of rhamnolipids. *Biores Technol*, 2016;212:144–150.

52 de Oliveira Schmidt VK, de Souza Carvalho J, de Oliveira D, de Andrade CJ. Biosurfactant inducers for enhanced production of surfactin and rhamnolipids: an overview. *World J Microbiol Biotechnol*. 2021;37(2):21.

53 Pathania AS, Jana AK. Improvement in production of rhamnolipids using fried oil with hydrophilic co-substrate by indigenous *Pseudomonas aeruginosa* NJ2 and characterizations. *Appl Biochem Biotechnol*. 2020;191:1223–1246.

54 Niu Y, Wu J, Wang W, Chen Q. Production and characterization of a new glycolipid, mannosylerythritol lipid, from waste cooking oil biotransformation by Pseudozyma aphidis ZJUDM34. *Food Sci Nutr*. 2019;7(3):937–948.

55 Deshpande M, Daniels L. Evaluation of sophorolipid biosurfactant production by Candida bombicola using animal fat. *Biores Technol*. 1995;54(2):143–150.

56 Bednarski W, Adamczak M, Tomasik J, Płaszczyk M. Application of oil refinery waste in the biosynthesis of glycolipids by yeast. *Biores Technol*. 2004;95(1):15–18.

57 Pekin G, Vardar-Sukan F, Kosaric N. Production of sophorolipids from *Candida bombicola* ATCC 22214 using Turkish corn oil and honey. *Eng Life Sci*. 2005;5(4):357–362.

58 Rufino RD, Sarubbo LA, Campos-Takaki GM. Enhancement of stability of biosurfactant produced by Candida lipolytica using industrial residue as substrate. *World J Microbiol Biotechnol*. 2007;23:729–734.

59 Shah V, Jurjevic M, Badia D. Utilization of restaurant waste oil as a precursor for sophorolipid production. *Biotechnol Prog.* 2007;23(2):512–515.

60 Rufino RD, Sarubbo LA, Neto BB, Campos-Takaki GM. Experimental design for the production of tensio-active agent by *Candida lipolytica*. *J Ind Microbiol Biotechnol.* 2008;35(8):907–914.

61 Sobrinho HB, Rufino RD, Luna JM, Salgueiro AA, Campos-Takaki GM, Leite LF. Utilization of two agroindustrial by-products for the production of a surfactant by Candida sphaerica UCP0995. *Proc Biochem.* 2008;43(9):912–917.

62 Thaniyavarn J, Chianguthai T, Sangvanich P, Roongsawang N, Washio K, Morikawa M et al. Production of sophorolipid biosurfactant by *Pichia anomala*. *Biosci, Biotechnol Biochem.* 2008;72(8):2061–2068.

63 Coimbra CD, Rufino RD, Luna JM, Sarubbo LA. Studies of the cell surface properties of Candida species and relation to the production of biosurfactants for environmental applications. *Curr Microbiol.* 2009;58:245–251.

64 Daverey A, Pakshirajan K. Production, characterization, and properties of sophorolipids from the yeast Candida bombicola using a low-cost fermentative medium. *Appl Biochem Biotechnol*, 2009;158:663–674.

65 de Gusmao CA, Rufino RD, Sarubbo LA. Laboratory production and characterization of a new biosurfactant from *Candida glabrata* UCP1002 cultivated in vegetable fat waste applied to the removal of hydrophobic contaminant. *World J Microbiol Biotechnol*, 2010;26:1683–1692.

66 Mohanty SS, Koul Y, Varjani S, Pandey A, Ngo HH, Chang JS et al. A critical review on various feedstocks as sustainable substrates for biosurfactants production: a way towards cleaner production. *Microb Cell Fact.* 2021;20(1):1–13.

67 Xie F, Pollet E, Halley PJ, Avérous L. Starch-based nano-biocomposites. *Prog Polymer Sci.* 2013;38(10–11):1590–1628.

68 Nasrollahzadeh M, Sajjadi M, Sajadi SM, Issaabadi Z. Green nanotechnology. In: Nasrollahzadeh M, Sajadi SM, Sajjadi M, Issaabadi Z, Atarod M editors. *An Introduction to Green Nanotechnology*. London: Elsevier, 2019; 145–198.

69 Almeida RRD Estudo do bagaço de mandioca (Manihot esculenta C.), nativo e tratado com α-amilase e amiloglucosidase, por meio de técnicas termoanalíticas. *Food Science and Technology* [dissertation]. Ponta Grossa (PR): Universidade Estadual de Ponta Grossa; 2009.

70 Sandhu KS, Singh N, Lim ST. A comparison of native and acid thinned normal and waxy corn starches: physicochemical, thermal, morphological and pasting properties. *LWT-Food Sci Technol.* 2007;40(9):1527–1536.

71 Woiciechowski AL, Nitsche S, Pandey A, Soccol CR. Acid and enzymatic hydrolysis to recover reducing sugars from cassava bagasse: an economic study. *Braz Arch Biol Technol.* 2002;45:393–400.

72 Sigüenza-Andrés T, Pando V, Gómez M, Rodríguez-Nogales JM. Optimization of a simultaneous enzymatic hydrolysis to obtain a high-glucose slurry from bread waste. *Foods.* 2022;11(12):1793.

73 Thompson DN, Fox SL, Bala GA Biosurfactants from potato process effluents. In: Finkelstein M, Davison BH, editors. *Twenty-First Symposium on Biotechnology for Fuels and Chemicals: Proceedings of the Twenty-First Symposium on Biotechnology for Fuels and*

Chemicals Held May 2–6, 1999 Fort Collins, Colorado. Totowa: Humana Press; 2000. 917–930.

74 Thompson DN, Fox SL, Bala GA. The effects of pretreatments on surfactin production from potato process effluent by *Bacillus subtilis*. In: Davison BH, McMillan J, Finkelstein M editors. *Twenty-second Symposium on Biotechnology for Fuels and Chemicals*. Totowa: Humana Press, 2001; 487–501.

75 Noah KS, Fox SL, Bruhn DF, Thompson DN, Bala GA. Development of continuous surfactin production from potato process effluent by *Bacillus subtilis* in an airlift reactor. In: Finkelstein M, Davison BH, McMillan J editors. *Twenty-Third Symposium on Biotechnology for Fuels and Chemicals*. Totowa: Humana Press, 2002; 803–813.

76 Noah KS, Bruhn DF, Bala GA. Surfactin production from potato process effluent by *Bacillus subtilis* in a chemostat. In: Davison BH, Evans BR, Finkelstein M, McMillan J, editors. *Twenty-Sixth Symposium on Biotechnology for Fuels and Chemicals*. Totowa: Humana Press, 2005; 465–473.

77 Jain RM, Mody K, Joshi N, Mishra A, Jha B. Effect of unconventional carbon sources on biosurfactant production and its application in bioremediation. *Int J Biol Macromol.* 2013;62:52–58.

78 He Q, McNutt J, Yang J. Utilization of the residual glycerol from biodiesel production for renewable energy generation. *Renewable Sustainable Energy Rev.* 2017;71:63–76.

79 Saimmai A, Rukadee O, Sobhon V, Maneerat S. Biosurfactant production by Bacillus subtilis TD4 and *Pseudomonas aeruginosa* SU7 grown on crude glycerol obtained from biodiesel production plant as sole carbon source. *Council of Scientific and Industrial Research–National Institute of Science Communication and Policy Research*, 2012;71:396–406.

80 Sousa M, Melo VMM, Rodrigues S, Sant'Ana HB, Gonçalves LRB. Screening of biosurfactant-producing *Bacillus* strains using glycerol from the biodiesel synthesis as main carbon source. *Bioprocess Biosystems Eng.* 2012;35:897–906.

81 Louhasakul Y, Cheirsilp B, Intasit R, Maneerat S, Saimmai A. Enhanced valorization of industrial wastes for biodiesel feedstocks and biocatalyst by lipolytic oleaginous yeast and biosurfactant-producing bacteria. *Int Biodeterioration Biodegradation.* 2020;148:104911.

82 Domingos AM, Pitt FD, Chivanga Barros AA. Purification of residual glycerol recovered from biodiesel production. *South Afr J Chem Eng.* 2019;29(1):42–51.

5

Biosurfactant Production by Solid-state Fermentation in Biorefineries

Daylin Rubio-Ribeaux[1],*, Rogger Alessandro Mata da Costa[1], Dayana Montero Rodríguez[2], Nathália Sá Alencar do Amaral Marques[2], Gilda Mariano Silva[1], and Silvio Silvério da Silva[1]

[1] *Department of Biotechnology, Engineering School of Lorena, University of São Paulo, 12.602-810, Brazil*
[2] *Catholic University of Pernambuco (UNICAP), Rua Nunes Machado, 42, Boa Vista, 50050-590*
* *Corresponding author*

5.1 Introduction

Biosurfactants are molecules fundamentally synthesized by microorganisms consisting of a hydrophilic and a hydrophobic portion. This characteristic gives these compounds the ability to divide between liquid interfaces with different degrees of polarity and consequently reduce surface tension and promote liquid emulsion [1, 2]. Biosurfactants are gaining more prominence over their chemically produced counterparts due to their notable advantages, such as: low toxicity, biodegradability, higher selectivity, improved environmental compatibility, and functionality over a wide range of pH, temperatures, and salinity among others [3].

These properties favor the potential application of these compounds in the pharmaceutical, food, cosmetic, textile, oil, and agricultural industries. In addition, the growth of biosurfactants in the global market is expected to reach 0.8% CAGR from $1.3754 billion in 2020 to $1.4427 billion in 2026 [2]. However, the presence of most of these biomolecules on the market remains limited due to the high production costs and the low yields regarding the production cost of chemical surfactants. Furthermore, raw materials account for 10–30% of the total expense of producing biosurfactants [4, 5]. Among the studied alternatives to overcome these obstacles, the use of renewable nutrient substrates can reduce production costs, making the process viable [6]. Furthermore, the selection of the fermentative process also contributes to this purpose. Thus, solid state fermentation (SSF) offers some advantages, such as the use of simple equipment and low volumes of water, higher concentration of products obtained, and low energy demand [7]. In this context, the biorefinery concept involves the use of facilities equivalent to oil refineries, but with the difference that biomass is used to generate a wide range of value-added products, such as biosurfactants [8]. On this basis, this work aims to provide a comprehensive view of the production of biosurfactants by SSF in the context of biorefineries as part of a sustainable future.

Biosurfactants and Sustainability: From Biorefineries Production to Versatile Applications, First Edition.
Edited by Paulo Ricardo Franco Marcelino, Silvio Silverio da Silva, and Antonio Ortiz Lopez.
© 2023 John Wiley & Sons Ltd. Published 2023 by John Wiley & Sons Ltd.

5.2 Advantages of Biosurfactant Production by Solid-State Fermentation

SSF is a fermentative process that consists of the microbial biotransformation of substrates, which takes place in the absence or near absence of free water. It can be stated that this technology simulates the microbial degradation process that occurs naturally in environments [9, 10]. The low water volume used in SSF brings several economic benefits, in comparison with submerged fermentation (SmF), such as the possibility of using smaller bioreactors and the lower energy demand for the sterilization of the substrates, once under solid-state conditions they are less susceptible to microbial contamination [11, 12]. The low activity of free water in the substrates in SSF tends to limit the growth of potential contaminants [13].

Besides, the scarce free water in the system avoids the serious foaming problem of biosurfactant production by conventional SmF and, consequently, the use of antifoam agents that increases the overall cost of the process [2, 14]. In addition, the lower effluent generation is considered an environmental benefit of SSF [13, 15]. The reduced downstream processing is another technical and economic advantage of SSF that can also boost large-scale production and commercialization of biosurfactants [11, 16]. It is widely known that the downstream processing costs account for approximately 60–80% of the total production costs of biosurfactants and they are still the main technological limitation for their industrial application [17, 18]. Once SSF offers the possibility of higher end-concentration of products, downstream processing can be simpler to that in SmF [15, 19].

In addition, yields of biosurfactants obtained by SmF are often low, when compared with chemical surfactants, hindering application at the industrial scale. In this context, SSF is an attractive alternative because of its higher productivity [2, 13]. Several researchers reported higher yields of biosurfactants using SSF in comparison to SmF, for example, El-Housseiny et al. [20], informed almost threefold rhamnolipids yield in solid-state culture. Surfactin production by *Bacillus pumilus* UFPEDA 448 in SSF was also reported at threefold higher than in SmF by Slivinski et al. [21]. Previously, Mizumoto et al. [22], achieved almost tenfold yield of iturin A when produced using okara, a by-product of tofu manufacture, in SSF.

Another positive aspect of SSF is the possibility of using solid agro-industrial by-products and wastes as alternative raw materials for biosurfactant production. The valorization of underutilized agricultural by-products such as potato peel, rice husk and corncob, allows the expensive conventional substrates to be replaced and reducing the production costs of biosurfactants [23, 24]. In addition, the reuse of these agrifood wastes contributes to their appropriate management, reducing their environmental impact and promoting a circular economy [18, 25].

5.3 Suitable Biomasses for Biosurfactant Production in Biorefineries

The biorefinery concept strictly follows the principle that the raw material used must be renewable to obtain energy and value-added products such as biosurfactants. Thus, a diversity of renewable and low-cost industry-based wastes have been investigated for their ability as nutritional sources for biosurfactant production [26, 27]. Carbon sources provided in the culture medium are of extreme importance to biosurfactant production since this compound is the main element for microbial growth followed by nitrogen [1, 3]. The selection of biomass is another key element due to its water holding capacity that maintains the moisture content of the solid particles. Solid substrates used in SSF are generally based on food and agro-industry crops and residues which are unprocessed and have different particle sizes. In addition, the nutrient composition and the quality of the biomass might vary from one batch to another which can affect the fermentation process of the biosurfactant [28, 29].

Among the biomass used for biosurfactant production, the agro-industrial by-products can be used as a source of carbohydrates, proteins, oils, and other micronutrients that are mainly obtained from the processing of various crops [30, 31]. In the context of biorefinery, this approach is useful in increasing the profitability of the process and improving the effective management of the wastes generated [32, 33]. Thus, starch and lignocellulosic by-products stand out due to their wide distribution, nutritional composition, and ability to be hydrolyzed from appropriate pretreatment techniques [13, 34].

Starchy biomass can be used in biorefineries due to its versatility and availability, which reduce the production cost. It is considered an important agricultural product mainly obtained from potatoes, corn, wheat, and tapioca although other sources include sweet sorghum, sugar beet, and sugar cane. Its basic structure comprises glucose monomers linked by α-1,4-glycosidic bonds with branches connected by α-1,6-glycosidic bonds [35, 36]. In addition, starchy residues are rich in carbohydrates, metal ions, nitrogen, and others which favor that nutritional supplementation is unnecessary into the culture medium. Among these wastes, potato peels account for 10% of the total potato waste, and 15–40% of the fruit, according to the selected peeling method [37, 38]. Thus, this by-product, of little value to the feed industry, represents an excellent and interesting raw material for the production of biosurfactants with different applications in human health and multiple industrial applications [2, 39]. In addition, cassava husk is another substrate rich in starch and mineral salts which is obtained from the processing of this common crop that can be used for the synthesis of microbial surfactants [40].

In contrast, lignocellulosic biomass is the most abundant and renewable organic carbon source. The main three types of conventional lignocellulosic materials can be divided into agricultural wastes, energy crops, and forestry residues. Among the agricultural residues are sugarcane bagasse, sugarcane straw, rice straw, rice husk, corn cobs, corn stover, wheat straw, wheat stover, and barley straw. Energy crops mainly include switchgrass, miscanthus, energy cane, and grass. Whereas woods, wood branches, woodchips, wood sawdust, and fruit bunch, among others are part of the forestry residues [41]. The composition of these feedstocks varies, however, the three major constituents involve cellulose, hemicellulose, and lignin. The structure of cellulose comprises a linear polymer of glucose unit which represents 23–53% of lignocellulosic biomass. The hemicellulose component accounts for 20–35% and can be identified as a heteropolysaccharide principally composed of xylose. The lignin fraction is a macromolecular complex of polyphenols molecules which are approximately 10–25% [31, 41]. Based on this, several studies have proved the potential of this biomass for the synthesis of biosurfactants due to its composition of sugars (C5, C6) that can be used by microorganisms as a carbon source. However, the use of this biomass for bioconversion into biosurfactants involves the pretreatment of waste [42–44].

5.4 Microorganisms Used in Biosurfactant Production by Solid-state Fermentation

Microorganisms can grow on solid particles and produce biosurfactants in SSF (Figure 5.1). The selection of microorganisms is generally dependent on the type of biomass, growth conditions, and targeted final product. Another important factor is the ability of microorganisms to synthesize enzymes that can hydrolyze polysaccharides into mono- and disaccharides from pretreated biomass. Thus, for example, α-amylase and glucoamylase are produced in the culture medium and convert starch into glucose and maltose. The action of endo-β-1,4-glucanases and microbial cellobiohydrolases allow the hydrolysis of cellulose to generate glucose. These general principles are important for the fermentation process and downstream methods [30].

Bacteria are more often referred to in the literature for the production of biosurfactans [11]. This is an interesting fact due to bacteria requiring high water activity (higher than 0.9) and for that reason, they are preferably cultivated in submerged fermentation. However, the production of biosurfactants is normally high when the bacteria are well adapted to solid biomass. For instance, bacteria are much less susceptible to mechanical damage in an agitated bed of solids. Another important fact is that there is the possibility of recovering the bacteria from the solids and trying to determine their quantity in some way. However, this will not be an exact determination of bacterial growth since it is a partial recovery [14, 45]. Among the bacteria, some of the most studied genera are *Bacillus* sp., *Pseudomonas* sp., *Serratia* sp [2, 45, 46].

On the other hand, both bacteria and yeast grow by adhering to the surfaces of solid substrate particles [15]. However, yeasts can capture the nutrients in the solid matrix with low water activity. These microorganisms evolved in conditions where oxygen availability was not a problem. Also, yeasts are less susceptible to mechanical shear stress and grow as single cells [47]. In addition, the use of yeasts as producers of biosurfactants became a viable alternative for the production of biosurfactants since many of them have GRAS (generally regarded as safe) status. Thus, they present a lower risk of toxicity and pathogenic reactions which are important factors during industrial fermentation processes [48, 49]. On the other hand, several yeast species such as *Starmerella bombicola*, *Yarrowia lipolytica*,

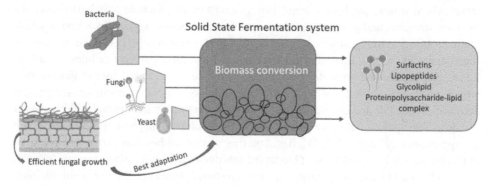

Figure 5.1 Microbial groups used in solid-state fermentation for biosurfactant production.

Rhodotorula bogoriensis, *Candida bombicola* ATCC 22214, and *Wickerhamiella domercqiae* among others, have been reported mainly as sophorolipid producers [50, 51].

Additionally, fungi are conventionally used in this kind of fermentative process, their metabolism is favored with a low water activity (Figure 5.1). This evidence is explained by the fact that higher fungi evolved and became more competitive in natural environments for bioconversion of solid substrates. In general, the hyphal structure gives a significant advantage to filamentous fungi over unicellular microorganisms in the colonization of solid biomass. Consequently, this structural advantage allows the active synthesis during the growth of several metabolites such as biosurfactant. This process plays a key role in the combination of the apical expansion of hyphal tips and the production of new hyphal tips through branching [52]. In addition, this process is also aided by hydrolytic enzymes that are excreted without large dilution like in the case of liquid fermentation, which makes the action of hydrolytic enzymes highly efficient. Some of the filamentous fungi used in SSF are *A. fumigatus* and *T. versicolor*, *Phialemonium* sp., *Fusarium oxysporum* LM 5634 [23, 53].

5.5 Raw Materials Used in Solid-state Fermentation for Biosurfactant Production

The difference in the economic development of different countries around the globe can be explained, currently, by the growth of economic activities linked to trade, industries, and technological sectors [54]. For example, the growth of agribusiness stands out in countries with large territorial extensions such as Brazil, Russia, India, China, and South Africa that produce coffee, corn, sugar cane, oranges, and cotton among others [55]. However, with the growing demand over the years, the number of agro-industrial by-products and residues with little or no industrial interest has also increased. This generated a problem from a logistical point of view due to disposal as well as environmental issues due to the emission of greenhouse gases [56]. In the search for solutions, the use of agro-industrial by-products as nutritional sources for microorganisms in biotechnological processes emerged as alternatives. Thus, in the context of the circular economy, this approach made it possible to reuse biomass for microbial synthesis of high value-added products, such as biosurfactants [57].

Among the various fermentation processes in biosurfactant production, SSF has several advantages already mentioned, mainly reduced consumption of water and electricity and no foam formation in the fermentation process [57, 58]. Several agro-industrial by-products and residues have already been investigated for their use as a carbon source in the production of biosurfactants. This is because the carbon source used in the process can determine the composition, concentration, yields, and physicochemical properties related to the biosurfactant [11]. Table 5.1 shows some biosurfactants reported in the literature that were produced through SSF in recent years, which demonstrates the valorization of different biomasses.

In addition, the production of biosurfactants is also favored with hydrophobic carbon sources supplemented into the culture medium. This hydrophobic source is usually associated with pure or waste fatty acids from residual vegetable, crude oils, and glycerol among

Table 5.1 Microorganisms, culture medium, type, and amount of biosurfactant from some works carried out in SSF.

Microorganism	Culture medium	Type of BS	BS produced	References
Pleurotus ostreatus	Sunflower seed shell and sunflower seed oil	Carbohydrate–peptide–lipid complex	4.69 g L^{-1}	[59]
Starmerella bombicola	Winterization oil cake and sugar beet molasses	Sophorolipids	0.179 g g^{-1} dry solids	[60]
Pseudomonas aeruginosa	Glycerol and rapeseed/wheat bran	Rhamnolipids	18.7 g L^{-1}	[61]
Pseudomonas aeruginosa	Sugarcane bagasse and fish oil	Rhamnolipid	—	[62]
Trametes versicolor	Two-phase olive mill waste, wheat bran and olive stones	—	373.6 mg g^{-1} substrates	[53]
Starmerella bombicola	Stearic acid and polyurethane foam	Sophorolipids	0.211 g g^{-1} substrates	[51]
Pseudomonas aeruginosa SS14	Rice distillers dried grain	Rhamnolipid	14.87 g L^{-1}	[63]
Pseudomonas aeruginosa 15GR	Sugarcane bagasse and sunflower seed meal	Rhamnolipid	83.7 g kg^{-1} dry solids	[20]
Aspergillus niger	Wheat bran, corncob and sugarcane molasse	—	—	[23]
Aspergillus niger	Banana stalk powder	—	2.3 g L^{-1}	[64]
Starmerella bombicola ATCC 22214	Winterization oil filtration cake and sugar beet molasses	Sophorolipids	0.20 g g^{-1} dry solids	[65]

others. For instance, hydrophobic substrates of plant origin used as a carbon source, was evaluated by Velioğlu and Ürek [59], in the production of a biosurfactant by *P. ostreatus* which reached a yield of 4.69 g L^{-1}. In addition, Lourenço et al. [53], analyzed the influence of olive mill waste in the biosurfactant production by *T. versicolor* that showed a maximum concentration of 373.6 mg g^{-1}. In addition, hydrophobic substrates of animal origin such as fish oil have been examined in the production of biosurfactants through SSF by *Pseudomonas aeruginosa* [62]. Whereas Martins et al. [66], investigated the use of biodiesel oil as carbon source to enhance a biosurfactant production by *Aspergillus fumigatus*.

On the other hand, raw materials can also serve as inert support in the biosurfactant production. Jiménez-Peñalver et al. [51] and Gong et al. [67] demonstrated in their studies that polyurethane foam can be used as physical support for the growth of *S. bambicola* and *P. aurigenosa* respectively for biosurfactant production. It is important to highlight that inert materials such as foams or clays are easier to handle and have fewer impurities when compared to agro-industrial by-products [67]. Another alternative is the use of biomass as support/substrate. From an economic and environmental point of view this is desirable for biosurfactant production due to the fact that the substrate can be use as support for microbial growth and culture medium. Although in the case of SSF the biomass requires, in most cases, preparation steps such as filtration (for oils) or pretreatments to remove lignin.

5.6 Pretreatment of Raw Materials for the Production of Biosurfactants in Solid-state Fermentation

The term biomass is not only used for plant-based materials, but also microorganisms. In the context of a biorefinery, the lignocellulosic materials used can be complex for the processes to obtain biosurfactants. Some elements such as the variation in composition from one species to another and the type of species present in different regions influence the availability of biomass. Despite this, the content of the main components in the biomass is similar, with about 50–60% of carbohydrates, i.e. cellulose and hemicelluloses and 20–30% of lignin. Whereas the rest consists of extractives, ash, and other compounds, etc. [68]. However, lignin is the main non-carbohydrate component of complex structure formed by aromatic compounds. Its association with cellulose and hemicellulose makes it difficult for microorganisms to access biomass bioconversion [69, 70].

Thus, pretreatment is a step that aims to reduce the content of the intrinsic recalcitrant components to facilitate the production of valuable chemicals. In general, in the pretreatment processes some requirements should be met, such as the high recovery of individual polymers and other compounds in the lignocellulosic material and the low formation of toxic compounds or inhibitors [71] (Table 5.2).

Likewise, the energy demand should be as low as possible. The general classification presents three categories which are: physical (crushing, grinding, irradiating, and sonication); chemical (organic solvents, alkalis, acids, ionic liquids, oxidizing agents, and deep eutectic solvents); physicochemical methods (wet oxidation and hydrothermolysis, pretreatment with steam w/w catalyst, among others) and biological methods [81]. Another important issue to be considered is that the biomass pretreatment will be selected according to the final product [82].

For instance, the first step of pretreatment for lignocellulosic biomass is particle size reduction which increases the total surface area, pore size, and microbial enzyme contact points. The different physical methods used for this purpose are ecological and rarely produce any toxic material [81, 83]. However, the energy consumption is high and will depend on the type of lignocellulosic material used [84]. This process is followed by the chemical or enzymatic hydrolyzed. Thus, chemical hydrolysis can be carried out under acidic, neutral, or alkaline conditions. It has been observed that the result of the pretreatment process is strongly influenced by pH. Thus, when the process is developed in the presence of low pH, the cellulose polymer is expected to remain intact, inhibitory or toxic compounds are not formed, and monomeric sugars are obtained from hemicellulose hydrolysis. Inorganic acids such as HCl and H_2SO_4 are used to treat lignocellulosic biomass for biosurfactant production. Although the results with concentrated acids are better, the process is more expensive. Hence, dilute-acid hydrolysis has been a successful alternative to address

Table 5.2 Common methods used for the pretreatments of lignocellulosic biomass.

Pretreatment	Advantage	Disadvantage	Improvements	References
Milling	Control of final particle size, reduction in cellulose crystallinity, increasing of available surface for enzymatic hydrolysis, improvement of material handling and the mass transfer due to fractioning of biomass	High energy requirement, non-removal of lignin during the process	Factors such as: initial feed of biomass, time of ball milling and ball to biomass mass ratio should be considered during the process	[72, 73]
Microwave	Rapid and steady heating, easy operation and energy efficient, high product yield, minimal degradation or formation of side products, short reaction time, eco-friendly	Increase of operating costs at high loads and high pressure, high-cost, energy consumption	Process efficiency can be improved through the knowledge of the dielectric properties of lignocellulosic biomass	[72, 74]
Acid pretreatment	Enzymatic hydrolysis is sometimes not required as the acid itself may hydrolyse the biomass to yield fermentable sugars	Possible formation of volatile degradation products, risk of vessels corrosion due to the high acid concentrations, high cost of the reactors, formation of inhibitory by-products	Optimization considering temperature and acid concentration, among other factors	[75]
Alkaline pretreatment	Removal of lignin content without losing reducing sugar and carbohydrates, increasing of porosity, enzymatic digestibility and surface area of biomass	Generation of inhibitors, high-cost associated with the neutralization of slurry after pretreatment, longer residence time, low efficiency to pretreat the biomass with high lignin content (wood)	Temperature optimization and pretreatment of biomass slurries combining the alkali with diluted acid	[76]
Ionic liquids	High yield of sugars and carbohydrates, low vapor pressure designer solvent, working under mild reaction conditions	Costly, complexity of synthesis and purification, toxicity, poor biodegradability and inhibitory effects on enzyme activity	Some factors such as: temperature and biomass loading should be considered during the process	[77]

Table 5.2 (Continued)

Pretreatment	Advantage	Disadvantage	Improvements	References
Supercritical fluids	High energy efficiency and low operational costs, no sugar degradation of sugars	Small residence time, formation of inhibitors, high temperature and pressure	Adjustments in temperature and pressure can be made to increase the efficiency of the process	[72, 78]
Ultrasound	High activation energy, short residence time, sufficient mass transfer	Energy intensive	Design of optimum reaction condition considering: selection of medium, frequency, sonication duration, temperature and sonication	[79]
Biological pretreatment	Selective degradation of lignin and hemicelluloses, environment friendly	Long treatment times, low yields	Combined approaches with other methods	[80]

this drawback. In the case of neutral or near neutral pH, the hemicellulose will remain in oligomeric or polymeric form because the conditions are not severe enough. Some of the alkali used for this purpose are sodium, potassium, calcium, and ammonium hydroxide which cause less sugar degradation. Whereas dissolution of the lignin fraction and preservation of hemicellulose will take place at high pH. On the other hand, for hydrolysis treatment can also utilize enzymes like β-glucosidase [85, 86]. The last step of the process involves drying the substrates (hydrolysates). Consequently, the pretreated substrate is incorporated into the medium to be used as the sugar source for growth and biosurfactant production [87].

5.7 Physicochemical Factors of Solid-state Fermentation

SmF has been the most commonly required technology in biotechnological processes involving the production of most microbial metabolism of industrial interest [88]. This is because this type of bioprocess has the advantage of providing a homogeneous distribution of nutrients in the environment, resulting in an efficient contact between nutrient and micro-organism, thus increasing its assimilation [89]. However, in the last decade, there has been a trend towards the increasing use SSF for the production of microbial biomolecules [90]. In SSF, the target microorganism is cultured in a medium with low water content or in its complete absence, using an inert and natural substrate as solid support [91, 92]. However, there are some disadvantages involving SSF, such as difficulties in scaling up and the large batch-to-batch variation of solid substrates [93]. Several aspects must be analyzed beforehand when developing and optimizing SSF bioprocesses. Castro and Sato [94] highlight the importance of paying special attention to the physicochemical factors of substrates. These factors correspond to the particle size, water absorption capacity, and the chemical composition of the substrates.

The selection of an appropriate substrate is an extremely important factor when it comes to SSF, as the solid material will act as physical support or/and nutritional source. The substrate must contain an adequate proportion of nutrients that will be used as carbon and nitrogen sources (C:N ratio) for adequate microbial growth during fermentation [95]. Various materials can be used as support for SSF, including inert supports such as vermiculite, perlite, and polyethene, which can be incorporated with a nutrient solution suitable for microbial growth [96]. On the other hand, there are natural supports such as agro-industrial residues that already have all the necessary characteristics to promote the growth of the cultivated micro-organism [94].

When selecting suitable substrates for the SSF process, it is important to ensure the availability and cost of substrates. They can provide adequate nutrients and physical support for the development of microorganisms [96, 97]. Organic residues from agro-industrial processing (rice straw, orange peel, soyabean oil cake, wheat straw, potato peel, and peanut oil cake) and household food residues are the most suitable substrates to be used in SSF due to their abundance, high availability, low or even no cost, and its chemical composition [96, 98]. However, in some cases, an additional supplement must be added to the organic waste. In other cases, a chemical or mechanical pretreatment (especially for lignocellulosic materials) is necessary due to the inaccessibility of certain nutrients to microorganisms [96].

On the other hand, the size of substrate particles is directly linked to the porosity of the solid substrate. This parameter can be analyzed by particle size distribution using sieves (mesh) or determining density. A more detailed analysis of this parameter includes the classification of particle properties based on substrate porosity, which will directly affect microbial growth and substrate bioconversion [94]. These properties can be classified as intraparticle (thermal, moisture, grain size, porosity, and kinetic property of the biological process); and extra-particles (heat transfer, permeability, and mass transfer condition properties). In this sense, the characteristics of the substrate particles can directly impact several other factors, since SSF is a system that contains a combination of three main phases: the substrate itself is the solid phase, the water trapped in the matrix and in the interparticle spaces is the liquid phase, and the gas present in spaces or pores is the gaseous phase [94, 99].

In general, small substrate particles provide a larger surface area for microbial attachment, which is desirable. However, very small particles can have the opposite effect as they clump together and reduce oxygen diffusion and, consequently, limit microbial growth. In contrast, larger particles provide larger spaces between them, enabling better conditions for heat and mass transfer, but this can provide a limited surface area for microbial fixation [99, 100]. In research carried out by Melikoglu et al. [101], it was suggested that solid substrates present heterogeneous particles or particles of intermediate size to carry out the production of biomolecules through SSF. As a result of these specific characteristics, there is a satisfactory distribution of oxygen, facilitating its diffusion process through a larger surface area and, therefore, a favorable environment for microbial growth.

The water-holding capacity is an extremely important factor for the suitability of a given substrate in SSF. This parameter indicates the capacity of substrate particles to absorb water and is directly related to the availability of binding hydrophilic groups with molecules [102]. The water absorption capacity of a substrate is essential for microbial growth and fermentation, as it directly impacts the physical characteristics of the solid substrate such as pore size, which can be changed by the swelling of solid particles after water absorption, making them favorable or unfavorable for the biodegradation and bioconversion of biomass [103]. The appropriate amount of water in the solid substrate is associated with the availability and diffusion of nutrients along with carbon dioxide and oxygen exchange mechanisms during fermentation [104]. The critical moisture point of a substrate is a measure that can be used at the same time as the estimated water absorption capacity of the substrate. This parameter can be calculated from the dehydration kinetics of the material, whose drying speed will be related to the amount of water removed as a function of time. This parameter explains the amount of water strongly bound to the support and, however, not accessible to microorganisms [94, 102].

5.8 Strategies for Scaling-up of Solid-state Fermentation for Biosurfactant Production

In an extensive review on the potential production of biosurfactants in SSF [11], the authors reported that 80% of the studies were conducted on a laboratory scale and only a small number of inquiries (2%) investigated SSF biosurfactants production in systems with 5–10 kg of culture medium. This limited number of investigations may be related to SSF issues that become even more critical in the scaling-up process: the establishment of water and gas gradients, the accumulation of heat, the monitoring or the fermentation parameters, and the downstream strategies for product recovery [60].

To address the main issues related to the scaling up of biosurfactants production by SSF, it is necessary to deepen the knowledge on bioreactor design and biosurfactant recovery procedures [105]. At industrial scales, it is usually necessary to equip bioreactors with an agitation system to equalize the distribution of nutrients and fluids, and to apply water-saturated air in order to disperse the heat and maintain the temperature level at its optimal value for the microbial metabolism [11, 57].

According to the type of aeration or mixed system there are four categories of bioreactors used in the SSF for the production of biosurfactants: 1) without aeration and mixing/agitation (tray), 2) forced aeration without mixing (packed bed), 3) without aeration and continuous or intermittent mixing/agitation (horizontal drum), and 4) forced aeration, with continuous or intermittent mixing/agitation (fluidized-bed, rocking-drum, and stirred-aerated bioreactors) (Figure 5.2). In the bioreactors of the first group the substrate remains on a tray or inside a plastic bag. Trays can be placed in rooms or equipment where air circulates around but not through the layer of biomass [106]. In the case of bags, the substrate can be left intact during the fermentation process or it can be mechanically mixed by hand [14].

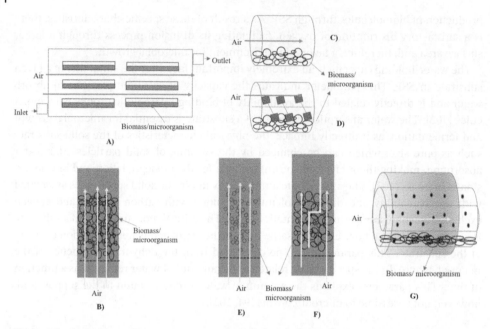

Figure 5.2 Main types of bioreactors used in SSF: A) tray bioreactor, B) fixed bed, C) rotated drum, D) agitated drum, E) fluidized bed, F) stirred aerated bioreactor, and G) rocking drum bioreactor.

Fixed bed bioreactors belong to the second group and they have the substrate placed on a perforated surface within a column and without agitation. The air is forcefully blown through the substrate inside the column. In this configuration, the growth and product formation can be influenced by the significant temperature and humidity gradients that occur within the system [60, 107]. In the third group the rotating or agitated drums are basically horizontally lying cylindrical containers that continuously rotate around their central axis to agitate the bed. The system is filled with substrate to about 20–30% of its volume and the air is supplied through the drum headspace [108].

A fluidized bed bioreactor contains air-fluidized substrate particles within a vertical vessel usually designed with sufficient height to facilitate separation of solids [14]. In the case of the stirred aerated bioreactor, it is a cylinder with a mechanical stirrer and a perforated plate at the bottom. Air is blown upwards from the bottom and the agitation can be intermittent or continuous [109, 110].

A rocking drum bioreactor consists of three drums (inner, middle, and outer) that are concentric. The substrate is placed between the inner and intermediate drums, which are perforated. Generally, air is supplied through the inner drum and moves through the perforated drums to reach the substrate layer and exit to the outer drum. The rotation of the two outer drums causes the substrate to mix [111].

Moreover, the choice of a bioreactor must take into consideration its operation costs and capital investment, product yield, and preservation of the microorganisms physical structure [112]. Mechanical agitation, for example, is known to be particularly deleterious to filamentous fungi because it provokes hyphae disintegration and the detachment of the cells from the solid medium. Although bacteria are less affected by mechanical forces, the disruption of biofilms by agitation could also compromise biosurfactant production [11, 57].

Regarding agitation, biosurfactants production can be influenced not only by the chosen system but also by its frequency. For instance, Jiménez-Peñalver et al. [57] demonstrated that intermittent mixing had a positive effect on sophorolipids yield by the yeast *Starmerella bombicola*, when compared to static conditions. The authors suggested that intermittent mixing increased the bioavailability of the nutrients, favoring higher rates of fat and oxygen consumption and that this strategy could successfully be exploited on the scaling up of biosurfactant production using rotating drum reactors since channeling and overheating in the system would be reduced, along with an increased product yield.

In contrast to submerged fermentation systems, cooling coils and water jackets are ineffective to stabilize the temperature in SSF cultures; for this reason, aeration is particularly important as a heat dispersion agent in large-scale cultivations. This parameter must be cautiously studied since high rates of air flow can compromise the humidity of the system and create gradients. As stated regarding agitation, rotating drums are indicated as a feasible bioreactor configuration to promote adequate aeration, along with agitated bioreactors [57].

It is worth noting that another important variable regarding scaling-up SSF production of biosurfactants is the support material. For instance, Gong et al. [67] identified that, when compared to agro-industrial by-products, high-density polyurethane foam (HPUF), presented fewer impurities and was a better support material for the production of rhamnolipids by *Pseudomonas aeruginosa* and could be re-used in the process three times without losing its integrity or compromising the yield of rhamnolipids.

It is generally recognized that the use of inert materials contributes to the stability of the physical structure of the solid phase because, when the solid material is subjected to microbial degradation, it faces compaction, a hindrance to mass transfer. Moreover, an inert support can facilitate the downstream processing of biosurfactants since the extraction of impurities together with the biosurfactant is reduced [57].

Another great challenge on the scaling-up process of biosurfactants production on SSF is the monitoring of fermentation parameters and product formation. Usually, metabolic activity is estimated by indirect measures such as the respiration parameters oxygen uptake rate (OUR) and the cumulative oxygen consumption (COC) [60]. It is considered that the control of SSF process is simplified when an inert support is used because nutrients can be more easily assessed, thus the optimization of the process is also simpler with this kind of support. Finally, scaling-up of SSF for biosurfactant production faces the difficulty to implement an environmentally safe and economically feasible method for the downstream processing of these compounds [4]. The downstream strategies suitable for submerged fermentation are less efficient for SSF since the solid phase, due to its complexity, generates a myriad of intricated interactions between the microorganisms, the medium, the product, and the solvent, hindering the biosurfactants extraction and compromising the final product purity [11]. Downstream steps can represent nearly 60% of a biosurfactants cost [48], therefore the feasibility of the large-scale production of biosurfactants on SSF is strictly dependent on the development of better downstream strategies specific to this cultivation system.

5.9 Conclusion

Biosurfactants can be produced in biorefineries through solid state fermentation and from renewable raw materials such as waste. However, the pretreatment of the substrates used is fundamental to ensure the best use for the microbial synthesis of these biomolecules. Several microorganisms are used to produce biosurfactants in SSF. Among them, filamentous fungi stand out due to the potential of secreted hydrolytic enzymes to degrade residual solids directly. Likewise, large-scale production is favored with the use of microorganisms with GRAS status. In addition, the adjustment of physicochemical factors allows higher yields of biosurfactants during SSF to be obtained. Another area of interest to make the process viable is production in bioreactor systems, as a variety of configurations need to be designed and implemented to suit the specific microorganism. However, studies are needed to verify the successful application to a commercially competitive system to achieve an economically feasible for biosurfactant production in biorefineries.

Acknowledgments

The authors express their gratitude to FAPESP (Fundação de Amparo à Pesquisa do Estado de São Paulo – Process number 2020/06323-0 and 2016/10636-8) and Conselho Nacional de Desenvolvimento Científico e Tecnológico (CNPq-Brazil) (DS 88882.379239/2019-01) for financial support.

References

1. Banat IM, Satpute SK, Cameotra SS, Patil R, Nyayanit NV. Cost effective technologies and renewable substrates for biosurfactants' production. *Front Microbiol*. 2014;5:697.
2. Costa JAV, Treichel H, Santos LO, Martins VG. Solid-state fermentation for the production of biosurfactants and their applications. In: Pandey A, Larroche C, Soccol CR (eds) *Current Developments in Biotechnology and Bioengineering*. 2018. pp. 357–372.
3. Santos DKF, Rufino RD, Luna JM, Santos VA, Sarubbo LA. Biosurfactants: multifunctional biomolecules of the 21st century. *Int J Mol Sci*. 2016;17:1–31.
4. Muthusamy K, Gopalakrishnan S, Ravi TK, Sivachidambaram P. Biosurfactants: properties, commercial production and application. *Cur Sci*. 2008;94:736–747.
5. Olasanmi IO, Thring RW. The role of biosurfactants in the continued drive for environmental sustainability. *Sustainability*. 2018;10(12):4817.
6. Al-Bahry SN, Al-Wahaibi YM, Elshafie AE, Al-Bemani AS, Joshi SJ, Al-Makhmari HS, et al. Biosurfactant production by Bacillus subtilis B20 using date molasses and its possible application in enhanced oil recovery. *Int Biodeterior Biodegrad*. 2012;81:1–6.
7. Akpinar M, Urek RO. Production of ligninolytic enzymes by solid-state fermentation using pleurotus eryngii. *Prep Biochem Biotechnol*. 2012;42(6):582–597.
8. Capolupo L, Faraco V. Green methods of lignocellulose pretreatment for biorefinery development. *Appl Microbiol Biotechnol*. 2016;100(22):9451–9467.

9 Soccol CR, da Costa ESF, Letti LAJ, Karp SG, Woiciechowski AL, Vandenberghe LPDS. Recent developments and innovations in solid state fermentation. *Biotechnol Res Innov*. 2017;1(1):52–71.

10 Barbosa JR, da Silva SB, de Carvalho RN. Biosurfactant production by solid-state fermentation, submerged fermentation, and biphasic fermentation. In: Inamuddin, Adetunji CO, Asiri AM (eds) *Green Sustainable Process for Chemical and Environmental Engineering and Science*, Elsevier Inc.; 2021. pp. 155–171.

11 Banat IM, Carboué Q, Saucedo-Castañeda G, de Jesús Cázares-marinero J. Biosurfactants: the green generation of speciality chemicals and potential production using Solid-state fermentation (SSF) technology. *Bioresour Technol*. 2021;320.

12 Chilakamarry CR, Mimi Sakinah AM, Zularisam AW, Sirohi R, Khilji IA, Ahmad N, et al. Advances in solid-state fermentation for bioconversion of agricultural wastes to value-added products: opportunities and challenges. *Bioresour Technol*. 2022;343.

13 Kumar V, Ahluwalia V, Saran S, Kumar J, Patel AK, Singhania RR. Recent developments on solid-state fermentation for production of microbial secondary metabolites: challenges and solutions. *Bioresour Technol*. 2021;323.

14 Krieger N, Neto DC, Mitchell DA. Production of microbial biosurfactants by solid-state cultivation. *Adv Exp Med Biol*. 2010;672:203–210.

15 Manan MA, Webb C. Design aspects of solid state fermentation as applied to microbial bioprocessing. *J Appl Biotechnol Bioeng*. 2017;4(1):91.

16 Ribeaux DR, Jackes CV, Medeiros ADM, Marinho J, Lins U, Nascimento I, et al. Innovative production of biosurfactant by Candida tropicalis UCP 1613 through solid-state fermentation. *Chem Eng Trans*. 2020;79:361–366.

17 Najmi Z, Ebrahimipour G, Franzetti A, Banat IM. In situ downstream strategies for cost-effective bio/surfactant recovery. *Biotechnol Appl Biochem*. 2018;65(4):523–532.

18 dos Santos RA, Rodríguez DM, Ferreira INDS, de Almeida SM, Takaki GMDC, de Lima MAB. Novel production of biodispersant by Serratia marcescens UCP 1549 in solid-state fermentation and application for oil spill bioremediation. *Environ Technol (United Kingdom)*. 2021;43(19):1–12.

19 Cano y Postigo LO, Jacobo-Velázquez DA, Guajardo-Flores D, Garcia Amezquita LE, García-Cayuela T. Solid-state fermentation for enhancing the nutraceutical content of agrifood by-products: recent advances and its industrial feasibility. *Food Biosci*. 2021;41.

20 El-Housseiny GS, Aboshanab KM., Aboulwafa MM, Hassouna NA. Rhamnolipid production by a gamma ray-induced Pseudomonas aeruginosa mutant under solid state fermentation. *AMB Express*. 2019;9(1).

21 Slivinski CT, Mallmann E, De Araújo JM, Mitchell DA, Krieger N. Production of surfactin by Bacillus pumilus UFPEDA 448 in solid-state fermentation using a medium based on okara with sugarcane bagasse as a bulking agent. *Process Biochem*. 2012;47(12):1848–1855.

22 Mizumoto S, Hirai M, Shoda M. Production of lipopeptide antibiotic iturin A using soybean curd residue cultivated with Bacillus subtilis in solid-state fermentation. *Appl Microbiol Biotechnol*. 2006;72(5):869–875.

23 Kreling NE, Simon V, Fagundes VD, Thomé A, Colla LM. Simultaneous production of lipases and biosurfactants in solid-state fermentation and use in bioremediation. *J Environ Eng*. 2020;146(9):04020105.

24 Rodríguez A, Gea T, Sánchez A, Font X. Agro-wastes and inert materials as supports for the production of biosurfactants by solid-state fermentation. *Waste and Biomass Valorization*. 2020.

25 Gaur VK, Sharma P, Sirohi R, Varjani S, Taherzadeh MJ, Chang JS, et al. Production of biosurfactants from agro-industrial waste and waste cooking oil in a circular bioeconomy: an overview. *Bioresour Technol*. 2022;343.

26 Zhao F, Shi R, Cui Q, Han S, Dong H, Zhang Y. Biosurfactant production under diverse conditions by two kinds of biosurfactant-producing bacteria for microbial enhanced oil recovery. *J Pet Sci Eng*. 2017;157:124–130.

27 Domínguez Rivera Á, Martínez Urbina MÁ, López y López VE. Advances on research in the use of agro-industrial waste in biosurfactant production. *World J Microbiol Biotechnol*. 2019;35(10):1–18.

28 Couto SR, Sanromán MÁ. Application of solid-state fermentation to food industry-A review. *J Food Eng*. 2006;76(3):291–302.

29 Lizardi-Jiménez MA, Hernández-Martínez R. Solid state fermentation (SSF): diversity of applications to valorize waste and biomass. *Biotech*. 2017;7(1):44.

30 nee'Nigam PS, Pandey A. Solid-state fermentation technology for bioconversion of biomass and agricultural residues. In: nee' Nigam PS, Pandey A (eds) *Biotechnology for Agro-Industrial Residues Utilisation*. Springer, Dordrecht; 2009. pp. 197–221.

31 Ascencio JJ, Chandel AK, Philippini RR, da Silva SS. Comparative study of cellulosic sugars production from sugarcane bagasse after dilute nitric acid, dilute sodium hydroxide and sequential nitric acid-sodium hydroxide pretreatment. *Biomass Convers Biorefinery*. 2019;10: 813–822.

32 Rene ER, Ge J, Kumar G, Singh RP, Varjani S. Resource recovery from wastewater, solid waste, and waste gas: engineering and management aspects. *Environ Sci Pollut Res*. 2020;27(15):17435–17437.

33 Varjani S, Lee DJ, Zhang Q. Valorizing agricultural biomass for sustainable development: biological engineering aspects. *Bioengineered*. 2020;11(1):522–523.

34 Galbe M, Wallberg O. Pretreatment for biorefineries: a review of common methods for efficient utilisation of lignocellulosic materials. *Biotechnol Biofuels*. 2019;12(1):1–26.

35 Preiss J, Sivak MN. Starch synthesis in sinks and sources. In: *Photoassimilate Distribution in Plants and Crops*. Routledge, New York; 2017. pp. 63–96.

36 Luchese CL, Benelli P, Spada JC, Tessaro IC. Impact of the starch source on the physicochemical properties and biodegradability of different starch-based films. *J Appl Polym Sci*. 2018;135(33):46564.

37 Liang S, McDonald AG. Chemical and thermal characterization of potato peel waste and its fermentation residue as potential resources for biofuel and bioproducts production. *J Agric Food Chem*. 2014;62(33):8421–8429.

38 Sepelev I, Galoburda R. Industrial potato peel waste application in food production: a review. *Res Rural Dev*. 2015;1:130–136.

39 Ansari FA, Hussain S, Ahmed B, Akhter J, Shoeb E. Use of potato peel as cheap carbon source for the bacterial production of biosurfactants. *Int J Biol Res*. 2014;2:27–31.

40 George S, Jayachandran K. Biosurfactants from processed wastes. In: Singhania R, Agarwal R, Kumar R, Sukumaran R (eds) *Waste to Wealth*. Springer, Singapore; 2018. pp. 45–58.

41 Varjani S, Shah AV, Vyas S, Srivastava VK. Processes and prospects on valorizing solid waste for the production of valuable products employing bio-routes: a systematic review. *Chemosphere*. 2021;282.

42 Das AJ, Kumar R. Utilization of agro-industrial waste for biosurfactant production under submerged fermentation and its application in oil recovery from sand matrix. *Bioresour Technol*. 2018;260:233–240.

43 Marcelino PRF, Peres GFD, Terán-Hilares R, Pagnocca FC, Rosa CA, Lacerda TM, et al. Biosurfactants production by yeasts using sugarcane bagasse hemicellulosic hydrolysate as new sustainable alternative for lignocellulosic biorefineries. *Ind Crops Prod*. 2019;129:212–223.

44 Sari SK, Trikurniadewi N, Ibrahim SNMM, Khiftiyah AM, Abidin AZ, Nurhariyati T. Bioconversion of agricultural waste hydrolysate from lignocellulolytic mold into biosurfactant by Achromobacter sp. BP (1) 5. *Biocatal Agric Biotechnol*. 2020;24:101534.

45 Nalini S, Parthasarathi R. Optimization of rhamnolipid biosurfactant production from Serratia rubidaea SNAU02 under solid-state fermentation and its biocontrol efficacy against Fusarium wilt of eggplant. *Ann Agrar Sci*. 2018;16(2):108–115.

46 Zhu Z, Li R, Yu G, Ran W, Shen Q. Enhancement of lipopeptides production in a two-temperature stage process under SSF conditions and its bioprocess in the fermenter. *Biores Technol*. 2013;127:209e215.

47 López-Pérez M, Viniegra-González G. Production of protein and metabolites by yeast grown in solid state fermentation: present status and perspectives. *J Chem Technol*. 2016;91(5):1224–1231.

48 Chaprão MJ, Soares da Silva R de CF, Rufino RD, Luna JM, Santos VA, Sarubbo LA. Formulation and application of a biosurfactant from Bacillus methylotrophicus as collector in the flotation of oily water in industrial environment. *J Biotechnol*. 2018;285:15–22.

49 Franco Marcelino PR, da Silva VL, Rodrigues Philippini R, Von Zuben CJ, Contiero J, Dos Santos JC, da Silva SS. Biosurfactants produced by Scheffersomyces stipitis cultured in sugarcane bagasse hydrolysate as new green larvicides for the control of Aedes aegypti, a vector of neglected tropical diseases. *PLoS One*. 2017;12(11):e0187125.

50 Van Bogaert IN, Zhang J, Soetaert W. Microbial synthesis of sophorolipids. *Proc Biochem*. 2011;46(4):821–833.

51 Jiménez-Peñalver P, Castillejos M, Koh A, Gross R, Sánchez A, Font X, et al. Production and characterization of sophorolipids from stearic acid by solid-state fermentation, a cleaner alternative to chemical surfactants. *J Clean Prod*. 2018;172:2735–2747.

52 De la Cruz Quiroz R, Roussos S, Hernández D, Rodríguez R, Castillo F, Aguilar CN. Challenges and opportunities of the bio-pesticides production by solid-state fermentation: filamentous fungi as a model. *Crit Rev Biotechnol*. 2015;35(3):326–333.

53 Lourenço LA, Alberton Magina MD, Tavares LBB, Guelli Ulson de Souza SMA, García Román M, Altmajer Vaz D. Biosurfactant production by Trametes versicolor grown on two-phase olive mill waste in solid-state fermentation. *Environ Technol (United Kingdom)*. 2018;39(23):3066–3076.

54 Lin JY. New structural economics: a framework for rethinking development. *World Bank Res Obs*. 2011;26(2):193–221.

55 Wu R, Geng Y, Liu W. Trends of natural resource footprints in the BRIC (Brazil, Russia, India and China) countries. *J Clean Prod*. 2017;142:775–782.

56 Guerrero AB, Muñoz E. Life cycle assessment of second generation ethanol derived from banana agricultural waste: environmental impacts and energy balance. *J Clean Prod.* 2018;174:710–717.

57 Jiménez-Peñalver P, Rodríguez A, Daverey A, Font X, Gea T. Use of wastes for sophorolipids production as a transition to circular economy: state of the art and perspectives. *Rev Environ Sci Biotechnol.* 2019;18(3):413–435.

58 Salim AA, Grbavčić S, Šekuljica N, Stefanović A, Jakovetić Tanasković S, Luković N, et al. Production of enzymes by a newly isolated Bacillus sp. TMF-1 in solid state fermentation on agricultural by-products: the evaluation of substrate pretreatment methods. *Bioresour Technol.* 2017;228:193–200.

59 Velioğlu Z, Öztürk Ürek R. Biosurfactant production by Pleurotus ostreatus in submerged and solid-state fermentation systems. *Turkish J Biol.* 2015;39(1):160–166.

60 Jiménez-Peñalver P, Gea T, Sánchez A, Font X. Production of sophorolipids from winterization oil cake by solid-state fermentation: optimization, monitoring and effect of mixing. *Biochem Eng J.* 2016;115:93–100.

61 Wu J, Zhang J, Wang P, Zhu L, Gao M, Zheng Z, et al. Production of rhamnolipids by semi-solid-state fermentation with Pseudomonas aeruginosa RG18 for heavy metal desorption. *Bioprocess Biosyst Eng.* 2017;40(11):1611–1619.

62 Narendrakumar G, Saikrishna NMD, Prakash P, Preethi TV. Production, characterization, and optimization of rhamnolipids produced by Pseudomonas aeruginosa by solid-state fermentation. *Int J Green Pharm.* 2017;11(2):92–97.

63 Borah SN, Sen S, Goswami L, Bora A, Pakshirajan K, Deka S. Rice based distillers dried grains with solubles as a low cost substrate for the production of a novel rhamnolipid biosurfactant having anti-biofilm activity against Candida tropicalis. *Colloids Surf B Biointerfaces.* 2019;182.

64 Asgher M, Arshad S, Qamar SA, Khalid N. Improved biosurfactant production from Aspergillus niger through chemical mutagenesis: characterization and RSM optimization. *SN Appl Sci.* 2020;2(5):1–11.

65 Rodríguez A, Gea T, Sánchez A, Font X. Agro-wastes and inert materials as the supports for production of biosurfactants by solid-state fermentation. *Waste and Biomass Valorization.* 2021;12(4):1963–1976.

66 Martins VG, Kalil SJ, Costa JAV. In situ bioremediation using biosurfactant produced by solid state fermentation. *World J Microbiol Biotechnol.* 2009;25(5):843–851.

67 Gong Z, He Q, Che C, Liu J, Yang G. Optimization and scale-up of the production of rhamnolipid by Pseudomonas aeruginosa in solid-state fermentation using high-density polyurethane foam as an inert support. *Bioprocess Biosyst Eng.* 2020;43(3):385–392.

68 Ståhl M, Nieminen K, Sixta H. Hydrothermolysis of pine wood. *Biomass Bioenergy.* 2018;109:100–113.

69 Hon DNS, Shiraishi N. *Wood and Cellulosic Chemistry, Revised, and Expanded.* 2nd ed. New York and Basel: CRC Press, 2000. 928.

70 Yoo CG, Meng X, Pu Y, Ragauskas AJ. The critical role of lignin in lignocellulosic biomass conversion and recent pretreatment strategies: a comprehensive review. *Biores Technol.* 2020;301:122784.

71 Jönsson LJ, Alriksson B, Nilvebrant N-O. New developments in microwave histoprocessing. *Biotechnol Biofuels.* 2013;10.

72. Mankar AR, Pandey A, Modak A, Pant KK. Pretreatment of lignocellulosic biomass: a review on recent advances. *Biores Technolo*. 2021;334:125235.
73. Maurya DP, Singla A, Negi S. An overview of key pretreatment processes for biological conversion of lignocellulosic biomass to bioethanol. *Biotechnol*. 2015;5:597–599.
74. Li H, Qu Y, Yang Y, Chang S, Xu J. Microwave irradiation - a green and efficient way to pretreat biomass. *Bioresour Technol*. 2016;199:34–41.
75. Rezania S, Oryani B, Cho J, Talaiekhozani A, Sabbagh F, Hashemi B, Parveen FR, Mohammadi AA. Different pretreatment technologies of lignocellulosic biomass for bioethanol production: an overview. *Energy*. 2020;199:117457.
76. Park YC, Kim JS. Comparison of various alkaline pretreatment methods of lignocellulosic biomass. *Energy*. 2012;47(1):31e5.
77. Elgharbawy AA, Alam MZ, Moniruzzaman M, Goto M. Ionic liquid pretreatment as emerging approaches for enhanced enzymatic hydrolysis of lignocellulosic biomass. *Biochem Eng J*. 2016;109:252e67.
78. Daza SLV, Orrego ACE, Alzate CAC. Supercritical fluids as a green technology for the pretreatment of lignocellulosic biomass. *Bioresour Technol*. 2016;199:113–120.
79. Baruah J, Nath BK, Sharma R, Kumar S, Deka RC, Baruah DC, Kalita E. Recent trends in the pretreatment of lignocellulosic biomass for value-added products. *Front Energy Res*. 2018;6:141.
80. Hassan SS, Williams GA, Jaiswal AK. Emerging technologies for the pretreatment of lignocellulosic biomass. *Biores Technol*. 2018;262:310–318.
81. Baruah J, Nath BK, Sharma R, Kumar S, Deka RC, Baruah DC, et al. Recent trends in the pretreatment of lignocellulosic biomass for value-added products. *Front Energy Res*. 2018;6:1–19 p.
82. Nauman Aftab M, Iqbal I, Riaz F, Karadag A, Tabatabaei M. Different pretreatment methods of lignocellulosic biomass for use in biofuel production. In: Abomohra AE (ed) *Biomass Bioenergy - Recent Trends Futur Challenges*. 2019;1–24.
83. Shirkavand E, Baroutian S, Gapes DJ, Young BR. Combination of fungal and physicochemical processes for lignocellulosic biomass pretreatment - a review. *Renew Sustain Energy Rev*. 2016;54:217–234.
84. Rajendran K, Drielak E, Sudarshan Varma V, Muthusamy S, Kumar G. Updates on the pretreatment of lignocellulosic feedstocks for bioenergy production – a review. *Biomass Convers Biorefinery*. 2018;8(2):471–483.
85. Tan YN, Li Q. Microbial production of rhamnolipids using sugars as carbon sources. *Microb Cell Fact*. 2018;17(1):1–13.
86. Xu Q, Liu X, Zhao J, Wang D, Wang Q, Li X, et al. Feasibility of enhancing short-chain fatty acids production from sludge anaerobic fermentation at free nitrous acid pretreatment: role and significance of Tea saponin. *Bioresour Technol*. 2018;254:194–202.
87. Mohanty SS, Koul Y, Varjani S, Pandey A, Ngo HH, Chang JS, et al. A critical review on various feedstocks as sustainable substrates for biosurfactants production: a way towards cleaner production. *Microb Cell Fact*. 2021;20(1):1–13.
88. Bhanja Dey T, Chakraborty S, Jain KK, Sharma A, Kuhad RC. Antioxidant phenolics and their microbial production by submerged and solid-state fermentation process: a review. *Trends Food Sci Technol*. 2016;53:60–74.

89 Wang F, Terry N, Xu L, Zhao L, Ding Z, Ma H. Fungal laccase production from lignocellulosic agricultural wastes by solid-state fermentation: a review. *Microorganisms*. 2019;7:12.

90 Nguyen KA, Wikee S, Lumyong S. Brief review: lignocellulolytic enzymes from polypores for efficient utilization of biomass. *Mycosphere*. 2018;9(6):1073–1088.

91 Jaramillo AC, Cobas M, Hormaza A, Sanromán MÁ. Degradation of adsorbed Azo Dye by solid-state fermentation: improvement of culture conditions, a kinetic study, and rotating drum bioreactor performance. *Water Air Soil Pollut*. 2017;228:6.

92 Jiang H, Wang W, Mei C, Huang Y, Chen Q. Rapid diagnosis of normal and abnormal conditions in solid-state fermentation of bioethanol using fourier transform near-infrared spectroscopy. *Energy and Fuels*. 2017;31(11):12959–12964.

93 Postemsky PD, Bidegain MA, González-Matute R, Figlas ND, Cubitto MA. Pilot-scale bioconversion of rice and sunflower agro-residues into medicinal mushrooms and laccase enzymes through solid-state fermentation with Ganoderma lucidum. *Bioresour Technol*. 2017;231:85–93.

94 de Castro RJS, Sato HH. Enzyme production by solid state fermentation: general aspects and an analysis of the physicochemical characteristics of substrates for agro-industrial wastes valorization. *Waste and Biomass Valorization*. 2015;6(6):1085–1093.

95 de Castro RJS, Nishide TG, Sato HH. Production and biochemical properties of proteases secreted by Aspergillus niger under solid state fermentation in response to different agroindustrial substrates. *Biocatal Agric Biotechnol*. 2014;3(4):236–245.

96 Sadh PK, Duhan S, Duhan JS. Agro-industrial wastes and their utilization using solid state fermentation: a review. *Bioresour Bioprocess*. 2018;5(1):1–15.

97 Yazid NA, Barrena R, Komilis D, Sánchez A Solid-state fermentation as a novel paradigm for organic waste valorization: a review. *Sustainability (Switzerland)*. 2017;9.

98 Leite P, Sousa D, Fernandes H, Ferreira M, Costa AR, Filipe D, et al. Recent advances in production of lignocellulolytic enzymes by solid-state fermentation of agro-industrial wastes. *Curr Opin Green Sustain Chem*. 2021;27.

99 Pandey A, Soccol CR, Mitchell D. New developments in solid state fermentation: i-bioprocesses and products. *Proc Biochem*. 2000;35(10):1153–1169.

100 Ruiz HA, Rodríguez-Jasso RM, Rodríguez R, Contreras-Esquivel JC, Aguilar CN. Pectinase production from lemon peel pomace as support and carbon source in solid-state fermentation column-tray bioreactor. *Biochem Eng J*. 2012;65:90–95.

101 Melikoglu M, Lin CSK, Webb C. Stepwise optimisation of enzyme production in solid state fermentation of waste bread pieces. *Food Bioprod Process*. 2013;91(4):638–646.

102 Mussatto SI, Aguilar CN, Rodrigues LR, Teixeira JA. Fructooligosaccharides and β-fructofuranosidase production by Aspergillus japonicus immobilized on lignocellulosic materials. *J Mol Catal B Enzym*. 2009;59(1–3):76–81.

103 Chen HZ, He Q. Value-added bioconversion of biomass by solid-state fermentation. *J Chem Technol Biotechnol*. 2012;87(12):1619–1625.

104 Orzua MC, Mussatto SI, Contreras-Esquivel JC, Rodriguez R, de la Garza H, Teixeira JA, et al. Exploitation of agro industrial wastes as immobilization carrier for solid-state fermentation. *Ind Crops Prod*. 2009;30(1):24–27.

105 Roelants S, Solaiman DKY, Ashby RD, Lodens S, Van Renterghem L, Soetaert W. Production and applications of sophorolipids. In: *Biobased Surfactants*, 2nd ed. Elsevier Inc. United Kingdom; 2019. pp. 65–119.

106 Velioglu Z, Urek RO. Optimization of cultural conditions for biosurfactant production by Pleurotus djamor in solid state fermentation. *J Biosci Bioeng*. 2015;120:526–1.

107 Castiglioni GL, Stanescu G, Rocha LAO, Costa JAV. Analytical modeling and numerical optimization of the biosurfactants production in solid-state fermentation by Aspergillus fumigatus. *Acta Sci Technol*. 2014;36(1):61–67.

108 Shima D, Ataei SA, Taheri A. Performance analysis of a laboratory scale rotating drum bioreactor for production of rhamnolipid in solid-state fermentation using an agro-industrial residue. *Biomass Convers Biorefinery*. 2021:1–8.

109 Bandelier S, Renaud R, Durand A. Production of gibberellic acid by fed-batch solid state fermentation in an aseptic pilot-scale reactor. *Process Biochem*. 1997;32:141e145.

110 Chamielec Y, Renaud R, Maratray J, Almanza S, Diez M, Durand A. Pilot-scale reactor for aseptic solid-state cultivation. *Biotechnol Tech*. 1994;8:245e248.

111 Ge X, Vasco-Correa J, Li Y. Solid-state fermentation bioreactors and fundamentals. In:Larroche C, Sanromán MA, Du G, Pandey A (eds) *Current Developments in Biotechnology and Bioengineering*, Elsevier; 2017. pp. 381–402.

112 Gurkok S. Important parameters necessary in the bioreactor for the mass production of biosurfactants. In: Inamuddin, Adetunji CO, Asiri AM (eds) *Green Sustainable Process for Chemical and Environmental Engineering and Science*, Elsevier Inc.; 2021. pp. 347–365.

6

An Overview of Developments and Challenges in the Production of Biosurfactant by Fermentation Processes

F.G. Barbosa[1], M.J. Castro-Alonso[1], T.M. Rocha[1], S. Sánchez-Muñoz[1], G.L. de Arruda[2], M.C.A. Viana[1], C.A. Prado[2], P.R.F. Marcelino[1], J.C. Santos[1,*], and Silvio S. Da Silva[1]

[1] Bioprocesses and sustainable products laboratory, Department of Biotechnology, Engineering School of Lorena, University of São Paulo (EEL-USP), 12.602.810. Lorena, SP, Brazil
[2] Bioprocesses, biopolymers, simulation and modeling laboratory, Department of Biotechnology, Engineering School of Lorena, University of São Paulo (EEL-USP), 12.602.810. Lorena, SP, Brazil
* Corresponding author

6.1 Introduction

Surfactants are molecules with amphipathic structure, that is, they have both hydrophobic and hydrophilic regions. Due to this characteristic, these molecules can reduce surface and interfacial tension, showing emulsifying action [1]. These compounds are usually synthesized by chemical routes from petroleum derivatives, and have long carbon chains, commonly branched by aromatic groups that prevent their biodegradation in the environment. They are present in different industrial products, and due to the indiscriminate use of chemically synthesized surfactants, as well as their disposal in rivers, seas, and lakes, these molecules lead to environmental problems, such as eutrophication of rivers and lakes, solubilization of toxic organic compounds, and formation of foam [2–4].

The interest in reducing the use of products harmful to human health and to the environment has increased the attention of researchers around the world on biological surfactants. Biosurfactants are synthesized by different organisms including plants, animals, and microorganisms. The microbial biosurfactants are produced by bacteria, fungi, and yeasts by the action of different biological enzymes. Studies have demonstrated the presence of biosurfactant-producing microorganisms in terrestrial habitats and aquatic ecosystems [5, 6].

In addition to synthesizing biosurfactants in their natural habitats, microorganisms can produce these molecules in fermentation processes using renewable substrates as raw materials [7, 8]. These biosurfactants are biodegradable molecules, with low toxicity, and are considered biocompatible and eco-friendly [3, 4, 9]. These biomolecules perform superior activities under extreme conditions, such as temperature, pH, and salinity, becoming potential substitutes for synthetic surfactants in several industrial sectors [9, 10].

Studies show an increase in the global market and in the search for biological substitutes for synthetic surfactants. The global market for biosurfactants is expected to exceed

$5.52 billion by 2022, and sales of these biomolecules are expected to reach 2.6 billion by 2023 [11]. However, biosurfactants still do not compete with their synthetic counterparts due to the high cost of raw materials, which represents about 50% of the final manufacturing cost, and to the low yields of product obtained at the end of the process [12].

Besides the high cost of raw material, there are other challenges to be faced in different stages of the bioprocess. They include foaming formation in different steps of the process, challenges in the scale-up of laboratory developed systems, and in the recovery and purification steps. In this chapter, the applications of biosurfactants are discussed, along with the main challenges involved in the fermentative process, and discussing sustainable alternatives for their production.

6.2 Current Market and Potential Applications of Biosurfactants

Biosurfactants have emerged as a promising alternative to replace synthetic surfactants, because of their advantageous characteristics, such as higher biodegradability, biocompatibility, low toxicity, specificity, and production from renewable sources [10, 13]. They are versatile molecules that can be applied in different areas.

The growing search for sustainable alternative molecules has resulted in the expansion of the market for this bioproduct. In 2020, the global biosurfactant market was about $1.7 billion and will grow with a CAGR of 13.38% from 2020 to 2027 [14]. It is expected that the high demand for this biomolecule induces industrial development and its use in the daily lives of the population [15, 16].

The market for biosurfactants includes a large number of medium and large companies distributed all over the world. Some important producers in 2021 include BASF SE (Germany), Ecover (Belgium), Jeneil Biotech, Inc (United States), Saraya Co. (Japan), AGAE Technologies (United States), Soliance (France), GlycoSurf (United States), and Kingorigin (China) (MarketWatch, 2021). Additionally, it is also worth mentioning Boruta Zachem AS (Poland), AkzoNobel NV (Netherlands), TeeGene Biotech Ltd. (UK), Biotensidon GmbH (Germany), Evonik Industries (Germany), Kao Corporation (Japan), Croda International (United Kingdom), Lion Corporation (Japan), Givaudan SA (Switzerland), and Henkel Corporation (Germany) [16, 17].

Biosurfactants have properties to promote the reduction of surface and interfacial tension, increase the solubility of immiscible compounds in water, and promote the stability of emulsions [18]. Thus, they have several applications, such as detergency, emulsification, lubrication, foaming capacity, wettability, and solubilization, which are already applied in several industrial sectors. Consequently, these substances can be used in the food industry, agriculture, oil recovery, cleaning solutions, skincare products, environmental bioremediation, and medical drugs (Figure 6.1) [3, 19, 20].

6.3 Biosurfactant as a Sustainable Alternative: Factors Influencing its Production

Despite the wide range of potential advantages and applications of biosurfactants in several sectors over synthetic ones, their commercialization is still low due to cost-attractive disadvantages. The large-scale outcomes of biosurfactants faces some challenges, such as the

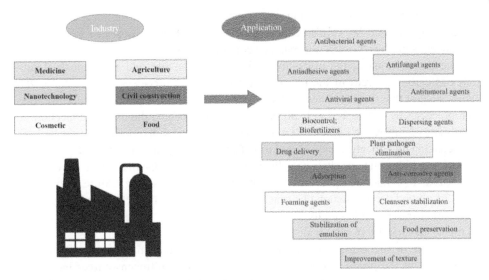

Figure 6.1 The main applications of biosurfactants.

high-cost of raw materials, low yield due to available downstream methods, and expenses involved in purification. Thus, researchers have been evaluating strategies to improve the production of biosurfactants by biotechnological routes, such as genetic modification of microorganisms, utilization of low-cost substrates (forest and agricultural residues and by-products), and optimization of bioprocesses via computational modeling [19, 21].

6.3.1 Factors Involved in the Biosurfactant Production

Biosurfactant concentration, types, and functionality are influenced by nutritional and physical factors, such as nutrients, inoculum concentration, pH, temperature, agitation, aeration, salinity, among others [3, 9, 22]. The main factors are discussed in the following sections.

6.3.1.1 Nutritional Parameters

The composition of the fermentation medium is a key parameter for the metabolism of microorganisms, influencing their cellular growth and the production of certain biomolecules [20]. Several studies show that different types and concentrations of carbon and nitrogen sources alongside metal ions directly affect the concentration, yield, quality, and cost of biosurfactant production in laboratory scale and in large-scale fermenters [20, 23, 24]. Carbon sources range from saccharine, starchy, and lignocellulosic substrates, in addition to lipophilic substrates [3].

For example, Jain et al. [23] showed that the different carbon sources influenced the production and quality of biosurfactants synthetized by the alkaliphilic bacterium *Klebsiella sp.* They observed that, when adding 30 g L^{-1} of carbon source to the culture medium, the amount of biosurfactant produced was the highest with starch, followed by sucrose, xylose, galactose, glucose, and fructose. This study also showed that the tested carbohydrates significantly influenced the physicochemical properties of biosurfactants. Biosurfactants obtained from the medium containing starch, sucrose, xylose, galactose, and glucose, demonstrated higher viscosity, emulsification activity, and reduction in surface tension than biosurfactants produced from culture media containing fructose and maltose.

Studies have evaluated the utilization of hydrocarbons, oils, and agro-industrial substrates as carbon sources in the biosurfactant production processes [7, 13, 22, 25, 26]. Studies suggest that the use of low-cost substrates or waste materials can reduce the costs of bioprocesses by 10% to 30% [27, 28]. Some examples include the use of hemicellulosic hydrolysate of sugarcane bagasse [7] and straw [13], corncob hemicellulosic liquor [22], waste cooking oil [29], glycerol [25], and tuna fish waste [30]. de Almeida et al. [29], using canola waste frying oil to produce glycolipid surfactant by *Pseudomonas cepacian*, reported that the cost to produce 40.5 g L^{-1} in a 50 L fermenter is around \$20 kg^{-1}. More recently, Hu et al. [30] studied the production of surfactin *Bacillus subtilis* ATCC 21332 on a large scale. They demonstrated that *B. subtilis* ATCC 21332 can use the fish waste peptones to produce 274 mg L^{-1} of surfactin in pilot-scale experiments.

In addition to the carbon source, the use of an adequate nitrogen source is also fundamental in the production of biosurfactants. Different nitrogen sources have been used, such as yeast extract, meat extract, tryptone, ammonium sulfate, ammonium nitrate, sodium nitrate, rice bran extract, soybean bran extract, among others [18, 31, 32]. Zargar et al. [33] showed glycolipid production by *Bacillus* sp. using ammonium sulfate and they observed that biosurfactant production of approximately 6.04 g L^{-1}. Furthermore, they found that the molecule produced was stable over a range of temperature from 30 °C to 70 °C, salinity from 0 to 150 g L^{-1}, and pH from 4 to 10. In another study, Gharaei et al. [34] optimized the production of glycolipid biosurfactant for *Shewanella algae* B12 by different nitrogen sources and found the best levels of $NaNO_3$ (0.2 g L^{-1}), NH_4Cl (0.7 g L^{-1}), and peptone (0.5 g L^{-1}). Under these conditions, the biosurfactant promoted the maximum reduction of the surface tension (31 mN m^{-1}).

Some compounds induce the production of biosurfactants, they act as supplements in the culture medium. These are added at low concentrations and are classified as hydrophobic and hydrophilic inducers [35, 36]. The hydrophobic inducers include hydrocarbons derived from petroleum and vegetable oils. While the hydrophilic inducers are metals and are involved in the biosynthetic reactions of microbial metabolism [36–38].

Several studies showed the effect of hydrophilic inducers such as Na^+, K^+, Ca^{2+}, Fe^{2+}, Mn^{2+}, and Mg^{2+} [18, 39] in microbial metabolism. The study of Gudina et al. [40] described the production of biosurfactants by *B. subtilis* in an alternative medium containing Mg^{2+}, Mn^{2+}, and Fe^{2+}. It was observed that supplementation with the three metals induced the production of about 3.6 times more biosurfactant than the control without the addition of trace elements. These minerals induce the production, acting as cofactors in the metabolic pathways involved in the synthesis of these biomolecules, and also in microbial growth [40–42]. More studies are necessary to elucidate the metabolic processes involved in the synthesis of biosurfactants and to improve the fermentation process on a large scale from the utilization of low-cost substrates.

6.3.1.2 Other Important Parameters
6.3.1.2.1 Inoculum Concentration

It is well documented that inoculum concentration proportionally affects the number of cells and biosurfactant production. Furthermore, the influence of inoculum concentration in the production of biosurfactants depends on the type and age of cells [43, 44]. Several studies have shown that the increase of inoculum concentration from 1.0% to 1.5% improved cell growth and biosurfactant production, whereas 2% of inoculum resulted in reduction of microbial activity due to nutritional limitations of *Bacillus sp* [45–47]. More recently, Sun et al. [44] optimized biosurfactant production using *Pseudomonas sp.* CQ2, adjusting several parameters, including inoculum concentration, pH, and temperature. They obtained a maximum biosurfactant production (40.7 g L^{-1}) with an inoculum ratio of 3%, using pH 7 at 35 °C.

6.3.1.2.2 pH

The influence of pH in biosurfactant production is related to the activity of enzymes involved in the microbial growth and biosynthesis of biosurfactants [1, 48]. For example, Chen et al. [49] evaluated different pH values (6–12) and nitrogen source concentrations (1.0–5.0 g L^{-1}) to optimize the production of rhamnolipid from kitchen waste oil using *Pseudomonas aeruginosa*. The authors demonstrated that the optimal pH and nitrogen source concentration for maximum biosurfactant production were 8 and 2 g L^{-1}, respectively. Sharma and Padey [50], optimized biosurfactant production and oil degradation by *B. subtilis* RSL 2 using response surface methodology. The maximum biosurfactant production was observed at pH 4.0, 25 °C, using 1 g L^{-1} crude oil as the only carbon source, in seven days.

Regarding the influence of pH on biosurfactant activity, De Freitas et al. [51] reported rhamnolipids as control agents of food pathogens. They observed that the antimicrobial activity of rhamnolipidic biosurfactant is pH dependent. The activity of rhamnolipid against Gram-positive pathogens (*Listeria monocytogenes, Bacillus cereus*, and *Staphylococcus. aureus*) was favored at more acidic conditions.

6.3.1.2.3 Temperature

The fermentation temperature has a great influence on enzymes catalytic activity in microorganisms. Najafi et al. [52] investigated the microbial-culture factors in the production of extracellular biosurfactants by a new strain of *Bacillus mycoides*. They showed that glucose and temperature were the parameters with greater effects on microbial growth and metabolism. The optimum temperature for microbial growth and biosurfactant production was 39 °C. The authors also showed that temperature is an important factor for microbial growth and biosurfactant production of *B. mycoides*. In other work, Hosseininoosheri et al. [53] studied the effect of salinity and temperature on microbial growth rate and the microbial enhanced oil recovery (MEOR) process. In MEOR processes, bacteria and nutrients are added to a reservoir. Bacteria use the nutrients present in the surroundings, and they can produce biosurfactants that act in the recovery of the remaining oil in reservoirs. The authors found that microbial growth rate is directly related to the effectiveness of the oil recovery process, and the enhancement of MEOR processes by biosurfactant-producing bacteria is highly dependent on temperature and salinity. They obtained an optimal oil recovery (50%) at 40 °C and 0.37 meq ml^{-1} of salinity. More recently, Sun et al. [44] enhanced the biosurfactant production yield of *Pseudomonas sp.* CQ2. Their results indicated the maximum biosurfactant production (40.7 g L^{-1}) was obtained at 35 °C using soybean oil and ammonium nitrate as carbon and nitrogen sources, respectively, at pH 7.

6.3.1.2.4 Salinity

Salt concentration in a fermentative medium affects the production, stability, and functions of biosurfactants in different ways. Hu et al. [54] showed that the absence of NaCl in culture medium resulted in poor microbial growth and low production of biosurfactant by marine *Vibrio sp.* strain 3B-2. In another study, Souza et al. [55] optimized the culture medium to improve the stability and production of biosurfactant in shaken flasks and bioreactor by *Wickerhamomyces anomalus* CCMA 0358. The authors showed that biosurfactant remained stable at high salinity (300 g L^{-1}) and pH values of 6–12.

Other biosurfactants produced by yeasts have been reported as stable in higher salinity up to 150 g L^{-1} [56] and 300 g L^{-1} [57]. In a recent work, Ibrahim et al. [58] evaluated the growth kinetics, biosurfactant production, and ability to degrade canola oil by a novel soil bacterium (*Rhodococcus erythropolis strain* AQ5-07). They showed that *Rhodococcus erythropolis* AQ5-07 is a potent biosurfactant producer and efficient canola oil degrader. Biosurfactant produced by this microorganism demonstrated high hydrophobicity, emulsification activity, and low surface tension value at optimum conditions, correspondent to 2% NaCl, pH 7.5, and 15 °C.

6.3.1.2.5 Aeration and Agitation

Aeration and agitation are the most crucial factors in bioprocesses, since these parameters provide the exchange of oxygen from the gas to aqueous phase in the fermentative medium [59–61]. For example, Nazareth et al. [60] evaluated important factors in the scale-up process, namely the impact of oxygen transfer coefficient (k_{La}) in volumetric productivity, and the effect of agitation (150–300 rpm) and aeration (0.5–1.5 vvm), on surfactin production by *B. subtilis* ATCC 6051 in a stirred tank bioreactor, using trub as a carbon source. They found that agitation presented a significant negative effect on surfactin concentration. Furthermore, the authors observed that biosurfactant volumetric productivity increased with k_{La} value and the aeration increases from 0.5 to 1.5 vvm, keeping the agitation constant (at 300 rpm). Zambry et al. [61] studied the lipopeptide biosurfactant production by *Streptomyces sp.* PBD-410L in batch and fed-batch fermentation in a 3 L stirred-tank reactor using palm oil as the sole carbon source. They observed that the optimum agitation speed and aeration rates for lipopeptide biosurfactant production were 200 rpm and 0.5 vvm, respectively. Applying fed-batch operation based on DO (dissolved oxygen) values at an initial feed rate of 0.6 mL h^{-1}, enhanced the lipopeptide biosurfactant production from 3.74 to 5.32 g L^{-1} compared to batch cultivation. Furthermore, the authors discussed the importance of oxygen for the oxidative metabolism of biosurfactant production by filamentous bacteria.

6.4 Strategies and Main Challenges for Biosurfactant Production

Increasing the production of biosurfactants is the main challenge for researchers and industrial sectors. Several microorganisms are capable of synthesizing a large number of biosurfactants. However, there are several problems still to be solved to favor the implementation of these molecules in the market [62].

The difficulties related to biosurfactant production can be attributed to the factors as limitations in microbial metabolism, the use of high-cost and non-renewable raw materials, and the cost of downstream processing [63]. Thus, this section has presented the obstacles associated with biosurfactant production aiming at its scale-up, discussing the use of bioreactor and the techniques and advances in separation and purification processes.

6.4.1 Process Configurations as Strategies for Biosurfactant Production

Due to the widespread utilization of biosurfactants, different strategies have been evaluated to make the production process feasible, more efficient, and cost-effective. Different strategies evaluated include: a) fermentative conditions (e.g. the choice of high-yielding microorganisms, adequate inoculum size, and appropriate C/N ratio in the medium), b) studies of downstream recovery processes, c) the use of cheap carbon and nitrogen substrates, d) the development of overproducing strains, and e) the evaluation of different operational modes [7, 64–67]. These strategies could be considered determinants for the economic and commercial establishment of biosurfactants [20].

6.4.1.1 Submerged and Solid-state Fermentation for Biosurfactant Production

There are two main possibilities to produce biosurfactants: submerged fermentation (SmF), also called liquid fermentation, and solid-state fermentation (SSF) [68, 69]. Most of biosurfactant production is performed in submerged fermentation. Bioprocesses conducted in submerged fermentation have some advantages such as online monitoring of parameters such as dissolved oxygen, temperature, and pH, homogenization of the culture media, biomass separation after fermentation, and high recovery of the products of interest [70, 71].

In this context, many reports have focused on the production of biosurfactants by SmF following different strategies. For example, Hentati et al. [72] studied the potential of a new hydrocarbonclastic marine bacterium, *Bacillus stratosphericus* FLU5, to produce an efficient surface-active agent using different carbon sources (crude oil, diesel fuel, motor oil, used motor oil, corn oil, olive oil, residual frying oil, and glycerol). They found a critical micelle concentration of the produced biosurfactant of 50 mg L^{-1}, and at this concentration, the surface tension of the water was reduced 2.57 times. The strain FLU5 showed its capacity to produce biosurfactants from residual frying oil (a cheap renewable carbon source), which could minimize the biosurfactant production cost. Chen et al. [49] also used a kitchen waste oil as a carbon source for biosurfactant production, and their results showed a critical micelle concentration of 55.87 mg L^{-1}. Both reports valued the use of cheap carbon sources and showed their potential to produced biosurfactants. In another work, Sun et al. [44] also studied the production of biosurfactants from different carbon sources (glucose, peanut oil, glycerol, soybean oil, olive oil, N-hexadecane, paraffin, and kerosene) by *Pseudomonas sp.* CQ2. Results showed that the maximum biosurfactant production (40.7 g L^{-1}) was obtained at 35 °C using soybean oil and ammonium nitrate as carbon and nitrogen sources, with pH 7, rotational speed of 175 rpm and inoculation ratio of 3%. This study also focused on the use of this surfactant for the remediation of heavy metal (Cd, Cu, and Pb) contaminated soil. As outlined in the examples, there is a prevailing concern to find low-cost carbon sources for the production of biosurfactants. SmF is a well-known strategy for the study and optimization of several parameters that could help to enhance the production of these high-value molecules.

On the other hand, solid-state fermentation (SSF) strategy is also an interesting alternative to produce biosurfactants. In SSF, microorganisms grow in their natural habitat and are able to produce various secondary metabolites (e.g. enzymes, pigments, organic acids, etc.) that are produced in low amounts or not produced in SmF [73]. As well as the microbiological advantages, SSF has other positive contributions for a cost-effective production of biomolecules, such as the use of low-cost substrates, simple equipment requirements, higher concentration of products, avoidance of foam formation (a usual drawback in SmF), and reduced water and electricity consumption [74, 75].

As well as SmF, SSF allows the use of alternative substrates such as industrial or agricultural by-products (e.g. lignocellulosic and starch-based biomass) to reduce costs [68, 76]. For example, Lourenço et al. [77] reported the production of biosurfactants by *Trametes versicolor*, which was grown in olive mill waste in SSF. Their results showed the highest biosurfactant production around 373.6 ± 19.4 mg in 100 g of culture medium, and it was also able to reduce the surface tension of an aqueous extract of the culture medium to 34.5 ± 0.3 mN m^{-1}. In other work, Rubio-Ribeaux et al. [78] evaluated the ability of *Candida tropicalis* UCP 1613 to produce biosurfactants (with values around of 25.8 mN m^{-1} for surface tension) through SSF using raw glycerol and instant noodle waste as substrates. Recently, Dos Santos et al. [79] obtained values around 28.4 mN m^{-1} when the surfactant produced from wheat bran by *Serratia marcescens* UCP 1549 was verified using SSF as the fermentative strategy.

Das and Mukherjee [80], compared the efficiency of lipopeptide biosurfactant production by *B. subtilis* DM-03 and DM-04 strains in SmF and SSF systems using dried potato peels as a cheap carbon source. They found that both strains produced a similar quantity of the lipopeptide surfactant in both fermentation strategies (*B. subtilis* DM-03: 80.0 ± 9 mg g^{-1} in SmF and 67.0 ± 6 mg g^{-1} in SSF; *B. subtilis* DM-04: 23.0 ± 5.0 mg g^{-1} in SmF and 20.0 ± 2.5 mg g^{-1} in SSF). Kreling et al. [81] used a mixture of wheat bran and corncob (80:20) for the simultaneous production of biosurfactants and lipases (6.67 ± 0.06 units of emulsifying activity and 10.74 ± 0.54 uits of lipase activity) by *Aspergillus niger* in SSF.

Other fermentative strategies have been used to enhance the production of surfactants, such as biphasic fermentation (SmF-SSF in a single system) and the intensification of the process to produce biosurfactants and other high-value molecules [81, 82].

6.4.1.2 Fermentative Configurations Applied to Biosurfactant Production

Biosurfactant production using low-cost substrates can be achieved by different configurations such as separated hydrolysis and fermentation, simultaneous saccharification and fermentation, and semi-simultaneous saccharification and fermentation that will be described in this section.

In the separate hydrolysis and fermentation (SHF) method, the first step is the enzymatic saccharification of the substrate (e.g. starch or lignocellulosic biomass). Subsequently, the fermentation process is performed by adding the fermenting microorganisms in the hydrolysate [83, 84]. On the other hand, in the simultaneous saccharification and fermentation (SSF) process, enzymatic hydrolysis and fermentation are implemented at the same time, usually in the same reaction vessel. Due to the lower temperature employed in the process, this method requires more enzymes than the SHF [84, 85]. The semi-simultaneous saccharification and fermentation (SSSF) process is a combination of SSF and SHF, with an initial hydrolysis step performed before fermentation, followed by the inoculation of microorganisms in the same reactor [86, 87]. In SSSF, the first phase (hydrolysis) and the second phase (simultaneous hydrolysis and fermentation) can be performed under different conditions, favoring optimum hydrolysis in first phase.

Jiménez et al. [88] studied different alkaline pretreatment strategies to produce sugarcane straw enzymatic hydrolysates. Those hydrolysates were used as the substrate in SHF strategy, producing surfactants with emulsification index of 54.33%, 48.66%, and 32.66% from sodium sulfite, sodium hydroxide, and ammonium hydroxide pretreated sugarcane straw, respectively. In another study, Panjiar et al. [89] studied the potential of *Serratia nematodiphila* to produce different biosurfactants from rice straw hydrolyzate. The results exhibited higher glycolipid production (4.5 ± 0.6 g L^{-1}) in xylose-rich hydrolysate than in glucose-rich enzymatic hydrolysate (3.1 ± 0.2 g L^{-1}) produced by SHF strategy.

Faria et al. [90] studied different strategies of configurations (SHF, SSF, and fed-batch SSF) for the production of mannosylerythritol lipids (MEL). They achieved a higher yield by using fed-batch SSF with pre-hydrolysis, obtaining 2.5 g L^{-1} (wheat straw) of MEL. Zheng et al. [91] also reported the use of SSF strategy to produce biosurfactants (rhamnolipids). In this study, cassava residues were used as substrate, and the maximum production of rhamnolipids (11.49 g L^{-1}) was achieved with 5% (w/v) of total solids in the reactor.

Many efforts have been made to produce biosurfactants at large scale; however, it has not reached a satisfactory cost-effective level due to their low yields for production and purification steps. Nevertheless, all the described strategies could enhance the high-cost input required for downstream processing to recover and purify biosurfactants [92–94]. As the strategies discussed in this section, the production and quality of biosurfactants can be improved with scaling up strategies at the bioreactor level and this will be discussed in the next section.

6.4.2 Bioreactors Used in the Biosurfactants Production: Types, Advantages, and Disadvantages

Bioreactors are fundamental for a fermentation process to obtain greater productivity [95, 96]. An empirical analysis of different bioreactors is very important to provide strong evidence of their use, such as design aspects, functionality, and kinetic behavior, which are specific for each bioreactor configuration. Furthermore, it is possible control several parameters such as dissolved oxygen concentration, temperature, pH, substrate feed rate, etc. by adding process sensors [97, 98].

Microbial fermentation can be divided into two main categories: using free cells or immobilized systems. In the case of suspended free cells, the types of bioreactors employed are stirred tank (STR), air-lift (ALR), and bubble column bioreactors (BCR) (Figure 6.2). For immobilized cells, membrane (MR), packed-bed (PBR), and fluidized-bed bioreactors (FBR) are usually applied [96, 98]. Non-conventional bioreactor configuration is also applied for the production of biosurfactants as reported by Chtioui et al. [99], who developed a rotating disc bioreactor to produce lipopeptides (Figure 6.2).

In the bioreactors used in biosurfactant production, foam production can be a limiting factor in agitated and aerated systems. The biosurfactants such as rhamnolipids, surfactin, and mannosylerythritol display a high-foaming capacity. On the other hand, sophorolipids are surfactants with low foaming capacity, which facilitate their use in standard-developed bioreactors [100]. Foam accumulation promotes a reduction in the work volume of the

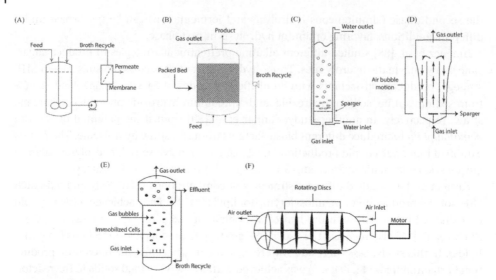

Figure 6.2 Types of bioreactors. (A) stirred tank (STR), (B) packed-bed (PBR), (C) bubble column bioreactors (BCR), (D) air-lift (ALR), (E) fluidized-bed bioreactors (FBR), and (F) rotating disc bioreactor.

fermenter, increases the risk of contamination, and reduces the production of biosurfactants with a decrease in the final yield [101–103].

Considerations about foam formation is crucial in the viability of the production process, and there are several alternatives evaluated to overcome this challenge. Among them are the use of anti-foaming agents, foam fractionation, foam adsorption, defoaming tandem system, as well as replacing the conventional submerged fermentation with solid fermentation [30, 104–106]. In the foam fractionation recovery process, the biosurfactant molecules are adsorbed on the surface of the air bubbles and subsequently collected [107]. Studies revealed that the concentration of recovered biosurfactants is approximately 50 times higher in the foam than in culture medium [108, 109]. Therefore, this technique can provide high efficiency of biosurfactant recovery and can reduce negative effects of this event [107, 110].

Thus, for the selection of the type of bioreactor to produce biosurfactant it is necessary to consider the main needs of the production processes [13, 100, 111]. Table 6.1 shows different studies covering the influence of bioreactor types, optimized conditions, biosurfactant concentration, and biosurfactant production.

The STR has advantages in the production of biosurfactants such as the facility in operation and the capacity to reach a high volumetric oxygen transfer coefficient (k_{La}) [67, 100]. Furthermore, STR can act in highly viscous environments up to 2 Pa s^{-1}. It is an important advantage as the viscosity of the medium increase with the fermentation time. The increment of viscosity promotes the reduction of oxygen mass transfer to the medium [123, 124].

On the other hand, ALR and BCR are less employed in biosurfactant production. The bubble system can induce foam formation and reduce the oxygen mass transfer. Consequently, there may not be enough oxygen available for the microorganism to synthesize the biosurfactant.

Table 6.1 Examples of biosurfactant production works using different bioreactors, mode of operation, and microorganisms.

Biosurfactant type/ microorganism	Bioreactor type/ mode of operation	Conditions	Concentration (yield)	References
Sophorolipid/*S. bombicola* ATCC 22214	STR/fed-batch ultrasound assisted	300 rpm	55.6 g L^{-1}	[112]
Sophorolipid/*S. bombicola* ATCC 22214	STR/fed-batch	130–250 rpm; 0.3–1.0 vvm	54.0 g L^{-1}	[113]
Sophorolipid/*S. bombicola* ATCC 22214	STR/fed-batch *in-situ* separation design	600 rpm; 2.67 vvm	240 g L^{-1} (0.73 g g^{-1})	[102]
Sophorolipid/*C. albicans* O-13-1	STR/semi-continuous (with modified ventilation design)	350–450 rpm; 8 air L min^{-1}	477 g L^{-1} (0.6 g g^{-1})	[114]
Rhamnolipid/*Pseudomonas aeruginosa* O-2-2	STR/fed-batch	500 rpm; pH-controlled 6.0	70.56 g L^{-1}	[115]
Rhamnolipid/*Pseudomonas aeruginosa* KT1115	STR/integrated foam-control fed-batch	300 rpm; 1.0 vvm	48.67 g L^{-1} (0.83 g g^{-1})	[116]
Rhamnolipid/*Pseudomonas aeruginosa* TIB-R02	STR/batch	200 rpm; 10–150 L air min^{-1}	60 g L^{-1} (0.80 g g^{-1})	[117]
Rhamnolipid/*Pseudomonas aeruginosa* HAK02	STR/Fed-batch	500–1000 rpm; > 40% dissolved oxygen	240 g L^{-1}	[104]
Rhamnolipid/*Pseudomonas aeruginosa*	STR/fed-batch (membrane assisted)	100 rpm; 4.0 mg O_2 L^{-1}	40–50 mg L^{-1} h	[118]
Lipopeptides/*B. subtilis* BBG21	FBR/batch with solid catalysts	0.015 s^{-1} (k_{La})	58 mg L^{-1}	[119]
Lipopeptides/*Bacillus* sp. GY19	STR/batch chitosan-immobilized cells	200 rpm; 0.5 mL air min^{-1}	10.9 g L^{-1} (foam fraction) + 7.12 g L^{-1} (culture media)	[120]
Surfactin/*Bacillus* sp. ITP-001	STR/semi-batch foam fractionation	200 rpm/1.0 vvm	4.54 g L^{-1}	[121]
Lipopeptides/*Streptomyces* sp. PBD-410L	STR/fed-batch	200 rpm/0.5 vvm	5.32 g L^{-1}	[61]
Lipid polyol/*Aureobasidium pullulans* LB83	STR/batch	0.1–1.1 vvm	1.52 g L^{-1}	[122]

In this sense, the increase in the viscosity of the medium could be disadvantageous to these systems. Shearing rates are lower and possibly lead to depletion of dissolved oxygen and cell death [125].

FBR are preferentially used in processes that employ microorganisms sensitive to shear forces [126]. A study showed the influence of some parameters in an inverse fluidized bed reactor for biosurfactant production by *B. subtilis*. The authors observed the positive influence of solid particles in the variation of k_{La}. The relation of this variable with other parameters could help with process control and scale-up considerations [119].

Other studies show novel strategies to overcome key challenges to biosurfactant production like the design of reactors. For instance, the rotating disc bioreactor was developed for surfactin production by *B. subtilis*. The authors observed biofilm-formation to build layers of biomass attached to the discs located through the reactor's length. A correlation of product yield per biomass-formed unit was found. Overall, as the number of discs increased (7–14), higher levels of lipopeptides per biomass unit were obtained [99].

The bubble-free membrane reactor is another non-usual alternative employed in the production of biosurfactants [127]. In this bioreactor, the biomass and biosurfactant are partially absorbed on the membrane surface. It can integrate fermentation-separation system, reducing the process cost [127, 128].

Aiming to improve productivity and reduce process challenges, changes in operating modes of fermentation in bioreactors can be performed, which include batch, repeated batch, fed-batch, continuous, and semi-continuous [95, 96, 98]. The better choice of operation mode is dependent on the target product, the selected microorganism, and its metabolic rates [129].

Studies show the production of sophorolipid by *Candida bombicola* by fed-batch operation. They obtained higher biosurfactant production (approximately 300–400 g L^{-1}) in comparison with conventional batch modes [130, 131]. The fed-batch approach is more convenient due to its ease of operation, low cost, and high yield [98].

Recently, efforts in the development of integrated processes have been performed to reduce the market prices for biosurfactants. Those processes integrate production and downstream processing operations. The objective is to create a semi-continuous process, increasing productivity and minimizing equipment costs [132]. Dolman et al. [133] showed the possibility of separating sophorolipids during the production process by density difference. When the biosurfactant is denser than the aqueous medium, by the action of gravity, it will settle at the bottom of the reactor. Thus, it can be pumped out of the vessel and transferred to a separatory funnel for further purification processes [134]. This opportunity may facilitate the scale-up of the process which is a limiting factor for biosurfactant industrial implementation.

Currently, the bioprocesses for obtaining biosurfactants are not commonly performed using integrated systems for their production and recovery. However, several methods are applied in the downstream stage for recovery and purification of these biomolecules. The main methods used and their main advantages and disadvantages are discussed below.

6.4.3 Biosurfactant Separation Processes

The downstream processes (separation and purification) of biosurfactants include fundamental unit operations after its production. These steps should be designed to ensure improved yield and greater competitiveness to the biosurfactant market [135, 136]. Challenges in downstream steps are due to the complexity of the fermented broth, scarce thermodynamic data, and low concentration of the product [137, 138].

The methods of separation and recovery of biosurfactants depend on the structural and physicochemical characteristics of these molecules, required degree of purity, and their application [132, 139]. The purification process of these biomolecules will also depend on their cellular location (intracellular or extracellular), as well as their affinity with organic solvents, interactions with cationic species, and molecular weight [140].

The main methods applied for separation of biosurfactants are gravity separation (no salt, salting out, and acidic), solvent extraction, foam fractionation, membrane separation, adsorption separation, and ion-exchange chromatography.

In the gravity separation method, the form of production and the physicochemical characteristics of the biosurfactants are determinants. The interruption of the fermentation process agitation and the structure of the biosurfactant facilitate phase separation by sedimentation. Due to the action of gravity, a second phase is formed with the precipitation of the product at the bottom of the flask or bioreactor [133, 141]. Separation by gravity can also be classified depending on the use or type of agent employed for decanting the product. This method can be carried out with or without the addition of salts or even in the presence of acids [142].

Gravity separation without the addition of salts occurs by the density difference between the biosurfactant and the aqueous medium. In the presence of salts, such as ammonium sulfate and zinc sulfate, changes occur in the surface charge of the molecule. There is then an attraction between the molecules of biosurfactants which self-assemble, accumulate, and subsequently precipitate [139, 140]. In acid precipitation, the reduction in pH (between 2 and 3) promotes the protonation of carboxyl groups, decreasing their solubility in aqueous solutions. This is considered an easy, inexpensive, and readily available method. However, this method has a limitation when the separation step occurs in the cultivation reactor. This is because this method has high operating times for the bioreactors because of the high operating times due to a non-operating period destined for the precipitation of biosurfactants.

The solvent extraction method occurs through the solubilization of biosurfactants by the interaction of portions of these biomolecules with the added solvent. The most commonly used organic solvents are acetone, alcohol, butanol, chloroform, methanol, ethyl acetate, dichloromethane, hexane, acetic acid, and isopropanol [7, 143, 144]. This method allows quick separation, the solvent is removed by evaporation and recovered, and the biosurfactant is obtained in its concentrated form. However, this method has some limitations, such as the high cost, toxicity, and unsustainable nature of these organic solvents [140, 145].

Another method to consider is foam fractionation, which is supported by the ability of biosurfactants to accumulate at the air–water interface. The foam formed can be collected during or after the fermentation process, allowing the integration of biological production and the separation of biosurfactants produced [126, 145].

The bubble formation kinetics in the foam fractionation is susceptible to saturation. In this condition the capacity of formation and concentration of biosurfactants in the foam is limited, regardless of the concentration [108]. Another limitation of this method is the need for a high degree of agitation in the bioreactor. In addition, the need for a minimum concentration of broth to promote foam formation [146].

The membrane separation method (filtration and ultrafiltration) uses membranes with selective permeability. This method is carried out to separate molecules of different sizes over a pressure gradient used to conduct the flow to the membrane [147]. In the purification of biosurfactants, the fermentation broth passes through different membranes. First, it performs filtration to separate cells and biosurfactants, then, carries the ultrafiltration of the cell-free broth to separate the biosurfactant [140, 148].

Separation by membranes allows the recovery of product independently of the fermentation process. This method allows easy separation of cells, enabling its reuse, and the biosurfactants are soluble in the aqueous medium [140, 149]. The main disadvantage of this method is related to incrustations and obstructions in the membrane wall during the process [138].

The adsorption method evaluates the adsorption and desorption capacity of biosurfactants to a polymeric matrix (hydrophobic substrate). This matrix is chosen according to the type of targeted molecule and can be reused. This process is always carried out with the fermented broth without the presence of cells, and elution is carried out with organic solvents to remove the retained biosurfactant [150, 151].

Chromatography ion-exchange explores the charge of biosurfactants. For a biosurfactant with negative charges, cationic resins are generally used and eluted with a buffer with greater ionic strength [149]. This method has a high specific capacity for interaction between resin and biomolecules, with good yields and high purity. Furthermore, it is an economical process since the resins can be reused [152]. The main features, advantages, and disadvantages of some separation methods are presented in Table 6.2.

The separation processes can also be combined to increase the degree of purity and allow applications of biosurfactants in medical and food areas. For example, the separation of the biosurfactant rhamnolipid was done by acid precipitation, followed by solvent extraction (ethyl acetate) [136]. The work of Félix et al. [155] describes the need to adopt several stages of separation of the surfactin biosurfactant, starting with obtaining the crude biosurfactant with acid precipitation (pH 2.0 and 3 M HCl) and centrifugation, followed by solvent extraction (dichloromethane) and the last purification step in reverse phase chromatography.

Furthermore, several strategies are being developed and used for the separation and purification of biosurfactants. These include high-performance liquid chromatography (HPLC), gel filtration, and activated carbon adsorption [156]. In addition, there is the development of new techniques and economic and sustainable strategies to obtain these biomolecules. For example, electrokinetic separation, a technique that applies an electric field in the reactor and allows the movement (electromigration) of the liquid phase via electro-osmosis of the colloidal particles present in the broth. This movement occurs due to the net surface charge of the particles acquired with the application of a magnetic field that promotes the accumulation and recovery of the bioproduct at the end of the anode [103].

Despite the advances, the separation and purification of biosurfactants are still challenging and not very competitive compared to methods for separating and obtaining synthetic surfactants. Most biosurfactant extraction methods still require a high degree of investment, use of organic solvents with high tixicity, more detailed studies of industrial application, in addition to high operating costs with energy and equipment usage time [133].

Table 6.2 Advantages and disadvantages of separation/purification methods.

Method		Description	Efficiency (%)	Advantages	Disadvantages	References
Gravity separation	no salting	Explores chemical and physical properties of biosurfactants as changes in the hydrophobic part and decreased solvation effects.	-	No expense with reagents, easy operation and handling.	Long standby and reactor occupancy and high expenses associated with energy consumption (reactor operation).	[133]
	salting out		70–80% (extraction of surfactin)	Low cost. Easy operation.	It can provide coprecipitation of other small molecules, decreasing purity and reductions.	[139, 153]
	acidic	Decreased solubility caused by low pH	-	Easy scale-up and low cost.	Degradation or inhibition of the final product, by strong aggregation.	[143]
Solvent extraction		Phase migration, solvent partition, and evaporation.	97% (extraction of rhamnolipids with ethyl acetate)	Ease of recovery and obtaining of the dry product. Solvent recovery.	High costs associated with the amount of solvents used, and the use and generation of toxic waste.	[140, 136]
Foam fractionation		Based on the production and removal of foam formed during fermentation.	80–97% (recover of rhamnolipids)	It can be carried out during and after fermentation. It can be integrated into a continuous process.	Need for high agitation. Requires minimal concentration of biosurfactant in the broth for bubble formation.	[120, 154, 110]
Membrane separation		Based on retention of particles (by size) in a semi-permeable membrane.	97% (extraction of surfactin, in pH 5.65)	Reuse of retained cells. The method is adjustable by membrane porosity.	Requirements of maintenance and problems with incrustations on the membrane.	[148, 145]
Adsorption		Retention of the molecule in a polymeric matrix.	-	Reusability and specificity of adsorbents.	High specificity and low scaling capacity. Cell separation can be an important stage.	[150]
Chromatography (ion-exchange)		Explores the ability to interact with the resin and the chemical affinity with the eluent using for biosurfactant concentration.	-	Obtaining of a product with high purity. Good reusability (resin can be used for several cycles).	High cost of ion-exchange columns.	[152]

6.5 Future Perspectives and Conclusion

Although interest in biosurfactants has been increasing in recent years, these compounds still do not compete with synthetic surfactants in the current market. As discussed, this is due to the high production cost, low yield, high cost of downstream processes, and purification of these biomolecules [90, 157]. Regarding production costs, some research demonstrates the use of alternative substrates, usually renewable resources, such as industrial waste and by-products [8, 158]. In recent decades, the use of industrial and agro-industrial waste for the development of sustainable biorefineries has been of great economic and environmental interest [159].

As a strategy to increase the production yield of biosurfactants, process optimization is carried out considering nutritional parameters, and bioprocess conditions such as temperature, pH, aeration, and agitation [63, 160]. Another approach is the use of carbon sources with different polarities. The combination of oleaginous (hydrophobic) substrates with hydrophilic substrates favors the production of biosurfactants [39]. Hydrophobic compounds promote the reduction of foam generated in the production process, increasing productivity and final yield. In addition, they form biomolecules with better physicochemical properties, and the combination of two substrates is advantageous [3, 9, 39, 161]. Moreover, with molecular biology technologies, genetic modification of producer strains to increase biosurfactant yields is another alternative [156, 162]. Studies revealing the biosynthesis pathway of biosurfactant production allows the improvement of their production through genetic manipulations [26, 45, 158].

Many tools are used to overcome the challenges involved in the production of biosurfactants, however, there are still few studies involving the scaling up and large-scale production of these biomolecules. Therefore, it is necessary to continue the studies looking for alternatives to be used in these bioprocesses to make biosurfactants competitive in the industrial market to synthetic surfactants.

References

1 Santos DKF, Rufino RD, Luna JM, Santos VA, Sarubbo LA. Biosurfactants: multifunctional biomolecules of the 21st century. *Int J Mol Sci*. 2016;17(3):401.
2 Marchant R, Banat IM. Microbial biosurfactants: challenges and opportunities for future exploitation. *Trends Biotechnol*. 2012;30(11):558–565.
3 Jimoh AA, Lin J. Biosurfactant: a new frontier for greener technology and environmental sustainability. *Ecotoxicol Environ Saf*. 2019;184:109607.
4 Farias CBB, Almeida FC, Silva IA, Souza TC, Meira HM, Rita de Cássia F, et al. Production of green surfactants: market prospects. *Electron J Biotechnol* 2021; 51:28–29.
5 Giri SS, Sen SS, Jun JW, Sukumaran V, Park SC. Role of Bacillus subtilis VSG4-derived biosurfactant in mediating immune responses in Labeo rohita. *Fish Shellfish Immunol*. 2016;54:220–229.
6 Guan R, Yuan X, Wu Z, Wang H, Jiang L, Li Y, Zeng G. Functionality of surfactants in waste-activated sludge treatment: a review. Science of the total environment. *Sci Total Environ*. 2017;609:1433–1442.

7 Marcelino PRF, Peres GFD, Terán-Hilares R, Pagnocca FC, Rosa CA, Lacerda TM, et al. Biosurfactants production by yeasts using sugarcane bagasse hemicellulosic hydrolysate as new sustainable alternative for lignocellulosic biorefineries. *Ind Crops Prod.* 2019;129:212–223.

8 Marcelino PRF, Gonçalves F, Jimenez IM, Carneiro BC, Santos BB, da Silva SS. Sustainable production of biosurfactants and their applications. In: Ingle AP, Chandel AK, da Silva SS, editors. *Lignocellulosic Biorefining Technologies.* 2020. pp. 159–183.

9 Jahan R, Bodratti AM, Tsianou M, Alexandridis P. Biosurfactants, natural alternatives to synthetic surfactants: physicochemical properties and applications. *Adv Colloid Interface Sci.* 2020;275:102061.

10 Drakontis CE, Amin S. Biosurfactants: formulations, properties, and applications. *Curr Opin Colloid Interface Sci.* 2020;48:77–90.

11 Markets and Markets. 2017 https://www.marketsandmarkets.com/PressReleases/biosurfactant.asp acesso de fevereiro de, 26 2022, às 14: 48.

12 Rodríguez-López L, Rincón-Fontán M, Vecino X, Cruz JM, Moldes AB. Extraction, separation and characterization of lipopeptides and phospholipids from corn steep water. *Sep Purif Technol.* 2020;248:117076.

13 da Silva Chaves F, Brumano LP, Marcelino PRF, da Silva SS, Sette LD, de Almeida Felipe MDG. Biosurfactant production by Antarctic-derived yeasts in sugarcane straw hemicellulosic hydrolysate. *Biomass Convers Biorefin.* 2021:1–11.

14 Research and Markets. Global Biosurfactants Market 2021–2025. https://www.researchandmarkets.com/reports/5125615/global-biosurfactants-market-2021-2025 access Feb 26 2022 at 14:48.

15 GVR (Grand view research). 2015 https://www.grandviewresearch.com/press-release/global-biosurfactants-market, acesso de fevereiro de, 26 2021, às 14: 48.

16 Global market insights. 2020, https://www.gminsights.com/industry-analysis/biosurfactants-market-report acesso de fevereiro de, 26 2022, às 14: 48.

17 Markets and Markets. 2020 https://www.marketsandmarkets.com/Market-Reports/biosurfactants-market-493.html acesso de fevereiro de 03, 2021, às 14: 48.

18 Fontes GC, Amaral PFF, Coelho MAZ. Production of biosurfactant from yeast. *Quim Nova.* 2008;31(8):2091–2099.

19 Sachdev DP, Cameotra SS. Biosurfactants in agriculture. *Appl Microbiol Biotechnol.* 2013;97:1005–1016.

20 Singh P, Patil Y, Rale V. Biosurfactant production: emerging trends and promising strategies. *J Appl Microbiol.* 2019;126(1):2–13.

21 Fenibo EO, Ijoma GN, Selvarajan R, Chikere CB. Microbial surfactants: the next generation multifunctional biomolecules for applications in the petroleum industry and its associated environmental remediation. *Microorganisms.* 2019;7(11):581.

22 Prado AAOS, Santos BLP, Vieira IMM, Ramos LC, de Souza RR, Silva DP, et al. Evaluation of a new strategy in the elaboration of culture media to produce surfactin from hemicellulosic corncob liquor. *Appl Biotechnol Rep.* 2019;24:e00364.

23 Jain RM, Mody K, Joshi N, Mishra A, Jha B. Production and structural characterization of biosurfactant produced by an alkaliphilic bacterium, Klebsiella sp.: evaluation of different carbon sources. *Colloids Surf B Biointerfaces.* 2013;108:199–204.

24 Ghasemi A, Moosavi-Nasab M, Setoodeh P, Mesbahi G, Yousefi G. Biosurfactant production by lactic acid bacterium Pediococcus dextrinicus SHU1593 grown on different carbon sources: strain screening followed by product characterization. *Sci Rep.* 2019;9(1):1–12.

25 Baskaran SM, Zakaria MR, Sabri ASMA, Mohamed MS, Wasoh H, Toshinari M, et al. Valorization of biodiesel side stream waste glycerol for rhamnolipids production by Pseudomonas aeruginosa RS6. *Environ Pollut.* 2021;276:116742.

26 Hu X, Subramanian K, Wang H, Roelants LKWS, To MH, Soetaert W, et al. Guiding environmental sustainability of emerging bioconversion technology for waste-derived sophorolipid production by adopting a dynamic life cycle assessment (dLCA) approach. *Environ Pollut.* 2021;269:116–121.

27 Zenati B, Chebbi A, Badis A, Eddouaouda K, Boutoumi H, El Hattab M, et al. A non-toxic microbial surfactant from Marinobacter hydrocarbonoclasticus SdK644 for crude oil solubilization enhancement. *Ecotoxicol Environ Saf.* 2018;154:100–107.

28 Liepins J, Balina K, Soloha R, Berzina I, Lukasa LK, Dace E. Glycolipid biosurfactant production from waste cooking oils by yeast: review of substrates, producers and products. *Fermentation.* 2021;7(3):136.

29 de Almeida DG, Brasileiro PPF, Rufino RD, de Luna JM, Sarubbo LA. Production, formulation and cost estimation of a commercial biosurfactant. *Biodegradation.* 2019;30(4):191–201.

30 Hu J, Luo J, Zhu Z, Chen B, Ye X, Zhu P et al. Multi-scale biosurfactant production by bacillus subtilis using tuna fish waste as substrate. *Catalysts.* 2021;11(4):456.

31 Nurfarahin AH, Mohamed MS, Phang LY. Culture medium development for microbial-derived surfactants production—an overview. *Molecules.* 2018;23(5):1049.

32 Franco Marcelino PR, da Silva VL, Rodrigues Philippini R, Von Zuben CJ, Contiero J, Dos Santos JC, et al. Biosurfactants produced by Scheffersomyces stipitis cultured in sugarcane bagasse hydrolysate as new green larvicides for the control of Aedes aegypti, a vector of neglected tropical diseases. *PLoS One.* 2017;12(11):e0187125.

33 Zargar AN, Lymperatou A, Skiadas I, Kumar M, Srivastava P. Structural and functional characterization of a novel biosurfactant from Bacillus sp. IITD106. *J Hazard Mater.* 2022;423:127201.

34 Gharaei S, Ohadi M, Hassanshahian M, Porsheikhali S, Forootanfar H. Isolation, optimization, and structural characterization of glycolipid biosurfactant produced by Marine Isolate Shewanella algae B12 and evaluation of its antimicrobial and anti-biofilm activity. *Appl. Biochem. Biotechnol.* 2022;194:1–20.

35 Hanson KG, Desai JD, Desai AJ. A rapid and simple screening technique for potential crude oil degrading microorganisms. *Biotechnol Tech.* 1993;7(10):745–748.

36 de Oliveira Schmidt VK, de Souza Carvalho J, de Oliveira D, de Andrade CJ. Biosurfactant inducers for enhanced production of surfactin and rhamnolipids: an overview. *World J. Microbiol Biotechnol.* 2021;37(2):1–15.

37 Pathania AS, Jana AK. Improvement in production of rhamnolipids using fried oil with hydrophilic co-substrate by indigenous Pseudomonas aeruginosa NJ2 and characterizations. *Appl Biochem Biotechnol.* 2020;191:1223–1246. 2020.

38 Niu Y, Wu J, Wang W, Chen Q. Production and characterization of a new glycolipid, mannosylerythritol lipid, from waste cooking oil biotransformation by Pseudozyma aphidis ZJUDM34. *Int J Food Sci Nutr.* 2019;7(3):937–948.

39 Amaral PFF, Da Silva JM, Lehocky BM, Barros-Timmons AMV, Coelho MAZ, Marrucho IM, et al. Production and characterization of a bioemulsifier from Yarrowia lipolytica. *Process Biochem*. 2006;41(8):1894–1898.

40 Gudiña EJ, Fernandes EC, Rodrigues AI, Teixeira JA, Rodrigues LR. Biosurfactant production by Bacillus subtilis using corn steep liquor as culture medium. *Front Microbiol*. 2015;6:59.

41 Nazareth TC, Zanutto CP, Maass D, de Souza AAU, Ulson SMDAG. Bioconversion of low-cost brewery waste to biosurfactant: an improvement of surfactin production by culture medium optimization. *Biochem Eng J*. 2021;172:108058.

42 Moshtagh B, Hawboldt K, Zhang B. Kinetic modeling of biosurfactant production by Bacillus subtilis N3-1P using brewery waste. *Chem Prod Process Model*. 2021;17.

43 Mnif I, Ellouze-Chaabouni S, Ghribi D. Optimization of inocula conditions for enhanced biosurfactant production by Bacillus subtilis SPB1, in submerged culture, using Box–Behnken design. *Probiotics Antimicrob Proteins* 2013;5(2):92–98.

44 Sun W, Zhu B, Yang F, Dai M, Sehar S, Peng C, et al. Optimization of biosurfactant production from Pseudomonas sp. CQ2 and its application for remediation of heavy metal contaminated soil. *Chemosphere*. 2021;265:129090.

45 Nalini S, Parthasarathi R. Production and characterization of rhamnolipids produced by Serratia rubidaea SNAU02 under solid-state fermentation and its application as biocontrol agent. *Bioresour Technol*. 2014;173:231–238.

46 Martins PC, Martins VG. Biosurfactant production from industrial wastes with potential remove of insoluble paint. *Int Biodeterior Biodegrad*. 2018;127:10–16.

47 Ostendorf TA, Silva IA, Converti A, Sarubbo LA. Production and formulation of a new low-cost biosurfactant to remediate oil-contaminated seawater. *J Biotechnol*. 2019;295:71–79.

48 Bjerk TR, Severino P, Jain S, Marques C, Silva AM, Pashirova T. Biosurfactants: properties and applications in drug delivery, biotechnology and ecotoxicology. *Bioengineering*. 2021;8(8):115.

49 Chen C, Sun N, Li D, Long S, Tang X, Xiao G, et al. Optimization and characterization of biosurfactant production from kitchen waste oil using Pseudomonas aeruginosa. *Environ Sci Pollut Res*. 2018;25(15):14934–14943.

50 Sharma S, Pandey LM. Production of biosurfactant by Bacillus subtilis RSL-2 isolated from sludge and biosurfactant mediated degradation of oil. *Bioresour Technol*. 2020;307:123261.

51 de Freitas Ferreira J, Vieira EA, Nitschke M. The antibacterial activity of rhamnolipid biosurfactant is pH dependent. *Int Food Res J*. 2019;116:737–744.

52 Najafi AR, Rahimpour MR, Jahanmiri AH, Roostaazad R, Arabian D, Ghobadi Z. Enhancing biosurfactant production from an indigenous strain of Bacillus mycoides by optimizing the growth conditions using a response surface methodology. *Chem Eng J*. 2010;163(3):188–194.

53 Hosseininoosheri P, Lashgari HR, Sepehrnoori K. A novel method to model and characterize in-situ bio-surfactant production in microbial enhanced oil recovery. *Fuel*. 2016;183:501–511.

54 Hu X, Wang C, Wang P. Optimization and characterization of biosurfactant production from marine Vibrio sp. strain 3B-2. *Front Microbiol* 2015;6:976.

55 Souza KST, Gudiña EJ, Schwan RF, Rodrigues LR, Dias DR, Teixeira JA. Improvement of biosurfactant production by Wickerhamomyces anomalus CCMA 0358 and its potential application in bioremediation. *J Hazard Mater* 2018;346:152–158.

56 Elshafie AE, Joshi SJ, Al-Wahaibi YM, Al-Bemani AS, Al-Bahry SN, Al-Maqbali D, et al. Sophorolipids production by Candida bombicola ATCC 22214 and its potential application in microbial enhanced oil recovery. *Front Microbiol.* 2015;6:1324.

57 Monteiro AS, Coutinho JO, Júnior AC, Rosa CA, Siqueira EP, Santos VL. Characterization of new biosurfactant produced by Trichosporon montevideense CLOA 72 isolated from dairy industry effluents. *J Basic Microbiol* 2009;49(6):553–563.

58 Ibrahim S, Abdul Khalil K, Zahri KNM, Gomez-Fuentes C, Convey P, Zulkharnain A, et al. Biosurfactant production and growth kinetics studies of the waste canola oil-degrading bacterium Rhodococcus erythropolis AQ5-07 from Antarctica. *Molecules.* 2020;25(17):3878.

59 Md F. Biosurfactant: production and application. *J Pet Environ Biotechnol.* 2012;3(4):124.

60 Nazareth TC, Zanutto CP, Maass D, de Souza AAU, Ulson SMDAG. Impact of oxygen supply on surfactin biosynthesis using brewery waste as substrate. *J Environ Chem Eng.* 2021;9(4):105372.

61 Zambry NS, Rusly NS, Awang MS, Noh NAM, Yahya ARM. Production of lipopeptide biosurfactant in batch and fed-batch Streptomyces sp. PBD-410L cultures growing on palm oil. *Bioprocess Biosyst Eng.* 2021;44(7):1577–1592.

62 Wang H, Tsang CW, To MH, Kaur G, Roelants SL, Stevens CV, et al. Techno-economic evaluation of a biorefinery applying food waste for sophorolipid production–a case study for Hong Kong. *Bioresour. Technol.* 2020;303:122852.

63 Roelants S, Solaiman DK, Ashby RD, Lodens S, Van Renterghem L, Soetaert W. Production and applications of sophorolipids. In: Hayes DG, Solaiman DKY, Ashby RD, editors. *Biobased Surfactants*, 2nd ed. 2019. pp. 65–119.

64 Reiling HE, Thanei-Wyss U, Guerra-Santos LH, Hirt R, Käppeli O, Fiechter A. Pilot plant production of rhamnolipid biosurfactant by Pseudomonas aeruginosa. *Appl Environ Microbiol.* 1986;51(5):985–989.

65 Thavasi R, Jayalakshmi S, Balasubramanian T, Banat IM. Production and characterization of a glycolipid biosurfactant from Bacillus megaterium using economically cheaper sources. *World J Microbiol Biotechnol.* 2008;24(7):917–925.

66 Bouassida M, Ghazala I, Ellouze-Chaabouni S, Ghribi D. Improved biosurfactant production by Bacillus subtilis SPB1 mutant obtained by random mutagenesis and its application in enhanced oil recovery in a sand system. *J Microbiol Biotechnol.* 2018;28(1):95–104.

67 Gurkok S. Important parameters necessary in the bioreactor for the mass production of biosurfactants. In: Inamuddin D, Boddula R, Asiri A, editors. *Green Sustainable Process for Chemical and Environmental Engineering and Science*, 1st ed. Elsevier; 2021. pp. 347–365.

68 Banat IM, Carboué Q, Saucedo-Castaneda G, de Jesús Cázares-Marinero J. Biosurfactants: the green generation of speciality chemicals and potential production using Solid-State fermentation (SSF) technology. *Bioresour Technol.* 2020:320;124222.

69 Rastogi S, Kumar R. Statistical optimization of biosurfactant production using waste biomaterial and biosorption of Pb2+ under concomitant submerged fermentation. *J Environ Manage*. 2021;295:113158.

70 Colla LM, Rizzardi J, Pinto MH, Reinehr CO, Bertolin TE, Costa JAV. Simultaneous production of lipases and biosurfactants by submerged and solid-state bioprocesses. *Bioresour Technol*. 2010;101(21):8308–8314.

71 Reihani SFS, Khosravi-Darani K. Influencing factors on single-cell protein production by submerged fermentation: a review. *Electron J Biotechnol*. 2019;37:34–40.

72 Hentati D, Chebbi A, Hadrich F, Frikha I, Rabanal F, Sayadi S, et al. Production, characterization and biotechnological potential of lipopeptide biosurfactants from a novel marine Bacillus stratosphericus strain FLU5. *Ecotoxicol Environ Saf*. 2019;167:441–449.

73 Al-Dhabi NA, Esmail GA, Valan Arasu M. Enhanced production of biosurfactant from Bacillus subtilis Strain Al-Dhabi-130 under solid-state fermentation using date Molasses from Saudi Arabia for bioremediation of Crude-oil-contaminated soils. *Int J Environ Res*. 2020;17(22):8446.

74 Akpinar M, Urek RO. Production of ligninolytic enzymes by solid-state fermentation using Pleurotus eryngii. *Prep Biochem Biotechnol*. 2012;42(6):582–597.

75 Nalini S, Parthasarathi R. Optimization of rhamnolipid biosurfactant production from Serratia rubidaea SNAU02 under solid-state fermentation and its biocontrol efficacy against Fusarium wilt of eggplant. *Ann Agrar Sci*. 2018;16(2):108–115.

76 Costa JA, Treichel H, Santos LO, Martins VG. Solid-state fermentation for the production of biosurfactants and their applications. In: Larroche C, Sanroman M, Du G, Pandey A, editors. *Current Developments in Biotechnology and Bioengineering*. Elsevier; 2018. pp. 357–372.

77 Lourenço LA, Alberton Magina MD, Tavares LBB, Guelli Ulson de Souza SMA, García Román M, Altmajer Vaz D. Biosurfactant production by Trametes versicolor grown on two-phase olive mill waste in solid-state fermentation. *Environ Technol*. 2018;39(23):3066–3076.

78 Rubio-Ribeaux D, De Oliveira CVJ, Marinho JDS, Lins UDBL, Do Nascimento IDF, Barreto GC, et al. Innovative production of biosurfactant by Candida tropicalis UCP 1613 through solid-state fermentation. *Chem Eng Trans*. 2020;79:361–366.

79 Dos Santos RA, Rodríguez DM, Ferreira INDS, de Almeida SM, Takaki GMDC, de Lima MAB. Novel production of biodispersant by Serratia marcescens UCP 1549 in solid-state fermentation and application for oil spill bioremediation. *Environ Technol*. 2021;43:431–12.

80 Das K, Mukherjee AK. Comparison of lipopeptide biosurfactants production by Bacillus subtilis strains in submerged and solid-state fermentation systems using a cheap carbon source: some industrial applications of biosurfactants. *Process Biochem*. 2007;42(8):1191–1199.

81 Kreling NE, Simon V, Fagundes VD, Thomé A, Colla LM. Simultaneous production of lipases and biosurfactants in solid-state fermentation and use in bioremediation. *J Environ Eng*. 2020;146(9):04020105.

82 Barbosa JR, da Silva SB, de Carvalho Junior RN. Biosurfactant production by solid-state fermentation, submerged fermentation, and biphasic fermentation. In: Inamuddin D, Boddula R, Asiri A, editors. *Green Sustainable Process for Chemical and Environmental Engineering and Science*, 1st ed. Elsevier; 2021. pp. 155–171.

83 Zaldivar J, Nielsen J, Olsson L. Fuel ethanol production from lignocellulose: a challenge for metabolic engineering and process integration. *Appl Microbiol Biotechnol* 2001;56(1):17–34.

84 Ishizaki H, Hasumi K. Ethanol production from biomass. In: Tojo S, Hirasawa T, editors. *Research Approaches to Sustainable Biomass Systems*. 2014. pp. 243–258.

85 Devos RJB, Colla LM. Simultaneous saccharification and fermentation to obtain bioethanol: a bibliometric and systematic study. *Bioresour Technol Rep*. 2021;17:100924.

86 Lu J, Li X, Yang R, Yang L, Zhao J, Liu Y, et al. Fed-batch semi-simultaneous saccharification and fermentation of reed pretreated with liquid hot water for bio-ethanol production using Saccharomyces cerevisiae. *Bioresour Technol* 2013;144:539–547.

87 Zeng G, You H, Wang K, Jiang Y, Bao H, Du M, et al. Semi-simultaneous saccharification and fermentation of ethanol production from Sargassum Horneri and biosorbent production from fermentation residues. In: Nzihou A, editors. *Waste and Biomass Valorization*, Vol. 11(9). Springer; 2020. 4743–4755.

88 Jiménez IM, Chandel AK, Marcelino PR, Anjos V, Costa CB, Bell MJV, et al. Comparative data on effects of alkaline pretreatments and enzymatic hydrolysis on bioemulsifier production from sugarcane straw by Cutaneotrichosporon mucoides. *Bioresour Technol*. 2020;301:122706.

89 Panjiar N, Mattam AJ, Jose S, Gandham S, Velankar HR. Valorization of xylose-rich hydrolysate from rice straw, an agroresidue, through biosurfactant production by the soil bacterium Serratia nematodiphila. *Sci Total Environ*. 2020;729:138933.

90 Faria NT, Santos M, Ferreira C, Marques S, Ferreira FC, Fonseca C. Conversion of cellulosic materials into glycolipid biosurfactants, mannosylerythritol lipids, by Pseudozyma spp. under SHF and SSF processes. *Microb Cell Factories*. 2014;13(1):1–13.

91 Zheng T, Lei F, Li P, Han C, Liu S, Jiang J. High rhamnolipid production from cassava residues by simultaneous saccharification and fermentation with low enzyme loading. *J Biobased Mater*. 2019;13(5):635–642.

92 Mukherjee S, Das P, Sen R. Towards commercial production of microbial surfactants. *Trends Biotech*. 2006;24(11):509–515.

93 Smyth T, Perfumo A, McClean S, Marchant R, Banat I. Isolation and analysis of lipopeptides and high molecular weight biosurfactants. In: Timmis KN, editors. *Handbook of Hydrocarbon and Lipid Microbiology*. Springer; 2010. pp. 3688–3704.

94 Karnwal A. Biosurfactant production using bioreactors from industrial byproducts. In: Sarma H, Prasad MNV, editors. *Biosurfactants for a Sustainable Future: Production and Applications in the Environment and Biomedicine*. Wiley; 2021. pp. 59–78.

95 Mustafa MG, Khan MGM, Nguyen D, Iqbal S. Techniques in biotechnology: essential for industry. In: Barh D, Azevedo V, editors. *Omics Technologies and Bio-Engineering*. Academic Press; 2018. pp. 233–249.

96 Jaibiba P, Vignesh SN, Hariharan S. Working principle of typical bioreactors. In: Yousuf A, Singh L, Mahapatra DM, editors. *Bioreactors*. Elsevier; 2020. pp. 145–173.

97 Gill NK, Appleton M, Baganz F, Lye GJ. Design and characterisation of a miniature stirred bioreactor system for parallel microbial fermentations. *Bioch Eng J*. 2008;39(1):164–176.

98 Zhou TC, Zhou WW, Hu W, Zhong JJ. Bioreactors, cell culture, commercial production. In: Flickinger MC, editors. *Encyclopedia of Industrial Biotechnology: Bioprocess, Bioseparation, and Cell Technology*. 2009. pp. 1–18.

99 Chtioui O, Dimitrov K, Gancel F, Dhulster P, Nikov I. Rotating discs bioreactor, a new tool for lipopeptides production. *Process Bioch*. 2012;47(12):2020–2024.

100 Beuker J, Syldatk C, Hausmann R. Bioreactors for the production of biosurfactants. In: Kosaric N, Vardar Sukan F, editors. *Biosurfactants: Production and Utilization; Processes, Technologies, and Economics*, 1st ed. 2014. pp. 117–128.

101 Jiang J, Zu Y, Li X, Meng Q, Long X. Recent progress towards industrial rhamnolipids fermentation: process optimization and foam control. *Bioresour Technol*. 2020;298:122394.

102 Wang H, Kaur G, To MH, Roelants SL, Patria RD, Soetaert W, et al. Efficient in-situ separation design for long-term sophorolipids fermentation with high productivity. *J Clean Prod*. 2020;246:118995.

103 Gidudu B, Chirwa EMN. Electrokinetic extraction and recovery of biosurfactants using rhamnolipids as a model biosurfactant. *Sep Purif Technol*. 2021;276:119327.

104 Bazsefidpar S, Mokhtarani B, Panahi R, Hajfarajollah H. Overproduction of rhamnolipid by fed-batch cultivation of Pseudomonas aeruginosa in a lab-scale fermenter under tight DO control. *Biodegradation*. 2019;30(1):59–69.

105 Zheng H, Fan S, Liu W, Zhang M. Production and separation of pseudomonas aeruginosa rhamnolipids using coupling technology of cyclic fermentation with foam fractionation. *Chem Eng Process: Process Intensif*. 2020;148:107776.

106 Gong Z, He Q, Che C, Liu J, Yang G. Optimization and scale-up of the production of rhamnolipid by Pseudomonas aeruginosa in solid-state fermentation using high-density polyurethane foam as an inert support. *Bioprocess Biosyst Eng*. 2020;43(3):385–392.

107 Gong Z, Yang G, Che C, Liu J, Si M, He Q. Foaming of rhamnolipids fermentation: impact factors and fermentation strategies. *Microb Cell Factories*. 2021;20(1):1–12.

108 Chen CY, Baker SC, Darton RC. Continuous production of biosurfactant with foam fractionation. *J Chem Technol*. 2006;81(12):1915–1922.

109 Chen CY, Baker SC, Darton RC. Batch production of biosurfactant with foam fractionation. *J Chem Technol*. 2006;81(12):1923–1931.

110 Blesken CC, Strumpfler T, Tiso T, Blank L. Uncoupling foam fractionation and foam adsorption for enhanced biosurfactant synthesis and recovery. *Microorganisms*. 2020;8:1–23.

111 Kronemberger FA, Borges CP, Freire DM. Fed-batch biosurfactant production in a bioreactor. *Int Rev Chem Eng*. 2010;2(4):513–518.

112 Maddikeri GL, Gogate PR, Pandit AB. Improved synthesis of sophorolipids from waste cooking oil using fed batch approach in the presence of ultrasound. *Chem Eng J*. 2015;263:479–487.

113 Morya VK, Park JH, Kim TJ, Jeon S, Kim EK. Production and characterization of low molecular weight sophorolipid under fed-batch culture. *Bioresour Technol*. 2013;143:282–288.

114 Zhang Y, Jia D, Sun W, Yang X, Zhang C, Zhao F, Lu W. Semicontinuous sophorolipid fermentation using a novel bioreactor with dual ventilation pipes and dual sieve-plates coupled with a novel separation system. *Microbial Biotech*. 2018;11(3):455–464.

115 Zhu L, Yang X, Xue C, Chen Y, Qu L, Lu W. Enhanced rhamnolipids production by Pseudomonas aeruginosa based on a pH stage-controlled fed-batch fermentation process. *Bioresour Technol*. 2012;117:208–213.

116 Xu N, Liu S, Xu L, Zhou J, Xin F, Zhang W, et al. Enhanced rhamnolipids production using a novel bioreactor system based on integrated foam-control and repeated fed-batch fermentation strategy. *Biotech Biofuels*. 2020;13(1):1–10.

117 Gong Z, Peng Y, Wang Q. Rhamnolipid production, characterization and fermentation scale-up by Pseudomonas aeruginosa with plant oils. *Biotech Lett.* 2015;37(10):2033–2038.

118 Pereira AG, Pacheco GJ, Tavares LF, Neves BC, Kronemberger FDA, Reis RS, et al. Optimization of biosurfactant production using waste from biodiesel industry in a new membrane assisted bioreactor. *Process Bioch.* 2013;48(9):1271–1278.

119 Fahim S, Dimitrov K, Vauchel P, Gancel F, Delaplace G, Jacques P, et al. Oxygen transfer in three phase inverse fluidized bed bioreactor during biosurfactant production by Bacillus subtilis. *Bioch Eng J.* 2013;76:70–76.

120 Khondee N, Tathong S, Pinyakong O, Muler R, Soonglerdsongpha S, Ruangchainikom O, et al. Lipopeptide biosurfactant production by chitosan-immobilized Bacillus sp. GY19 and their recovery by foam fractionation. *Biochem Eng J.* 2015;93:47–54. mudar ano no texto tabela thiago.

121 da Silva MTS, Soares CMF, Lima AS, Santana CC. Integral production and concentration of surfactin from Bacillus sp. ITP-001 by semi-batch foam fractionation. *Bioch Eng J.* 2015;104:91–97.

122 Brumano LP, Antunes FAF, Souto SG, Dos Santos JC, Venus J, Schneider R, et al. Biosurfactant production by Aureobasidium pullulans in stirred tank bioreactor: new approach to understand the influence of important variables in the process. *Bioresour Technol.* 2017;243:264–272.

123 Fernandes B, Mota A, Vicente A. Fundamentals of bio-reaction engineering. In: Pandey A, Sanromán MA, Du G, Soccol CR, Dussap CG, editors. *Current Developments in Biotechnology and Bioengineering.* Elsevier; 2017. pp. 153–185.

124 Guilmanov V, Ballistreri A, Impallomeni G, Gross RA. Oxygen transfer rate and sophorose lipid production by Candida bombicola. *Biotech Bioeng.* 2002;77(5):489–494.

125 Sastaravet P, Bun S, Wongwailikhit K, Chawaloesphonsiya N, Fujii M, Painmanakul P. Relative effect of additional solid media on bubble hydrodynamics in bubble column and airlift reactors towards mass transfer enhancement. *Processes.* 2020;8(6):713.

126 Cerri MO, Futiwaki L, Jesus CDF, Cruz AJG, Badino AC. Average shear rate for Non-Newtonian fluids in a concentric-tube airlift bioreactor. *Bioch Eng J.* 2008;39(1):51–57.

127 Coutte F, Lecouturier D, Yahia SA, Leclère V, Béchet M, Jacques P, et al. Production of surfactin and fengycin by Bacillus subtilis in a bubbleless membrane bioreactor. *App Microbiol Biotech* 2010;87(2):499–507.

128 Bongartz P, Bator I, Baitalow K, Keller R, Tiso T, Blank LM, et al. A scalable bubble-free membrane aerator for biosurfactant production. *Biotech Bioeng.* 2021;118(9):3545–3558.

129 Wehrs M, Tanjore D, Eng T, Lievense J, Pray TR, Mukhopadhyay A. Engineering robust production microbes for large-scale cultivation. *Trends Microbiol.* 2019;27(6):524–537.

130 Daniel HJ, Reuss M, Syldatk C. Production of sophorolipids in high concentration from deproteinized whey and rapeseed oil in a two stage fed batch process using Candida bombicola ATCC 22214 and Cryptococcus curvatus ATCC 20509. *Biotech Lett.* 1998;20(12):1153–1156.

131 Davila AM, Marchal R, Vandecasteele JP. Kinetics and balance of a fermentation free from product inhibition: sophorose lipid production by Candida bombicola. *Appl Microbiol Biotechnol.* 1992;38(1):6–11.

132 Ambaye TG, Vaccari M, Prasad S, Rtimi S. Preparation, characterization and application of biosurfactant in various industries: a critical review on progress, challenges and perspectives. *Environ Technol Innov.* 2021;24:102090.

133 Dolman BM, Kaisermann C, Martin PJ, Winterburn JB. Integrated sophorolipid production and gravity separation. *Process Bioch*. 2017;54:162–171.

134 Roelants SLKW, Van Renterghem L, Maes K, Everaert B, Redant E, Vanlerberghe B, et al. Microbial biosurfactants: from lab to market. In: Banat IM, Thavasi R, editors. *Microbial Biosurfactants and Their Environmental and Industrial Applications*, Vol. 1. 2019. pp. 340–362.

135 Helmy Q, Kardena E, Funamizu N. Strategies toward commercial scale of biosurfactant production as potential substitute for it's chemically counterparts. *Int J Biotechnol*. 2011;12(1–2):66–86.

136 Invally K, Sancheti A, Ju L. A new approach for downstream purification of rhamnolipid biosurfactants. *Food Bioprod Process*. 2019;114:122–131.

137 Weber A, May A, Zeiner T, Górak A. Downstream processing of biosurfactants. *Chem Eng Trans*. 2012;27:115–120.

138 Najmi Z, Ebrahimipour G, Franzetti A, Banat IM. In situ downstream strategies for cost-effective bio/surfactant recovery. *Biotechnol Appl Biochem*. 2018;65(4):523–532.

139 Sahah MUH, Sivapragasam M, Moniruzzaman M, Yusup SB. A comparison of recovery methods of rhamnolipids produced by Pseudomonas aeruginosa. *Procedia Eng*. 2016;148:494–500.

140 Satpute SK, Banpurkar AG, Dhakephalkar PK, Banat IM, Chopade BA. Methods for investigating biosurfactants and bioemulsifiers: a review. *Crit Rev Biotechnol*. 2010;30(2):127–144.

141 Wang H, Roelants SLKW, To MH, Patria RD, Kaur G, Lau NS, et al. Starmerella bombicola: recent advances on sophorolipid production and prospects of waste stream utilization. *J Chem Technol Biotechnol*. 2018;94:999–1007.

142 Varjani SJ, Upasani VN. Critical review on biosurfactant analysis, purification and characterization using rhamnolipid as a model biosurfactant. *Bioresour Technol*. 2017;232:389–397.

143 Banat IM, Franzetti A, Gandolfi I, Bestetti G, Martinotti MG, Fracchia L, et al. Microbial biosurfactants production, applications and future potential. *Appl Microbiol Biotechnol*. 2010;87:427–444.

144 Pardhi DS, Panchal RR, Raval VH, Rajput KN. Statistical optimization of medium components for biosurfactant production by Pseudomonas guguanensis D30. *Prep Biochem Biotechnol*. 2021;52:1–10.

145 Dimitrov K, Gancel F, Montrastruc L, Nikov I. Liquid membrane extraction of bio-active amphiphilic substances: recovery of surfactin. *Biochem Eng J*. 2008;42:248–253.

146 Dolman BM, Wang F, Winterburn JB. Integrated production and separation of biosurfactants. *Process Bioch*. 2019;83:1–8.

147 Decesaro A, Machado TS, Cappellaro ÂC, Rempel A, Margarites AC, Reinehr CO. Biosurfactants production using permeate from whey ultrafiltration and bioproduct recovery by membrane separation process. *J Surfactants Deterg*. 2020;23(3):539–551.

148 Rentergehm LV, Roelants SLKW, Baccile N, Uyttersprot K, Taelman MC, Everaert B, et al. From lab to market: an integrated bioprocess design approach for new-to-nature biosurfactants produced by Starmerella bombicola. *Biotechnol Bioeng*. 2017;115:1195–1206.

149 Thavasi R, Banat IM. Downstream processing of microbial biosurfactants. In: Thavasi R, Banat IM, editors. *Microbial Biosurfactants and Their Environmental and Industrial Applications*, 1st ed. CRC Press; 2019.

150 Haba E, Bouhdid S, Torrego-Solana N, Marqués AM, Espuny MJ, García-Celma MJ. Rhamnolipids as emulsifying agents for essential oil formulations: antimicrobial effect against Candida albicans and methicillin-resistant Staphylococcus aureus. *Int J Pharm.* 2014;467:134–141.

151 Syldatk C, Wagner F. Production of biosurfactants. In: Kosaric N, Caims WL, Gray NCC, editors. *Biosurfactants and Biotechnology*, 1st ed. Routledge; 2017. pp. 89–120.

152 Matsufuji M, Nakata K, Yoshimoto A. High production of rhamnolipids by Pseudomonas aeruginosa growing on ethanol. *Biotechnol Lett* 1997;19(12):1213–1215.

153 Chen H, Chen Y, Juand R. Recovery of surfactin from fermentation broths by a hybrid salting-out and membrane filtration process. *Sep Purif Technol* 2008;59:244–252.

154 Sarachat T, Pornsunthorntawee O, Chavadej S, Rujiravanit R. Purification and concentration of a rhamnolipid biosurfactant produced by Pseudomonas aeruginosa SP4 using foam fractionation. *Bioresour Technol* 2010;101:324–330.

155 Félix AKN, Martins JJL, Almeida JGL, Giro MEA, Cavalcante KF, Melo VMM, et al. Purification and characterization of a biosurfactant produced by Bacillus subtilis in cashew apple juice and its application in the remediation of oil-contaminated soil. *Colloids Surf B*. 2019;175:256–263.

156 Mnif I, Ghribi D. Glycolipid biosurfactants: main properties and potential applications in agriculture and food industry. *J Sci Food Agric*. 2016;96:4310–4320.

157 Elshikh M, Moya-Ramírez I, Moens H, Roelants S, Soetaert W, Marchant R, et al. Rhamnolipids and lactonic sophorolipids: naturas antimicrobial surfactants for oral hygiene. *J Appl Microbiol*. 2017;123:1364–5072.

158 Markande AR, Patel D, Varjani S. A review on biosurfactants: properties, applications and current developments. *Bioresour Technol*. 2021;330:124963.

159 Reddy AS, Chen CY, Chen CC, Jean JS, Fan CW, Chen HR, et al. Synthesis of gold nanoparticles via an environmentally benign route using a biosurfactant. *J Nanosci Nanotechnol*. 2009;9(11):6693–6699.

160 Castelein M, Verbruggen F, Van Renterghem L, Spooren J, Yurramendi L, Du Laing G, et al. Bioleaching of metals from secondary materials using glycolipid biosurfactants. *Miner Eng*. 2021;163:106665.

161 Zinjarde SS, Pant A. Emulsifier from a tropical marine yeast, Yarrowia lipolytica NCIM 3589. *J Basic Microbiol*. 2002;42(1):67–73.

162 Knepper TP, de Voogt P, Barcelo D. *Analysis and Fate of Surfactants in the Aquatic Environment*. Elsevier, 2003.

7

Enzymatic Production of Biosurfactants

Ana Karine F. de Carvalho[1,5,*], *Heitor B.S. Bento*[2], *Felipe R. Carlos*[5], *Vitor B. Hidalgo*[1,6], *Cintia M. Romero*[3], *Bruno C. Gambarato*[4], *and Patrícia C.M. Da Rós*[1,6]

[1] *Department of Basic Sciences and Environmental Engineering, School of Engineering of Lorena, University of São Paulo, Lorena – SP, Brazil*
[2] *Department of Bioprocess Engineering and Biotechnology - School of Pharmaceutical Sciences – São Paulo State University-UNESP, Araraquara – SP, Brazil*
[3] *Planta Piloto de Procesos Industriales Microbiológicos, Consejo Nacional de Investigaciones Científicas y Técnicas – CONICET-Tucuman-Argentina*
[4] *Department of Material Science - University Center of Volta Redonda Volta Redonda – RJ, Brazil*
[5] *Postgraduate Program in Biotechnology, Institute of Chemistry, Federal University of Alfenas, Alfenas – MG, Brazil*
[6] *Department of Chemical Engineering, School of Engineering of Lorena, University of São Paulo, Lorena – SP, Brazil*
* *Corresponding author*

7.1 Introduction

For industrial enzymatic reactions, enzymes are preferably used in an immobilized state in order to easily separate the catalyst from the product [1, 2]. Using immobilized enzymes, improved stability, reuse, continuous operation, the possibility of better control of reactions, and hence more favorable economic factors can be expected [3]. Immobilization is the practice responsible for the initial breakthrough innovation that allowed efficient reutilization of enzymes, thus reducing the cost per batch [4]. Several methods of enzyme immobilization are available, each involving a singular degree of complexity and efficiency [1–3].

Increased attention to the environment has led to growing interest in the field of natural biosurfactants, since they have different structural characteristics and physical properties, which make them comparable or superior to synthetic surfactants in terms of efficiency [5, 6]. In addition, natural biosurfactants present amphiphilic properties of interest for the formulation of several products in the fields of detergents, foods, medicines, pharmaceuticals, agriculture, and cosmetics [7], due to their non-toxic, non-irritant, odorless, tasteless, biodegradable, and biocompatible characteristics [8]. Another advantage is that they are compounds that are not petroleum derivatives, this is an important aspect as there is a worldwide trend of increasing oil prices. Indeed, interest in these compounds stems from the natural origin of the raw materials, the synthetic processes involved, and the performance of the final product.

Biosurfactants and Sustainability: From Biorefineries Production to Versatile Applications, First Edition.
Edited by Paulo Ricardo Franco Marcelino, Silvio Silverio da Silva, and Antonio Ortiz Lopez.
© 2023 John Wiley & Sons Ltd. Published 2023 by John Wiley & Sons Ltd.

7.2 What are the Biosurfactants Produced Enzymatically? Esterification Reactions of Sugars and Fatty Acids Catalyzed by Enzymes

Biosurfactants are compounds produced on living surfaces, mostly microbial cell surfaces or excreted extracellularly, that reduce surface tension and interfacial tension between individual molecules at the surface and interface respectively [9, 10]. There are various types of biosurfactants produced microbially or enzymatically, including peptides, glycolipids, glycopeptides, fatty acids, and phospholipids. Fatty acid derivatives (fatty acid ester, fatty acid ester of sugar, and sugar alcohols) are important for cosmetic and industrial purposes [9]. Biosurfactants have been traditionally classified into low molecular weight biosurfactants (BS) and high molecular weight bioemulsifiers (BE) [10].

Therefore, BS or BE are considered the most versatile chemical groups used in various industrial processes. As they have to compete in the market, however, it is essential to produce BS or BE that are eco-friendly and to do so economically [11, 12]. The roles and applications of surface-active agents have been reported to have microbial physiology and industrial application [10], such as:

- **Interaction of biosurfactants with hydrophobic moieties**: The ability of biosurfactants to form stable micelles has increased their applications in formulations of nanoemulsions and other drug-delivery systems used against major diseases like thrombosis, Alzheimer's, and cancer [10, 13].
- **Adherence and de-adherence**: The amphipathic nature of biosurfactants enables them to interact with both polar and non-polar surfaces as well as charged surfaces forming the first layer and acting as wetting agents.
- **Biofilm formation and removal**: The amphipathic nature of biosurfactants also gives them anti-biofilm applications. Many unwanted biofilms formed on surfaces of human interest (environmental, industrial, or even biomedical) have been reported to be effectively removed after the use of biosurfactants.
- **Antimicrobial activity**: many biosurfactants have been reported for antifungal, antibacterial and anti-helminth activities, recent studies on these activities in the situation of widespread antibiotic resistance reports have increased their potential.

The characteristics and micelle formation kinetics of biosurfactants and their possibility of various applications include the formulations of surface-active-agents exploited in food, cosmetics, and medicine [14, 15].

7.2.1 Esterification Reactions of Sugars and Fatty Acids Catalyzed by Enzymes

Enzyme-catalyzed esterification has attracted increasing attention in many applications, due to the significance of the derived products. Within the biocatalysts, which can be used in the synthesis of esters, the most common are lipases [1, 8, 16] which have attracted research interest over the past decade, due to an increased use of organic esters in biotechnology and the chemical industry. Microbial lipases have been employed in esterification reactions using either primary or secondary alcohols, or both, free-solvent systems, or organic solvents [13, 17].

Sugar based surfactants (SBS – e.g. long- or medium-chain fatty acids of glucose, fructose, or sucrose) syntheses can be carried out by chemical or enzymatic processes. Chemical syntheses are often carried out at high temperature using acid or alkaline catalysts but are non-regio-selective and lead to the formation of undesirable products. In addition, the chemical catalysts are mostly alkaline and the reaction often uses dimethylformamide (DMF), pyridine, or dimethylsulfoxide (DMSO) that are toxic solvents. This implies high energy costs, the production of undesired by-products, and low selectivity [1, 18].

On the other hand, enzymatic syntheses are performed under mild reaction conditions using enzyme as a catalyst and are highly selective [8, 19]. Moreover, the most challenging features of this process are the control of the degree of esterification and the position of acylation, and both are difficult to achieve using alkaline catalysis. Therefore, enzymatic synthesis of sugar esters might be a useful alternative, which is already used on industrial scales for some applications. In these reactions, in order to dissolve the carbohydrate, the use of very polar solvents is required [1]. In addition, the solvent must not present reactive hydroxyl groups able to compete with the sugars, carboxylic groups or ester bonds, which could be substrates of the enzyme and reduce the product yield and/or produce side-products. In general, tertiary alcohols or ionic liquids meet these requirements [1, 20, 21].

Butylene glycol laurate is a surfactant that finds application in the production of synthetic detergents. The synthesis of butylene glycol laurate ester with an enzymatic preparation of pancreatic lipase in an organic solvent medium was studied to determine conditions for esterification, and the yield of esters under optimum conditions was 78% in 24 h. Esters of fatty acids and polyethyleneglocol-400 (PEG) are valuable surfactants, and, the possibility of replacing the chemical synthesis of polyethylene glycol esters with enzymatic synthesis was demonstrated in the literature [12].

The search for ways of using enzymes, especially lipases, to synthesize esters with different structures is important and highly relevant for biotechnology, and industry in which esters are also used.

Lipases are biocatalysts of great importance in different areas, and can catalyze reactions both in aqueous and organic medium, with restricted water content. This phenomenon is primarily due to their ability to use a wide range of substrates, stability against temperature, pH, organic solvents, and maintenance of their chemo-, region-, and enantio-selectivity. Lipases have been used in a variety of biotechnological areas, such as food industries, detergents, pharmacological, agrochemical, and oleo chemical plants [4].

Among the important factors that influence the ester yield are the concentrations of enzyme and substrates, their molar ratio, the reaction pH-value and temperature, the mixing rates, and the water content [17].

7.3 Enzymes and Methods for Biosurfactant Production: Bioreactors and Ways of Conducting Enzymatic Processes

Enzymatic reactions may be conducted either in batch or continuous processes. However, only solid support-immobilized enzymes can be applied in continuous processes. Despite the other advantages of the enzyme immobilization technique, such as higher thermal and storage stabilities, the reusability and possibility of its application in continuous processes are essential tools to make a feasible industrial bioprocess [22, 23].

An industrial biocatalyst must present some desired characteristics to meet the economic feasibility requirements for implementation in the market. High stability, reusability, and elevated enzymatic activity are among the ideal parameters. Once a potential biocatalyst is selected, choosing the most suitable bioreactor for the industrial scale implementation is the next key step [24]. Figure 7.1 shows the main types of reactors feasible to biosurfactant production.

Batch processes based on a stirred tank reactor usually present the advantages of being the easiest implemented and simple methods, combined with biocatalyst recuperation and reusability and can be selected as a potential low-cost method of biosurfactant enzymatic production. However, the mechanical agitation may cause enzyme lixiviation and damages to the biocatalyst solid matrix, resulting in productivity losses and higher maintenance costs. In that sense, continuous processes may show more interesting characteristics for industrial application [22, 25]. The main advantages and disadvantages of applying theses batch and continuous enzymatic processes are summarized in Table 7.1.

Continuous processes usually show higher productivities with lower biocatalyst losses. The packed bed reactor system is the most frequently used configuration in continuous

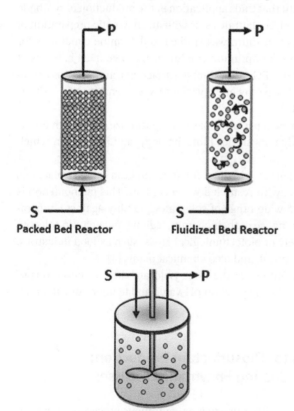

Figure 7.1 Main types of reactors used for biosurfactant production.

Table 7.1 Main advantages and disadvantages of applying batch and continuous enzymatic processes.

	Batch process	Continuous processes	
	Stirred tank reactor (STR)	Packed bed reactor (PBR)	Fluidized bed Reactor (FBR)
Advantages	Low-cost implementation Simple operation	High productivities Reduced enzyme stress and support damaging	High productivities Reduced enzyme stress and support damaging Facilitate substrate flow Better mixing and matrix distribution
Disadvantages	Low productivity May harm the biocatalyst by the agitation	Sugars substrate may form crystals reducing substrate flow and causing clogging of the column	Complex operational parameters involved Difficulties in maintaining bed stability

enzymatic reactions due to its relatively simple design. Some limitations of the method may be related to substrate diffusional difficulties through the packed bed, especially when there are sugars involved that may form solid crystals. Thus, the fluidized bed reactor may stand out as a promising alternative for biosurfactant enzymatic production [22].

The fluidized bed continuous process presents several advantages regarding mass transfer and substrate diffusional limitations. However, the selection of the biocatalyst needs to be more intensely judicious in order to allow a stable fluidized bed. Also, several complex operational parameters need to be deeply evaluated, as medium viscosity, density, flow rate, and others. In this sense, some engineered tools have been highlighted as potential complements for fluidized bed reactor applications, e.g. a magnetized stabilized fluidized bed reactor using magnetic enzymes supports [25–27].

In order to perform lipase-mediated glycolipid synthesis, it is necessary to consider some factors that influence the process. When it comes to a reverse hydrolysis (condensation) reaction, water activity is one of the most important factors to be considered when choosing the reaction conditions, since in this process, water acts as a solvent, but also as a reagent and substrate. Therefore, the enzymatic synthesis of glycolipids depends on finding the appropriate combination of the enzyme, the solvent and the consequent water activity. The ideal conditions involve strategies to reduce the water content in order to prevent hydrolysis. Among the ways to conduct the process in order to reduce water are the use of molecular sieves, membrane reactors and the use of solvents containing salt solutions. However, it is important to bear in mind that a low water activity is essential to keep the water layer bound to the enzyme, ensuring its activity [28].

Another way to carry out the synthesis of glycolipids without water formation is through the enzymatic reaction of transesterification instead of reverse hydrolysis using fatty acid esters or free fatty acids. In this process, a factor to be considered is the molar ratio sugar/fatty acid, since an excess of long chain fatty acids promotes a decrease in the conversion rate, mainly due to the inhibiting effect caused by the blockage of the catalytic site of some lipases [29].

The solubility of the reagents in the reaction medium is directly related to the process yield and to other operating parameters. Typically, the greatest possible solubility of the chemical species in the solvent promoted by an increase in temperature is desired. However, despite improving solubility, high temperatures can affect stability and enzymatic activity. Therefore, when considering the aspects related to water activity, sugar/fatty acid molar ratio, as well as the solubility, it is possible to optimize the conditions for a synthesis of different glycolipids [19].

The use of ionic liquids (IL) or deep eutectic solvents (DES) are reported ways to overcome the problem related to solubility. Qin et al. [30] found that ionic liquids containing N,N-dialkylimidazolium cations and different anions improve lipase activity and stability. It was also found [31] that it is possible to carry out the lipase-catalyzed synthesis of glucose-6-O-hexanoate using two deep eutectic solvents composed of choline chloride and urea (CC:U) and choline chloride and glucose (CC:G). Although there are not many studies related to processes catalyzed by lipases, DES have the advantage of being easily produced from non-toxic resources.

7.4 Advantages and Disadvantages of Enzymatic Biosurfactant Production

More developed countries tend to have greater sophistication in the global surfactant market. The most modern industries that use surfactants as inputs prefer simpler products, reinforcing the search for cheaper raw materials. The modernization of the surfactant industry has led to the gradual replacement of chemically-derived surfactants with those obtained via enzymatic reactions [32].

Current forecasts estimate a 20% annual growth rate in the bioproduct markets over the next decade and a market share for bioproducts and biomaterials of around 17% to 30% by 2050 [33]. According to the Center for Strategic Management and Studies, the production of numerous products through the development of bioprocesses was encouraged, becoming an option for products traditionally marketed and produced according to the concepts of "green chemistry, environmentally correct chemistry, making the sustainable development, an emerging culture that aims to conduct scientific actions and/or ecologically correct industrial processes". Among the diverse products that can be obtained from sustainable sources, the production and use of natural surfactants, also known as biosurfactants, as an alternative to synthetic surfactants, is a good example of bio-processes stimulated by consumers and distributors of consumer goods [34].

Most surfactants developed today are commonly obtained from chemical processes that involve catalysts with high acidity or alkalinity and/or use organic solvents, requiring large amounts of energy due to the use of high temperatures. On the other hand, there are the surfactants obtained through bioprocesses that involve a more ecological approach using enzymes, and therefore, they are receiving attention as examples of "green manufacturing" that can lead to a more suitable sustainability profile [35]. These biosurfactants have lower toxicity, biodegradability, and minimal impact on the ecosystem. Due to these qualities, biosurfactants are highly suitable for various applications in industries such as the environment, food and beverage, pharmaceuticals, and cosmetics [15, 36].

7.5 Potential Use of Enzymes for the Production of Biosurfactants

The application of purified enzymes in the production of surfactant compounds has been shown to be a viable alternative to synthetic methods. Processes such as the enzymatic synthesis of monoglycerides, sugar fatty acid esters, phospholipids, anomerically pure alkyl glycosides, and amino acid-based surfactants are discussed and have been researched by different groups. Table 7.2 shows enzymes that can be applied in industrial biosurfactant production.

The surfactant biosynthesis reaction can be carried out through processes that use enzymes as catalysts and comply with the concepts of green chemistry, in addition to ensuring lower energy costs, since the enzymes work under mild conditions of temperature and pressure, which also contributes to the prevention of accidents. However, the enzymatic route is not yet available in the market, since the industrial production mainly focuses on synthetic surfactants. This section shows the advantages and disadvantages of using enzymes for the synthesis of surfactants.

We can list the disadvantages of enzymatic processes, mainly in relation to free enzymes. Such disadvantages include: the enzyme's susceptibility to inhibition, the high cost of the commercial enzyme, instability in the reaction medium, loss of catalytic activity throughout the process, among others. These drawbacks derail the current use of enzymatic processes as they require pre-purification processes and create operational limitations such as milder temperatures to prevent loss of enzyme catalytic activity [37].

There is a basic difference between enzymatic and microbial syntheses of surfactants. The first is essentially an organic synthesis, where hydrolytic enzymes are used as biological alternatives to traditional catalysts. The second is a biosynthetic process catalyzed by a cascade of enzymes in metabolically active cells. Consequently, microorganisms can produce

Table 7.2 Enzymes applied in biosurfactant production.

Enzyme	Surfactant or precursor
Alcohol dehydrogenase	Aldehyde or ketone (from fatty alcohol)
α-Amylase	Alkyl polyglucosides
Glucosidade	Alkyl glucosides
Glucosyl transferases	Alkyl polyglucosides
Lipases	Fatty acid esters (polyol, polyglycerol), fatty amide-based, lysophospholipids, carbonates, amino acid-based
Papain	Amino acid surfactants
Phospholipases	Tailor-made phospholipids

Source: With [37] / The American Oil Chemists' Society.

much smaller surfactants than it would be possible on a practical scale by any other means. On the other hand, the use of isolated enzymes allows the preparation of a wide range of surfactants which, although simpler in structure, can be designed for different industrial purposes. Table 7.3 shows the advantages and disadvantages of applying enzymes for surfactants production.

It would be optimistic to assume that the use of biocatalysts in the surfactant industry will become significant in the short term, as the cost of purified enzymes still remains the main obstacle to the practical realization of their potential, in addition to the long reaction times. However, the enzymatic route can gain numerous advantages over conventional processes by developing strategies to mitigate these drawbacks. For example, the possibility of immobilizing the enzymatic system, increasing its stability, and allowing the result to be processed. Enzymatic reactions can be accelerated by using unconventional heating, such as ultrasound and microwaves, which meet the concept of green energy [38]. We can also count on advances in enzyme immobilization technology, recombinant DNA technology, and the discovery of extremophile microorganisms that are producing more active and stable enzyme preparations, enabling them to extend their operational lifetime in batch and continuous processes. In addition, the integration of different industrial segments in the context of biorefineries can reduce the costs associated with the production of enzymes and, therefore, can reduce the costs associated with the production of biosurfactants by enzymatic routes.

7.6 Production of Biosurfactants by the Enzymatic Route in Biorefineries: Demand for More Modern Production Processes

Since their introduction in the early twentieth century, the production of surfactants from petrochemical sources has steadily increased to 18.5 million tons per year and it is forecasted to grow at an annual compound growth rate of 5% per year over the period 2018–2023. Combined with the high growth rate, there is an awareness that more ecologically and economically viable surfactants are needed, preferably derived from renewable resources. Economically, the global biosurfactant market was expected to reach $39.86 billion in the year 2021, from $30.64 billion in 2016 [38].

Table 7.3 Advantages and disadvantages of using biocatalysts for the synthesis of surfactants.

Advantages	Disadvantages
Lower energy use (lower temperature) → reduced CO_2 production	Limitations on operating conditions, such as temperature range and pH
Lower amounts of waste products and by-products	Concerns for the supply of enzymes on a large-scale
Regioselectivity	High cost of enzymes
Absence of toxic metal catalysts or acids/bases	Low rate of reaction, frequently
Can result in lower solvent usage	

Source: Adapted from [37].

The generation of waste is becoming a global concern due to its adverse effects on the environment and human health. The use of waste as raw material to produce value-added products opened new pathways contributing to sustainable development. It is known that a complete valorization of agro-industrial residues can be achieved when these serve as substrates for microorganisms to produce a variety of high-value chemical products. Using waste as a substrate significantly reduces the cost of the overall process. In this section, we discuss the production of biosurfactants from several agro-industrial residues biotransformed by enzyme-mediated reactions, thus integrating industries from different sectors with the enzyme and surfactant production industries (Figure 7.1).

The reactions for obtaining surfactants by the enzymatic route are: enzymatic synthesis of monoglycerides, sugar fatty acid esters, modification of phospholipids, and pure alkyl glycosides. Figure 7.2 shows the production of some enzymes that can mediate different surfactant biosynthesis reactions.

The most widely used bio-based resource to make surfactants is the fatty acyl group derived from seed oils, which are becoming more readily available to manufacturers due to the growth of biodiesel, which provides an abundant and consistent supply of fatty acid methyl esters as chemical intermediates. The hydrophilic portion of surfactants can also be derived from renewables that readily occur in a biorefinery operation, such as sugars, glycerol (a low-value co-product derived from biodiesel), amino acids and their derivatives. Other biorefinery derived chemicals such as phospholipids, sterols, and glycols can also be used as base materials [39].

Due to the emergence of new technologies based on renewable resources, several countries have considered that an ideal biorefinery can be linked to the concept of an integrated biorefinery. By definition, biorefineries can be classified by phases [40]. Phase I biorefineries are capable of processing only one raw material in fixed proportions and without

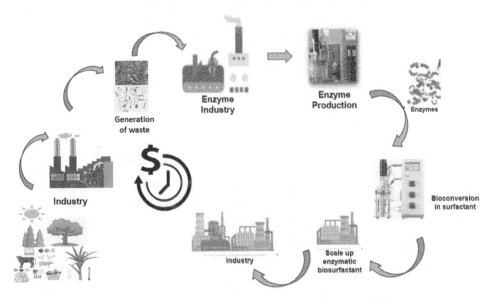

Figure 7.2 Usual stages in industrial use of enzymes for biosurfactant production.

processing flexibility; phase II biorefineries also process one type of biomass, but they have the capacity to produce a greater variety of products; phase III biorefineries are considered to be the most advanced type of biorefinery, as they are capable of processing different types of biomasses and can generate a wide variety of products. The actual participation in the bioproducts market will depend primarily on technological development and real policies that support a shift from the fossil fuel economy to bioeconomy.

Figure 7.3 shows how the process integration in a biorefinery context may lead to the diminishing of the number of industrial process steps in order to achieve elevated productivity of different products and lowering the production costs at the same time.

Europe is the biggest emerging market, followed by the United States. The growing infrastructure and awareness in Asian countries has increased the demand for biosurfactant [41]. Industries such as Ecover (Belgium) and BASF Cognis, Urumqui Unite, Evonik Industries, Akzo Nobel and Saraya (Germany), and Jeneil Biotech Inc. (United States) are the main players in the biosurfactant sector [41]. To overcome the problem of economic biosurfactant production, scientific investigations have shifted to the use of cheaper/renewable sources, such as agricultural or agro-industrial residues (for example: sugarcane bagasse and straw, rice husk, wheat straw, industrial wastewater, dairy by-products), taking into account the context of advanced biorefinery [42]. The lipase-mediated production of biosurfactants from agro-industrial co-products may contribute to making the process cheaper, while providing process integration within the context of an advanced biorefinery.

Surfactant production by enzymatic synthesis from lipases has provided a new approach to biosurfactant production, especially in the application of biosurfactants in the cosmetics, foods, and drugs industries. With increased research into developing technologies, lipase producer species enhancement, and biosurfactant production processes around the world, biosurfactants are currently considered very versatile and can be used in different industrial areas. Nowadays, biosurfactant production is relatively low in the global market, mainly due to its high raw material and further processing cost [39]. Expenses with raw materials represent more than 50% of the total cost of producing the biosurfactant [36]. In fact, enzymes are recognized as powerful synthetic tools, demanding increasing attention due to their potential use in the manufacture of surfactants.

In addition to the regio- and stereo-selectivity and mild reaction conditions associated with the enzyme-mediated process, the increasing attention paid to enzymes as biocatalysts for surfactant synthesis can be explained by two other noticeable trends in the

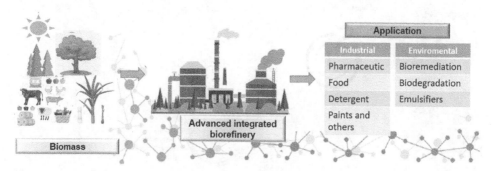

Figure 7.3 Industrial process steps integratated in a biorefinery context.

surfactant industry. They are: (1) the need to replace petrochemicals, when possible; (2) associate the biosurfactant production industry with other industrial areas, for example the biofuel industries; (3) use agro-industrial waste to produce enzymes and these being applied in biosurfactants production, we would have enzymatic processes advantageous, with synthetic green route and high reaction yield.

There is now an awareness that we need more environmentally-friendly and economically viable surfactants preferably derived from renewable resources. Glycerol is an underexploited by-product of biodiesel production, in particular from the hydrolysis of biomass derived triglycerides (such as palm oil, sunflower oil, or rapeseed oil) which results in valuable methyl esters. Between 2007 and 2016, biodiesel production increased by 83% in the European Union. Thus, glycerol is widely available, and its price is inversely proportional to the increase in biodiesel production. The world production of biodiesel has reached 46.8 billion in 2020, generating large volumes of glycerol at an affordable price. It is believed that full utilization of the by-product can be accomplished through microbial biotransformation, for example, glycerol can be used as a carbon source for enzyme production and also in lipase-mediated oil glycerolysis processes to produce surfactants [43].

Although the production costs of enzymatic biosurfactants are still higher than those produced by synthetic routes, the use of low-cost substrates in the fermentation and techniques as enzyme immobilization for reutilization may diminish the global process cost and increase the feasibility of these new bioprocesses [44, 45].

7.7 Conclusion

Briefly, in this chapter we have emphasized the versatility of biosurfactant production and the advantages of using the enzymatic route. Moreover, we demonstrated that biosurfactants from the enzymatic route have the potential to increase the sustainability factor of the process and even reduce the global cost by the use of wastes and by-products in a biorefinery context.

References

1 Lima LN, Mendes AA, Lafuente RF, Tardioli PW, Giordano RLC. Performance of different immobilized lipases in the syntheses of short- and long-chain carboxylic acid esters by esterification reactions in organic media. *Molecules*. 2018;23:766.

2 Santos LFS, Silva MRL, Ferreira EEA, Gama RS, Carvalho AKF, Barbosa JC et al. Decyl oleate production by enzymatic esterification using *Geotrichum candidum* lipase immobilized on a support prepared from rice husk. *Biocatalysis and Agricultural Biotechnology*. 2021;36:102–142.

3 Pereira EB, Zanin GM, de Castro HF. Immobilization and catalytic properties of lipase on chitosan for hydrolysis and esterification reactions. *Brazilian Journal of Chemical Engineering*. 2003;20(4):343–355.

4 Cortez DV, Reis C, Perez VH, de Castro HF. The realm of lipases in biodiesel production. In: Singh OV, Chandel AK, editors. *Sustainable Biotechnology- Enzymatic Resources of Renewable Energy*, 1st ed. Springer International Publishing; 2018: 247–288.

5 Habulin M, Sabeder S, Knez Z. Enzymatic synthesis of sugar fatty acid esters in organic solvent and in supercritical carbon dioxide and their antimicrobial activity. *J Supercrit Fluids*. 2008;45:338–345.

6 Rudakova MA, Galitskaya PY, Selivanovskaya SY Biosurfactants: current application trends. Uchenye Zapiski Kazanskogo Universiteta. Seriya Estestvennye Nauki. 2021; 163(2):177–208.

7 Ibinga SKK, Fabre JF, Bikanga R, Mouloungui Z. A typical reaction media and organized systems for the synthesis of low-substitution sugar esters. *Front Chem*. 2019;7:587.

8 Shin DW, Mai NL, Bae SW, Koo YM. Enhanced lipase-catalyzed synthesis of sugar fatty acid esters using supersaturated sugar solution in ionic liquids. *Enzyme Microb Technol*. 2019;126:18–23.

9 Gautam KK, Tyagi VK. Microbial surfactants: a review. *J Oleo Sci*. 2006;55(4):155–166.

10 Markande AR, Patel D, Varjani S. A review on biosurfactants: properties, applications and current developments. *Bioresour Technol*. 2021;330:124963.

11 Bhattacharya B, Ghosh TK, Das N. Application of bio-surfactants in cosmetics and pharmaceutical industry. *Sch Acad J Pharm*. 2017;6(7):320–329.

12 Gamayurovaa VS, Zinov'evaa ME, Shnaidera KL, Davletshina GA. Lipases in esterification reactions: a review. *Catal Ind*. 2021;13(1):58–72.

13 Calabrese I, Gelardi G, Merli M, Liveri MLT, Sciascia L. Clay-biosurfactant materials as funtional drug delivery systems: slowing down effect in the in vitro release of cinnamic acid. *Appl Clay Sci*. 2017;135:567–574.

14 Liu K, Sun Y, Cao M, Wang J, Lu J, Xu H. Rational design, properties, and applications of biosurfactants: a short review of recent advances. *Curr Opin Colloid Interface Sci*. 2020;45:57–67.

15 Varjani S, Rakholiya P, Ng HY, Taherzadeh MJ, Ngo HH, Chang JS, et al. Bio-based rhamnolipids production and recovery from waste streams: status and perspectives. *Bioresour Technol*. 2021;319:124213.

16 Vescovi V, Giordano RLC, Mendes AA, Tardioli PW. Immobilized lipases on functionalized silica particles as potential biocatalysts for the synthesis of fructose oleate in an organic solvent/water system. *Molecules*. 2017;22:212.

17 Stergiou PY, Foukis A, Filippou M, Koukouritaki M, Parapouli M, Theodorou LG, et al. Advances in lipase-catalyzed esterification reactions. *Biotechnol Adv*. 2013;31:1846–1859.

18 Zhang X, Nie K, Wang M, Liu L, Li K, Wang F, et al. Site-specific xylitol dicaprate ester synthesized by lipase from *Candida sp.* 99-125 with solvent-free system. *J Mol Catal B: Enzym*. 2013;89:61–66.

19 Grüninger J, Delavaultm A, Ochsenreither K. Enzymatic glycolipid surfactant synthesis from renewables. *Process Biochem*. 2019;87:45–54.

20 Fallavena LP, Antunes FHF, Alves JS, Paludo N, Ayub MAS, Fernandez-Lafuente R, et al. Ultrasound technology and molecular sieves improve the thermodynamically controlled esterification of butyric acid mediated by immobilized lipase from *Rhizomucor miehei*. *RSC Adv*. 2014;4:8675.

21 Mai NL, Ahn K, Bae SW, Shin DW, Morya VK, Koo YM. Ionic liquids as novel solvents for the synthesis of sugar fatty acid ester. *Biotechnol J*. 2014;1:1–8.

22 Hidayat C, Fitria K, Hastuti P. Enzymatic synthesis of bio-surfactant fructose oleic ester using immobilized lipase on modified hydrophobic matrix in fluidized bed reactor. *Agric Agric Sci Procedia*. 2016;9:353–362.

23 Watanabe Y, Miyawaki Y, Adachi S, Nakanishi K, Matsuno R. Continuous production of acyl mannoses by immobilized lipase using a packed-bed reactor and their surfactant properties. *Biochem Eng J*. 2001;8(3):213–216.

24 Mijone PD, Bôas RNV, Bento HB, Reis CER, de Castro HF. Coating and incorporation of iron oxides into a magnetic-polymer composite to be used as lipase support for ester syntheses. *Renew. Energy*. 2020;149:1167–1173.

25 Bento H, Reis CE, Pinto PA, Cortez DV, Vilas Bôas RN, Costa-Silva TA, et al. Continuous synthesis of biodiesel from outstanding kernel oil in a packed bed reactor using *Burkholderia cepacia* lipase immobilized on magnetic nanosupport. *Catalysis Lett*. 2021;1:1–11.

26 Morte EFB, Marum DS, Saitovitch EB, Alzamora M, Monteiro SN, Rodriguez RJS. Modified magnetite nanoparticle as biocatalytic support for magnetically stabilized fluidized bed reactors. *J Mater Res Technol*. 2021;14:1112–1125.

27 Bento HBS, De Castro HF, De Oliveira PC, Freitas L. Magnetized poly (STY-co-DVB) as a matrix for immobilizing microbial lipase to be used in biotransformation. *J Magn Magn Mater*. 2017;426:95–101.

28 Chamouleau F, Coulon D, Girardin M, Ghoul M. Influence of water activity and water content on sugar esters lipase-catalyzed synthesis in organic media. *J Mol Catal – B Enzym*. 2001;1:949–954.

29 Kumar A, Dhar K, Kanwar SS, Arora PK. Lipase catalysis in organic solvents: advantages and applications. *Biol Proceed Online*. 2016;18:2.

30 Qin J, Zou X, Lv S, Jin Q, Wang X. Influence of ionic liquids on lipase activity and stability in alcoholysis reactions. *RSC Adv*. 2016;6:87703–87709.

31 Pöhnlein M, Ulrich J, Kirschhöfer F, Nusser M, Muhle-Goll C, Kannengiesser B, et al. Lipase-catalyzes syntese of flucose-6-O-hexanoate in deep eutectic solvents. *Eur J Lipid Sci Technol*. 2015;117:161–166.

32 ACMITE MARKET INTELLIGENCE. Global surfactant market. 2016.

33 Budzianowski WM. High-value low-volume bioproducts coupled to bioenergies with potential to enhance business development of sustainable biorefineries. *Renew Sust Energy Rev*. 2017;70:793–804.

34 Gargalo CL, Udugama I, Pontius K et al. Towards smart biomanufacturing: a perspective on recent developments in industrial measurement and monitoring technologies for bio-based production processes. *J Ind Microbiol Biotechnol*. 2020;47:947–964.

35 Joseph KE, Krumm CA. Bio-renewable surfactant. *Inform*. 2027;28:5.

36 Kee SH, Chiongson JBV, Saludes JP, Vigneswari S, Ramakrishna S, Bhubalan K. Bioconversion of agro-industry sourced biowaste into biomaterials via microbial factories–a viable domain of circular economy. *Environ Pollut*. 2020;271:116311.

37 Hayes DG. Using enzymes to prepare biobased surfactants. *Int News Fats Oils Relat Mater*. 2012;23:477.

38 Gameiro MDA, Goddard A, Taresco V, Howdle SM. Enzymatic one-pot synthesis of renewable and biodegradable surfactants in supercritical carbon dioxide (scCO 2). *Green Chem*. 2020;22(4):1308–1318.

39 Adesra A, Srivastava VK, Varjani S. Valorization of dairy wastes: integrative approaches for value added products. *Indian J Microbiol*. 2021;61:270–278.

40 Kamm B, Kamm M. Principles of biorefineries. *Appl Microbiol Biotechnol* 2004;6:137–145.

41 Singh P, Patil Y, Rale V. Biosurfactant production: emerging trends and promising strategies. *J Appl Microbiol* 2019;126:2–13.

42 Sharma P, Gaur VK, Kim SH, Pandey A. Microbial strategies for bio-transforming food waste into resources. *Bioresour Technol*. 2020;299:122580.

43 Murdock HE, Gibb D, André T, editors. Market and industry trends. In: *Renewables 2021 Global Status Report*. National Technical University of Athens; (Paris: REN21 Secretariat); ISBN 978-3-948393-03-8; 2021:371.

44 Ostendorf TA, Silva IA, Converti A, Sarubbo LA. Production and formulation of a new low-cost biosurfactant to remediate oil-contaminated seawater. *J Biotechnol*. 2019;295:71–79.

45 Jimoh AA, Lin J. Biotechnological applications of *Paenibacillus* sp. D9 lipopeptide biosurfactant produced in low-cost substrates. *Appl Biochem Biotechnol*. 2020;191(3):921–941.

8

Co-production of Biosurfactants and Other Bioproducts in Biorefineries

Martha Inés Vélez-Mercado, Carlos Antonio Espinosa-Lavenant, Juan Gerardo Flores-Iga, Fernando Hernández Teran, María de Lourdes Froto Madariaga, and Nagamani Balagurusamy**

Laboratorio de Biorremediación, Facultad de Ciencias Biológicas, Ciudad Universitaria de la Universidad Autónoma de Coahuila, Carretera Torreón-Matamoros km. 7.5, Torreón, México
* *Correspondence author*

8.1 Introduction

Hydrocarbon pollution from fossil fuel and heavy metals have various disadvantages due to the increasing demand for crude oil and their related products in various fields of application [1]. Some of the pollutants are petroleum hydrocarbons, heavy metals (such as thallium, copper, zinc, mercury, arsenic, iron, titanium, cadmium, nickel) [2], pesticides, herbicides, bisphenol, sulfonamides, volatile organic compounds (e.g. benzene, toluene, chloroform, ethylbenzene, xylenes), nitroaromatics, organophosphorus compounds, trichlorethylene, perchlorethylene, solvents, and chlorinated hydrocarbons which can be extremely harmful and cause significant damage such as corrosive wounds, toxicity, general pathologies, and ecological damage in aquatic and terrestrial ecosystems [3]. Different techniques have been proposed to integrate chemical surfactants and complicated technologies for the remediation of the environment, however, these chemical compounds have harmful effects on the environment [4].

Surfactant compounds obtained by natural processes have become quite prominent in recent years and for this reason the number of publications based on the containment, characterization, and improvement of biosurfactant producers is steadily increasing [5]. Surfactants are molecules with amphipathic structures (one polar and one apolar region) [6] and for this reason, they are responsible for two extremely important phenomena: reduction of surface and interfacial tensions and emulsification of liquids with different polarities [7]. In comparison with the chemically produced surfactants, biosurfactants have several advantages like extensive foaming activities, environmentally friendliness non-toxicity, biodegradability, bioavailability, biocompatibility, ecological acceptability, high selectivity, effectiveness at extreme environments and active in a wide range of pH, temperature and salt concentrations, and long storage time [4, 7].

Biosurfactants and Sustainability: From Biorefineries Production to Versatile Applications, First Edition.
Edited by Paulo Ricardo Franco Marcelino, Silvio Silverio da Silva, and Antonio Ortiz Lopez.
© 2023 John Wiley & Sons Ltd. Published 2023 by John Wiley & Sons Ltd.

Diverse groups of microorganisms are capable of producing biosurfactants, bacteria and yeasts are the microorganisms generally used for the production of biosurfactant [8]. The bacterial members of the genus *Bacillus, Pseudomonas genus, Thiobacillus thiooxidans, Gluconobacter* spp.*, Paenibacillus polymyxa, Flavobacterium* spp.*, Rhodococcus erythropolis, Brevibacterium brevis, Agrobacterium* spp.*, Acinetobacter calcoaceticus, Leuconostoc mesenteroides,* and *Lactobacillus fermentum* are responsible for the production of lipopeptides, fatty acids, polymeric surfactants, phospholipids, glycolipids, neutral lipids, and biosurfactants [7]. Although bacteria are well reported as biosurfactant producers, some authors consider yeasts the best fermentation agents to produce these biomolecules, in comparison with bacteria yeast have cell structures that makes them more resistant, the yeasts used in biosurfactant production in majority have GRAS status (generally regarded as safe). The most frequently used yeasts in biosurfactant production include members of the genus *Candida, Starmerella,* and *Pseudozyma* (currently known as *Moesziomyces*) [9].

With the arrival of the concept of a biorefinery, many renewable sources are being used as raw materials in the production processes. More specifically, lignocellulosic sources have been used. Some studies report the use of lignocellulosic biomass as a source of carbon in the production of biosurfactants, for example the use of a hemicellulosic hydrolyzate of sugarcane bagasse as a source of carbon for the production of glycolipids [7]. Depending on the nature of the contaminant, the removal of the petroleum oil from the soil environment by surfactant occurs using two primary mechanisms: solubilization and mobilization, for aquatic environments the removal of contaminants, which are trapped in the aqueous zone, occurs by mobilization via formation of microemulsion [3].

8.2 Microbial Surfactant Production

Nowadays biosurfactant production is somehow limiting this due the utilization of non-renewable and expensive raw material together with the fact that, in most cases, processes are carried out using some bacteria (e.g. *Pseudomonas* and *Bacillus genus*) [10]. Organisms that have cell structures with membranes are more sensitive to secrete biosurfactants in the culture, although the use of lignocellulosic material for biosurfactant production is increasing as an alternative route for its production, in this context it is possible to use lignocellulosic biomass from agricultural. Lignocellulosic biomass is rich in carbohydrates such as cellulose and hemicellulose, and their acid or enzymatic hydrolysates are rich in fermentable sugars (e.g. glucose, xylose, and arabinose) [11, 12] that have the possibility of being used as substrates in fermentation processes, as they have the possibility of being metabolized by microorganisms to offer products of interest such as biosurfactants. Lignocellulosic biomass is conveniently classified into three main types (i.e. sugars, starches, and cellulose materials) [13] and then biomass can be further segmented into different types, such as forestry (birch, oak, pine, poplar), forestry wastes (the residues of trees and shrubs, sawdust), energy crops (sorghum, corn, sugarcane), agricultural residues (corn stover, wheat straw), algae (directly used as feedstock), industry and domestic wastes (waste paper, fruits or vegetables waste), and any other animal manure (cattle, swine, poultry) [14].

In the green technology and biorefineries context the lignocellulosic biomass is a suitable feedstock for biosurfactant production, one of the biggest problems is the selection of a microorganism that is capable of producing biosurfactants with a high production yield using a determinate feedstock. Table 8.1 show some reports of biosurfactant production by different microorganisms.

Some bacteria from the *Pseudomonas*, *Bacillus*, and *Rhodococcus* genera are the most widely implicated in the production of different types of biosurfactants [20], although bacteria are well reported as biosurfactants producers, yeast is considered to be the best fermentative microorganism for biosurfactant production when bacteria are compared with yeast, yeast have several advantages, e.g. cellular resistance to biosurfactant excretion in the culture, they are GRAS and the most frequently used yeasts for biosurfactant production are mainly from the *Candida* genus [6].

The biosurfactants are generally classified by their chemical composition, the major classes of biosurfactants include glycolipids, lipopeptides/lipoproteins, phospholipids/fatty acids, polymeric surfactants, and particulates. Some bacteria like *Bacillus subtilis*, *Pseudomonas. aeruginosa*, *Rhodococcus erythropolis*, and *Acinetobacter calcoaceticus* are capable of synthetizing lipopeptides, glycolipids, polymeric surfactants, fatty acids, phospholipids, neutral lipids, and particulate biosurfactants. The rhamnolipids are biosurfactants from the glycolipid group and they are the best known biosurfactants but their applications still have some restrictions because it is produced by *Pseudomonas aeruginosa* which is a pathogenic bacteria [6], iturin is another type of biosurfactant and is a cyclic lipoheptapeptide containing an amino fatty acid in its side chain and it can be produced by *B. subtilis* and form part of lipopeptides group [21, 22]. Another type of biosurfactant, also from the phospholipid group, can be produced from some microorganism from the *Acinetobacter* genus like *A. radioresistens* [23]. The last group particulate biosurfactants are

Table 8.1 Feedstock biorefinery for biosurfactant production by different microorganisms.

Microorganism	Biosurfactant	Substrate	Production yield	Reference
Bacillus Subtilis N3-1P	Not specified	Mash and lauter tun of beer brewery in a 7% (v/v) concentration	63.11 %	[15]
Bacillus subtilis ATCC 6633	Lipopeptides BS	Crude glycerol, a by-product of biodiesel production. 5% (v/v) of glycerol and 0.05 mM	793 mg L^{-1}	[16]
Pseudomonas aeruginosa HAK01	Glycolipid BS	Sunflower oil	1.55 g L^{-1}	[17]
Rhodococcus erythropolis AQ5–07	Not specified	Canola oil 2% (v/v)	1.6 mg ml^{-1}	[18]
Candida tropicalis UCP0996	Not specified	Residual frying oil	9.0 g L^{-1}	[19]

formed as extracellular membrane vesicles, creating a microemulsion that exerts an influence on the absorption of alkanes in microbial cells [24].

8.3 Co-production of Biosurfactants in a Biorefinery

There are cases where microbial production of precious metabolites and biological elements i.e. biosurfactants is a current challenge due inefficient and expensive processes that do not allow the systems to produce a reliable industry level yield [20, 25–27]. Meanwhile, present day biosurfactant production is a high-cost and long-time process production in comparison with chemical surfactants, although the biosurfactant market is estimated to have a revenue of approximately US$ 2.0 billion in 2023, this market does not have wide industrial production mainly due to the inability to develop profitable bioprocesses [20, 26, 28]. Even though they are naturally environmentally friendly and laws force their use due their biodegradability, companies still prefer to utilize the cheaper oil-based toxic surfactants compounds. Hence, the development of alternatives such as cost-effective single process coproduction strategies with biofuels and value-added products under laboratory and pilot scales to reduce the productivity cost balance are excellent proposals for industrial bioprocesses [20, 25–27].

One of the major problems of coproduction strategies is the downstream process [20, 26]. The simultaneous recovery of metabolites with similar chemical characteristics or identical cell compartments tend to utilize high-priced purification methods, making this an infeasible scheme. The previous design of the bioprocess before starting out is essential to yield considerable amounts of the desirable metabolites since the best production conditions of one metabolite can affect the generation of the other. Reaction kinetics and redox potential are two examples that take into account the microbial coproduct harness and its growth [20]. Interestingly, most of the simultaneous coproduced metabolites in biosurfactants refineries i.e. enzymes, PHA, lipids, etc. do not have this issue. For instance, while PHA is produced intracellularly, biosurfactants are excreted from the cell. Hence, this permits easy separation through centrifugation [29, 30]. However, there are certain occasions where the co-production possess the same intermediate metabolites that can lead to an inhibition of the generation of one product at the expense of the other. In addition, there are some systems and molecular tools that have been utilized to monitor the biosurfactant co-production process that can further utilize the data to optimize the performance of the process [26, 29, 31, 32].

A wide variety of microorganisms are utilized in biorefineries to produce biosurfactants [20]. However, in this synergistic biosurfactant production with other profitable products, there are some bacterial genera that predominate. For example, Bacillus sp. and *Pseudomonas* sp. species have been widely studied in a biorefinery context to co-synthetize biosurfactants with enzymes and biopolymers, respectively [29–32]. For instance, pectinase is a key produced enzyme for energy production from biomass oxidation, while biopolymers such as PHA are carbon-storing microbial polymers that are promising replacements for non-degradable oil-based plastics [20, 30–32]. Moreover, coproduction is stimulated by value-added products with the same microbial utilities that function synergistically. Examples are biosurfactants and lipases, used by microorganisms i.e. *Aspergillus sp.* to gain benefits from those hydrophobic substrates that do not solubilize in hydrophilic media [32]. Finally, metabolic engineering and synthetic strains have been used to enhance metabolic pathways, giving results that enable the better performance of the biosurfactant co-production with valued metabolites [28].

8.3.1 Co-production of Biosurfactants and Polyhydroxyalkanoates

Polyhydroxyalkanoates (PHA) are intracellular storage biopolymers synthetized by microbes to cope with energy and carbon nutrient imbalances. The PHA family includes short-chain length PHA (3–5 carbon atoms monomers), medium-chain length PHA (6–14 carbon atoms), and short-chain length co-medium chain length PHA (units of 3–5 and 6–14 carbon atoms repeated) that have been described as a potential bioplastic to replace traditional fossil-fuel based polymers [33]. PHA accumulation and monomer composition depend on the carbon source, key PHA synthesis enzymes present in the strain and environmental conditions [34]. Currently, PHA mass production has not attracted companies because of its high production cost in comparison with petrochemical plastics [35]. PHA simultaneous production with other substances have been reported as a profitable approach in industrial production [29]. PHA market growth rate rose to 93.5 million in 2021 with higher expectations in the future [36]. The co-production of PHA with biosurfactants is a potential line of investigation to harness the microbial metabolism of some bacterial genera that perform this output in a cost effective manner. Thus, enabling the concomitant biosurfactant and PHA to yield a competitive bioprocess performing the strategy of sharing the culture cost for an intracellular product, PHA and an extracellular biosurfactant product commonly found in the fermentation broth, as most of the microorganisms synthesize with a further easy separation performed by precipitation methods without interference in each product processing. As shown in Figure 8.1, PHA synthesis has been described as closely related to a biosurfactant (e.g. rhamnolipid synthesis), since (R)-3-hydroxyfatty acids are the common intermediates of both pathways, present in B-oxidation and *de novo* biosynthesis [37], PHA and rhamnolipid synthesis lines respectively. Thus, the related metabolism can promote competition for this intermediate metabolite [38]. Production and degradation of PHA can simultaneously occur in a cell, therefore leading to a possible rhamnolipid synthesis since (R)-3-hydroxyfatty acid, the intermediate, is produced [39] during the stationary growth phase [36]. It has been shown that nitrogen limiting conditions increase the production of polymer accumulation and rhamnolipid synthesis [36].

The genus *Pseudomonas* sp., and species *Thermus thermophilus*, *Burkholderia thailandensis* [40], and *Klebsiella aerogenes* [36] were reported for the synthesis of PHA and biosurfactants promising scalable production. Hydrophobic substrates generally reported better yields of PHA and biosurfactant co-production since oily carbon sources are thought to reach the cell, first produce PHA stimulated by low carbon source and further biosurfactant synthesis to allow the solubility of the non-hydrophilic substrate. However, hydrophilic substrates do not generate downstream separation problems. In addition, feedstock such as waste biomass from food industry and agriculture are potential low-cost substrates for fermentation processes that can be used in surfactants and PHA production [40]. Wastewater from oil and gas drilling have been tested as a substrate for this co-production bioprocess with promising performance [41].

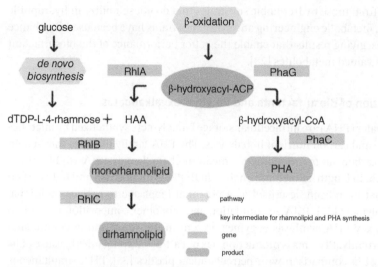

Figure 8.1 PHA and rhamnolipid biosynthetic pathway from *P. aeruginosa* proposed firstly by Soberón-Chávez et al. [37] and afterwards renovated by Nitschke et al. [33]. The key intermediate β-hydroxyacyl-CoA goes through a competition between RhlA (rhamnosyltransferase 1) and PhaG(3-hyroxyacyl-CoA-ACP transferase) genes for rhamnolipid and PHA production, respectively. HAA (β-hydroxydecanoyl-β-hydroxydecanoic acid), RhlB (rhamnosyltransferase 1), RhlC (rhamnosyltransferase 2), PhaC (PHA synthase).

8.3.2 Co-production of Biosurfactants and Enzymes

Recently, there was an increase in interest in microorganisms capable of producing enzymes and biosurfactants (Table 8.2) mainly due to enzymes being compatible with surfactants and detergent compounds. *Bacillus* spp is one of the most important microorganisms used for the simultaneous production of enzymes and biosurfactants under variable conditions [31]. These strains can produce a variety of enzymes such as pectinases, proteases, cellulases, and amylases in addition to large amounts of biosurfactants using different carbon sources including agro-industrial residues. Bhange et al. [42] optimized the production of protease, amylase, and biosurfactant from *Bacillus subtilis* PF1 using potato peel, rape seed cake, and feather meal. They also evaluated the washing efficiency of the combination of both enzymes, biosurfactant, and SDS to remove blood, chocolate, and beetroot juice.

In another study, Hmidet et al. [46] evaluated the simultaneous production of amylase with biosurfactant by *Bacillus methylotrophicus* DCS1 under different concentrations of potato starch as a carbon source and different nitrogen sources. This resulted in higher production of amylase and lipopeptides (i.e. biosurfactant) using organic nitrogen sources over inorganic sources. In this study, a stable amylase was found which has optimal activity at 60–65 °C and pH 8 and high stable biosurfactant at the same conditions.

Moreover, *Ochrobactrum intermedium* MZV101 was reported to produce thermostable lipase and biosurfactant at pH 10–13 and 70–90 °C. In addition, *O. intermedium* MZV101 has been reported as a good antimicrobial and efficient oil removal, with active dependence on metal ions such as Ca^{2+} and Co^{2+} [47].

Table 8.2 Co-production of biosurfactants and enzymes from various microorganisms using different substrates.

Microorganism	Substrate	Maximum concentration of biosurfactant	Units of emulsifying activity	Enzyme	Maximum activity	Reference
Aspergillus niger	Wheat bran:corncob 80:20	NIA	6.67 UE	Lipase	10.74 $\mu mol\ min^{-1}\ g^{-1}$	[43]
Aspergillus niger	Malt extract (75 g/L)	NIA	2.24 UE	Lipase	3.89 ($\mu mol\ mL^{-1}\ min^{-1}$)	[32]
Bacillus subtilis BKDS	Yeast extract pectin (YEP) agar medium	1.45 g L^{-1}	NIA	Pectinase	1.288 U mg^{-1} at 72 h	[44]
Bacillus subtilis A1	Sucrose and yeast extract	4.85 g L^{-1}	NIA	Alkane hydrolase	188 $\mu mol\ min^{-1}\ mg^{-1}$	[45]
				Alcohol Dehydrogenase	88 $\mu mol\ min^{-1}\ mg^{-1}$	

NIA: No information available; UE: units of emulsifying activity

The coproduction of enzymes is a promising strategy due to some of them (e.g. protease, amylase, lipase, cellulase, and mannanase) having the property of removing stains; for that reason they have been used in the formulation of detergents for better efficiency and as an eco-friendly alternative. For example, proteases can remove protein stains, lipases lipid stains, and cellulases are used for color clarification [48]. Table 8.3 shows a variety of enzymes with promising uses in the detergent industry.

Some conditions are needed to use these enzymes as additives such as stability under different inhibitors (e.g. detergents, solvents, chemicals, and metals), high catalytic activity, cost-effective production, and thermostability preferably [53].

The application of enzymes in detergent formulations began in 1913 with Rohm's patent for the use of pancreatic enzymes in presoak solutions [54]. Furthermore, in the 1960s microbial the first proteases were from *Bacillus* spp, especially from the subtilisin superfamily, they were more extensively used due to their unique functions, stability, and substrate specificity [55].

Moreover, cold-active enzymes are a development of biotechnology that have been used in various industries including the detergent industry due their high catalytic activity and stability under extreme conditions. Cold-active enzymes are mainly isolated from microorganisms that live in cold regions such as *Planococcus halocryophilus* Or1 which can produce extracellular enzymes at −15 °C [56]. However, more studies are needed to identify and optimize the microbial production of cold-active enzymes.

Additionally, some microorganisms such as *Pseudomonas*, *Burkholderia*, *Sphingomonas*, *Rhodococcus*, and *Bacillus* sp. can produce biosurfactants at low temperatures which can interact with more physical phases like ice, gases, and hydrophobic compounds. Most of

Table 8.3 Enzymes with detergent additive capability.

Enzyme activity	Microorganism source	Activity with detergents and solvents	Reference
Protease	Bacillus cereus	100% with SDS (0.5%) and Triton-X-100 (0.1% and 0.5%)	[49]
		80% with benzene, toluene, xylene, acetone, and isopropanol (0.1%, 0.5% v/v)	
		10% EDTA	
Protease	Bacillus pumilus CBS	119% with SDS (0.1%)	[50]
		107.5 ± 1.5 with Triton with X-100 (1% and 5%)	
		80% EDTA (5mM)	
β-keratinase	Brevibacillus sp. AS-S10-II	97% with EDTA (2mM)	[51]
		113% with SDS (20mM)	
		100% with Triton-X-100	
Lipase	Pseudomonas aeruginosa	100% with SDS (2%)	[52]
		113% with Triton-X-100 (2%)	
		100% with Tween 80 (2%)	

SDS: sodium dodecyl sulfate; EDTA: ethylenediamine tetra-acetic acid

them can produce biosurfactants at 4 °C. For example, Bacillus sp. and *Alcanivorax borkumensis* have the capacity to produce unidentified biosurfactant and glucose lipids respectively, in the range 4–32°C [57]. Furthermore, the yeast isolated from a polar habitat *Moesziomyces antarcticus* can produce extracellular biosurfactants and lipase that degrade thermoplastics such as poly-butylene succinate and polybutylene succinate-co-adipate [58].

8.3.3 Co-production of Biosurfactants and Lipids

Lipids can be coproduced with a variety of bioproducts such as enzymes, PHAs, lipoproteins and pigments [59]. Microalgae are a promising coproducer of lipids with other bioproducts due their capacity of over producing and accumulating lipids under stress conditions. Additionally, microalgae are the main producers of glycolipids, phospholipids, and polysaccharides, some of which can be used as biosurfactants [60].

Moreover, microalgae cells when erupted during lipid extraction have biosurfactant activity due the liberation of cell wall debris, protein-rich serum, and the extracted lipid which adheres to hydrophobic molecules, stabilizes the emulsion, and has surfactant activity [61].

The protein-rich extract from disrupted microalgae can be divided into water soluble and insoluble proteins. A 49.9% content of soluble protein extract and 56.5% of insoluble protein extract can be obtained from *Chlorella sorokiniana* using ethanol:acetone (1:1 v/v) as solvent for the extraction [62]. This protein-rich extract, particularly the soluble fraction, can be used as a moderately effective biosurfactant by the formation of rigid interfacial films and can stabilize the interface between hydrophobic and hydrophilic compounds such as oil and water [61]. Moreover, it has been reported that protein extract with microalgal polysaccharides could form a complex that results in a better emulsifying activity [63].

Cell debris is an organic fraction of the cell wall that was liberated when a cell ruptures, dies or is damaged. This fraction can stabilize emulsion and is a highly effective surfactant especially in the presence of the lipid fraction that is the most surface-active component leading to the best surfactant performance of the cell components [61, 64].

Additionally, *Bacillus spp* also coproduces microbial lipids and biosurfactants. It has been reported that *Bacillus velezensis* is able to produce 31% of lipid content (1.24 g L^{-1} yield) and 0.818 g L^{-1} of biosurfactant using waste office paper hydrolysates as substrate and NH4Cl as nitrogen source after 72 h [65]. Also, *Bacillus amyloliquefaciens* was evaluated for the production of some lipopeptide biosurfactant (surfactin, bacillomycin F, and fengycin) and lipids using glycerol or glucose as a carbon source. A better lipid production with glucose as substrate and a higher biosurfactant production with glycerol was shown[66].

8.3.4 Co-production of Biosurfactants and Ethanol

Alternatives to petroleum-derived fuels are being sought in order to reduce dependence on non-renewable resources [67]. The production of liquid biofuels from lignocellulosic biomass is the most promising renewable energy to reduce the environmental crisis [68]. Some microorganisms like *Pseudomonas aeruginosa, Saccharomyces cerevisiae* and *Bacillus subtilis* have been used to produce biofuels and biosurfactants simultaneously. Different strategies have been implemented to achieve this co-production such as co-fermentation, enzyme addition, and metabolic engineering.

Pseudomonas aeruginosa produced large amounts of biosurfactant and can utilized cheap substrates and lignocellulosic residues as substrate. De Sousa et al. [69], evaluated the biosurfactant production of *Pseudomonas aeruginosa* using different carbon and nitrogen sources, finding that *P. aeruginosa* was capable of producing 1269.79 mg L^{-1}, 898.26 mg L^{-1} and 503.13 mg L^{-1} of rhamnolipids with hydrolyzed glycerin as substrate using $NaNO_3$, $(NH_4)_2SO_4$ and peptone as nitrogen sources respectively. De Sousa et al. [69] also reported a rhamnolipids production of 354.63 mg L^{-1} using crude glycerin as the carbon source and $NaNO_3$ as the nitrogen source and associated the low production with the high salt concentration caused for the transesterification and pretreatment which result in an inhibitory effect. In addition, *P. aeruginosa* can grow under high ethanol concentration conditions [70], for that reason it has been used in co-cultures to produce ethanol and rhamnolipids simultaneously. Together *P. aeruginosa* with *Saccharomyces cerevisiae* achieved a production of 8.4 g L^{-1} of ethanol and 9.1 g L^{-1} of rhamnolipids after 72 h with the addition of crude enzyme complexes (obtained from the solid-state fermentation of *Aspergillus niger*) [71]. The ideal production of ethanol may be around 40 g L^{-1}, however, it is expected that co-production reduces the efficiency of the process, also, *P. aeruginosa* is able to utilize ethanol as a substrate which can affect ethanol concentration [72]. Nevertheless, Lopes et al. [71] reported that the presence of *P. aeruginosa* did not affect ethanol concentration.

In the other hand, *Bacillus subtilis* is another microorganism that produce a lipopeptide biosurfactants using different substrate including corn steep liquor [73], molasses [74], glycerol [75], and algae biomass [76]. *B. subtilis* is a GRAS microorganism able to produce a variety of value-added products such as lactate, acetate, butanediol, and a low concentration of ethanol. Because ethanol is the bioproduct of more interest, *B. subtilis* has been modified to produce ethanol as the main fermentation product achieving a production of 8.9 g L^{-1} of ethanol with a theoretical yield of 87% [77].

8.4 Conclusions

Production of biosurfactants and other bioproducts simultaneously is an important achievement to continue to develop for a better performance and increase in product yield. It is important to understand how microorganisms are capable of producing the product of interest and how it can be regulated for a higher production. Moreover, co-cultures and metabolic engineering are promising strategies to improve the co-production of biosurfactants. Furthermore, *Bacillus subtilis* and *Pseudomonas aeruginosa* seems to be two of the most important microorganisms for the coproduction of biosurfactant and other value-added products due the fact that they can participate in all the process discussed in this chapter.

References

1 Karlapudi AP, Venkateswarulu TC, Tammineedi J, Kanumuri L, Ravuru BK, Dirisala V, et al. Role of biosurfactants in bioremediation of oil pollution-a review. *Petroleum*. 2018;4(3):241–249.
2 Da Rocha Junior RB, Meira HM, Almeida DG, Rufino RD, Luna JM, et al. Application of a low-cost biosurfactant in heavy metal remediation processes. *Biodegradation*. 2018;30:215–233.
3 Karthick A, Roy B, Chattopadhyay P. A review on the application of chemical surfactant and surfactant foam for remediation of petroleum oil contaminated soil. *J Environ Manage*. 2019;243:187–205.
4 Jimoh AA, Lin J. Biosurfactant: a new frontier for greener technology and environmental sustainability. *Ecotoxicol Environ Saf*. 2019;184:109607.
5 Rawat G, Dhasmana A, Kumar V. Biosurfactants: the next generation biomolecules for diverse applications. *Environ Sustain*. 2020;3:353–369.
6 Ingle AP, Chandel AK, da Silva SS. Biorefining of lignocellulose into valuable products. In: *Lignocellulosic Biorefinig Technologies*; Eds. Avinash P. Ingle,Anuj Kumar Chandel,Silvio Silvério da Silva. Wiley-Blackwell, West Sussex, UK.2020. pp. 1–5.
7 Marcelino PRF, Gonçalves F, Jimenez IM, Carneiro BC, Santos BB, Silva SS. Sustainable production of biosurfactants and their applications. In: *Lignocellulosic Biorefining Technologies*; Eds. Avinash P. Ingle,Anuj Kumar Chandel,Silvio Silvério da Silva. Wiley-Blackwell, West Sussex, UK.2020. pp. 159–183.
8 Thakur IS, Kumar M, Varjani SJ, Wu Y, Gnansounou E, Ravindran S. Sequestration and utilization of carbon dioxide by chemical and biological methods for biofuels and biomaterials by chemoautotrophs: opportunities and challenges. *Bioresour Technol*. 2018;256:478–490.

9 Amaral PFF, Coelho MAZ, Marrucho IMJ, Coutinho JAP. Biosurfactants from yeasts: characteristics, production and application. *Adv Exp Med Biol*. 2010;672:236–249.

10 Marcelino PRF, Peres GFD, Terán-Hilares R, Pagnocca FC, Rosa CA, Lacerda TM, et al. Biosurfactants production by yeasts using sugarcane bagasse hemicellulosic hydrolysate as new sustainable alternative for lignocellulosic biorefineries. *Ind Crops Prod*. 2019;129:212–223.

11 De Assis T, Huang S, Driemeier CE, Donohoe BS, Kim C, Kim SH, et al. Toward an understanding of the increase in enzymatic hydrolysis by mechanical refining. *Biotechnol Biofuels*. 2018;11(1):1–11.

12 Hernández C, Escamilla-Alvarado C, Sánchez A, Alarcón E, Ziarelli F, Musule R, et al. Wheat straw, corn stover, sugarcane, and agave biomasses: chemical properties, availability, and cellulosic-bioethanol production potential in Mexico. *Biofuel Bioprod Biorefin*. 2019;13(5):1143–1159.

13 Saini JK, Saini R, Tewari L. Lignocellulosic agriculture wastes as biomass feedstocks for second-generation bioethanol production: concepts and recent developments. *3 Biotech*. 2015;5:337–353.

14 Liu Y, Nie Y, Lu X, Zhang X, He H, Pan F, et al. Cascade utilization of lignocellulosic biomass to high-value products. *Green Chem*. 2019;21(13):3499–3535.

15 Moshtagh B, Hawboldt K, Zhang B. Optimization of biosurfactant production by *Bacillus subtilis* N3-1P using the brewery waste as the carbon source. *Environ Technol (United Kingdom)*. 2019;40(25):3371–3380.

16 Cruz JM, Hughes C, Quilty B, Montagnolli RN, Bidoia ED. Agricultural feedstock supplemented with manganese for biosurfactant production by *Bacillus subtilis*. *Waste Biomass Valori*. 2018;9:613–618.

17 Khademolhosseini R, Jafari A, Mousavi SM, Hajfarajollah H, Noghabi KA, Manteghian M. Physicochemical characterization and optimization of glycolipid biosurfactant production by a native strain of *Pseudomonas aeruginosa* HAK01 and its performance evaluation for the MEOR process. *RSC Adv*. 2019;9(14):7932–7947.

18 Ibrahim S, Khalil KA, Nabilah K, Zahri M, Gomez-fuentes C, Convey P, et al. Biosurfactant production and growth kinetics studies of the waste canola oil-degrading bacterium rhodococcus erythropolis AQ5-07 from Antarctica. *Molecules*. 2020;25(17):3878.

19 Almeida DG, Soares da Silva RDCF, Meira HM, Pinto P, Brasileiro PPF, Silva EJ, et al. Production, characterization and commercial formulation of biosurfactant from *Candida tropicalis* UCP0996 and its application in decontamination of petroleum pollutants. *Processes*. 2021;9(885):1–18.

20 Singh P, Patil Y, Rale V. Biosurfactant production: emerging trends and promising strategies. *J Appl Microbiol*. 2019;126(1):2–13.

21 Hentati D, Chebbi A, Hadrich F, Frikha I, Rabanal F, Sayadi S, et al. Production, characterization and biotechnological potential of lipopeptide biosurfactants from a novel marine *Bacillus stratosphericus* strain FLU5. *Ecotoxicol Environ Saf*. 2019;167:441–449.

22 Nelson J, El-Gendy AO, Mansy MS, Ramadan MA, Aziz RK. The biosurfactants iturin, lichenysin and surfactin, from vaginally isolated lactobacilli, prevent biofilm formation by pathogenic *Candida*. *FEMS Microbiol Lett*. 2020;367(15):fnaa126.

23 Mujumdar S, Joshi P, Karve N. Production, characterization, and applications of bioemulsifiers (BE) and biosurfactants (BS) produced by *Acinetobacter* spp.: a review. *J Basic Microbiol*. 2019;59(3):277–287.

24 Domínguez Rivera Á, Martínez Urbina MÁ, López y López VE. Advances on research in the use of agro-industrial waste in biosurfactant production. *World J Microbiol Biotechnol*. 2019;35(10):1–18.

25 Byun J, Han J. Sustainable development of biorefineries: integrated assessment method for co-production pathways. *Energy Environ Sci*. 2020;13(8):2233–2242.

26 Gaur VK, Sharma P, Gupta S, Varjani S, Srivastava JK, Wong JWC, et al. Opportunities and challenges in omics approaches for biosurfactant production and feasibility of site remediation: strategies and advancements. *Environ Technol Innov*. 2022;25:102132.

27 Liang Q, Qi Q. From a co-production design to an integrated single-cell biorefinery. *Biotechnol Adv*. 2014;32(7):1328–1335. doi: 10.1016/j.biotechadv.2014.08.004.

28 Gaur VK, Sharma P, Sirohi R, Varjani S, Taherzadeh MJ, Chang JS, Yong Ng H, Wong JWC, Kim SH. Production of biosurfactants from agro-industrial waste and waste cooking oil in a circular bioeconomy: an overview. *Bioresour Technol*. 2022;343. Elsevier Ltd. doi: 10.1016/j.biortech.2021.126059.

29 Li T, Elhadi D, Chen GQ. Co-production of microbial polyhydroxyalkanoates with other chemicals. *Metab Eng*. 2017;43(A):29–36.

30 Yadav B, Talan A, Tyagi RD, Drogui P. Concomitant production of value-added products with polyhydroxyalkanoate (PHA) synthesis: a review. *Bioresour Technol*. 2021;337:125419.

31 Ramnani P, Suresh Kumar S, Gupta R. Concomitant production and downstream processing of alkaline protease and biosurfactant from *Bacillus licheniformis* RG1: bioformulation as detergent additive. *Process Biochem*. 2005;40(10):3352–3359.

32 Sperb JGC, Costa TM, Bertoli SL, Tavares LBB. Simultaneous production of biosurfactants and lipases from *Aspergillus niger* and optimization by response surface methodology and desirability functions. *Braz J Chem Eng*. 2018;35(3):857–868.

33 Nitschke M, Costa SGVAO, Contiero J. Rhamnolipids and PHAs: recent reports on *Pseudomonas*-derived molecules of increasing industrial interest. *Process Biochem*. 2011;46(3):621–630.

34 Costa SGVAO, Lépine F, Milot S, Déziel E, Nitschke M, Contiero J. Cassava wastewater as a substrate for the simultaneous production of rhamnolipids and polyhydroxyalkanoates by *Pseudomonas aeruginosa*. *J Ind Microbiol Biotechnol*. 2009;36(8):1063–1072.

35 Sabapathy PC, Devaraj S, Meixner K, Anburajan P, Kathirvel P, Ravikumar Y, et al. Recent developments in Polyhydroxyalkanoates (PHAs) production – a review. *Bioresour Technol*. 2020;306:123132.

36 Arumugam A, Furhana Shereen M. Bioconversion of calophyllum inophyllum oilcake for intensification of rhamnolipid and polyhydroxyalkanoates co-production by Enterobacter aerogenes. *Bioresour Technol*. 2020;296:122321.

37 Soberón-Chávez G, Lépine F, Déziel E. Production of rhamnolipids by *Pseudomonas aeruginosa*. *Appl Microbiol Biotechnol*. 2005;68(6):718–725.

38 Pantazaki AA, Papaneophytou CP, Lambropoulou DA. Simultaneous polyhydroxyalkanoates and rhamnolipids production by *Thermus thermophilus* HB8. *AMB Express*. 2011;1:17.

39 Kourmentza C, Costa J, Azevedo Z, Servin C, Grandfils C, De Freitas V, et al. Burkholderia thailandensis as a microbial cell factory for the bioconversion of used cooking oil to polyhydroxyalkanoates and rhamnolipids. *Bioresour Technol*. 2018;247:829–837.

40 Savich V, Novik G. Waste biodegradation and utilization by *Pseudomonas* species. *J Microbiol Biotechnol Food Sci*. 2016;6(2):851–857.

41 Koutinas M, Kyriakou M, Andreou K, Hadjicharalambous M, Kaliviotis E, Pasias D, et al. Enhanced biodegradation and valorization of drilling wastewater via simultaneous production of biosurfactants and polyhydroxyalkanoates by *Pseudomonas citronellolis* SJTE-3. *Bioresour Technol*. 2021;340:125679.

42 Bhange K, Chaturvedi V, Bhatt R. Simultaneous production of detergent stable keratinolytic protease, amylase and biosurfactant by *Bacillus subtilis* PF1 using agro industrial waste. *Biotechnol Rep*. 2016;10:94–104.

43 Kreling N, Simon V, Fagundes V, Thomé A, Colla L. Simultaneous production of lipases and biosurfactants in solid-state fermentation and use in bioremediation. *J Environ Eng*. 2020;146(9):04020105.

44 Kavuthodi B, Thomas S, Sebastian D. Co-production of pectinase and biosurfactant by the newly isolated strain *Bacillus subtilis* BKDS1. *Br Microbiol Res J*. 2015;10(2):1–12.

45 Parthipan P, Preetham E, Machuca L, Rahman P, Murugan K, Rajasekar A. Biosurfactant and degradative enzymes mediated crude oil degradation by bacterium *Bacillus subtilis* A1. *Front Microbiol*. 2017;8.

46 Hmidet N, Jemil N, Nasri M. Simultaneous production of alkaline amylase and biosurfactant by *Bacillus methylotrophicus* DCS1: application as detergent additive. *Biodegradation*. 2018;30:247–258.

47 Zarinviarsagh M, Ebrahimipour G, Sadeghi H. Lipase and biosurfactant from *Ochrobactrum intermedium* strain MZV101 isolated by washing powder for detergent application. *Lipids Health Dis*. 2017;16:177.

48 Kirk O, Borchert TV, Fuglsang CC. Industrial enzyme applications. *Curr Opin Biotechnol*. 2002;13(4):345–351.

49 Doddapaneni KK, Tatineni R, Vellanki RN, Rachcha S, Anabrolu N, Narakuti V, et al. Purification and characterization of a solvent and detergent-stable novel protease from *Bacillus cereus*. *Microbiol Res*. 2009;164(4):383–390.

50 Jaouadi B, Ellouz-Chaabouni S, Rhimi M, Bejar S. Biochemical and molecular characterization of a detergent-stable serine alkaline protease from *Bacillus pumilus* CBS with high catalytic efficiency. *Biochimie*. 2008;90(9):1291–1305.

51 Rai SK, Mukherjee AK. Optimization of production of an oxidant and detergent-stable alkaline β-keratinase from *Brevibacillus sp.* strain AS-S10-II: application of enzyme in laundry detergent formulations and in leather industry. *Biochem Eng J*. 2011;54(1):47–56.

52 Ruchi G, Anshu G, Khare SK. Lipase from solvent tolerant *Pseudomonas aeruginosa* strain: production optimization by response surface methodology and application. *Bioresour Technol*. 2008;99(11):4796–4802.

53 Gulmez C, Atakisi O, Dalginli KY, Atakisi E. A novel detergent additive: organic solvent- and thermo-alkaline-stable recombinant subtilisin. *Int J Biol Macromol*. 2018;108:436–443.

54 Christakopoulos P, Topakas E. Editorial note: advances in enzymology and enzyme engineering. *Comput Struct Biotechnol J*. 2012;2(3):e201209001.

55 Maurer KH. Detergent proteases. *Curr Opin Biotechnol*. 2004;15(4):330–334.

56 Al-Ghanayem AA, Joseph B. Current prospective in using cold-active enzymes as eco-friendly detergent additive. *Appl Microbiol Biotechnol*. 2020;104(7):2871–2882.

57 Perfumo A, Banat IM, Marchant R. Going green and cold: biosurfactants from low-temperature environments to biotechnology applications. *Trends Biotechnol.* 2018;36(3):277–289.

58 Torres Faria N, Marques S, Castelo Ferreira F, Fonseca C. Production of xylanolytic enzymes by *Moesziomyces* spp. using xylose, xylan and brewery's spent grain as substrates. *N Biotechnol.* 2018;49:137–143.

59 Chen L, Qian X, Zhang X, Zhou X, Zhou J, Dong W, et al. Co-production of microbial lipids with valuable chemicals. *Biofuel Bioprod Biorefin.* 2021;15(3):945–954.

60 Radmann EM, de Moris EG, de Oliveira CF, Zanfonato K, Vieira Costa JA. Microalgae cultivation for biosurfactant production. *Afr J Microbiol Res.* 2015;9(47):2283–2289.

61 Law SQK, Mettu S, Ashokkumar M, Scales PJ, Martin GJO. Emulsifying properties of ruptured microalgae cells: barriers to lipid extraction or promising biosurfactants? *Colloids Surf B Biointerfaces.* 2018;170:438–446.

62 Grossmann L, Ebert S, Hinrichs J, Weiss J. Production of protein-rich extracts from disrupted microalgae cells: impact of solvent treatment and lyophilization. *Algal Res.* 2018;36:67–76.

63 Schwenzfeier A, Helbig A, Wierenga PA, Gruppen H. Emulsion properties of algae soluble protein isolate from *Tetraselmis* sp. *Food Hydrocoll.* 2013;30(1):258–263.

64 Talukder MA, Menyuk CR, Kostov Y. Distinguishing between whole cells and cell debris using surface plasmon coupled emission. *Biomed Opt Express.* 2018;9(4):1977–1991.

65 Nair AS, Al-Bahry S, Sivakumar N. Co-production of microbial lipids and biosurfactant from waste office paper hydrolysate using a novel strain *Bacillus velezensis* ASN1. *Biomass Convers Biorefin.* 2019;10:383–391.

66 Etchegaray A, Coutte F, Chataigné G, Béchet M, dos Santos RHZ, Leclère V, et al. Production of *Bacillus amyloliquefaciens* OG and its metabolites in renewable media: valorisation for biodiesel production and p-xylene decontamination. *Can J Microbiol.* 2016;63(1):46–60.

67 Gray KA, Zhao L, Emptage M. Bioethanol. *Curr Opin Chem Biol.* 2006;10(2):141–146.

68 Correa DF, Beyer HL, Fargione JE, Hill JD, Possingham HP, Thomas-Hall SR, et al. Towards the implementation of sustainable biofuel production systems. *Renew Sustain Energy Rev.* 2019;107:250–263.

69 De Sousa JR, da Costa Correia JA, de Almeida JGL, Rodrigues S, Pessoa ODL, Melo VMM, et al. Evaluation of a co-product of biodiesel production as carbon source in the production of biosurfactant by *P. aeruginosa* MSIC02. *Process Biochem.* 2011;46(9):1831–1839.

70 Görisch H. The ethanol oxidation system and its regulation in *Pseudomonas aeruginosa*. *Biochim Biophys Acta – Proteins Proteom.* 2003;1647(1–2):98–102.

71 Lopes VDS, Fischer J, Pinheiro TMA, Cabral BV, Cardoso VL, Coutinho Filho U. Biosurfactant and ethanol co-production using *Pseudomonas aeruginosa* and *Saccharomyces cerevisiae* co-cultures and exploded sugarcane bagasse. *Renew Energy.* 2017;109:305–310.

72 Mern DS, Ha SW, Khodaverdi V, Gliese N, Gorisch H. A complex regulatory network controls aerobic ethanol oxidation in *Pseudomonas aeruginosa*: indication of four levels of sensor kinases and response regulators. *Microbiology.* 2010;156(5):1505–1516.

73 Gudiña EJ, Fernandes EC, Rodrigues AI, Teixeira JA, Rodrigues LR. Biosurfactant production by *Bacillus subtilis* using corn steep liquor as culture medium. *Front Microbiol.* 2015;6.

74 Al-Bahry SN, Al-Wahaibi YM, Elshafie AE, Al-Bemani AS, Joshi SJ, Al-Makhmari HS, et al. Biosurfactant production by *Bacillus subtilis* B20 using date molasses and its possible application in enhanced oil recovery. *Int Biodeterior Biodegradation*. 2013;81:141–146.
75 Yun J-H, Cho D-H, Lee B, Kim H-S, Chang YK. Application of biosurfactant from *Bacillus subtilis* C9 for controlling cladoceran grazers in algal cultivation systems. *Sci Rep*. 2018;8:5365.
76 Yun J-H, Cho D-H, Lee B, Lee YJ, Choi D-Y, Kim H-S, et al. Utilization of the acid hydrolysate of defatted chlorella biomass as a sole fermentation substrate for the production of biosurfactant from *Bacillus subtilis* C9. *Algal Res*. 2020;47:101868.
77 Romero S, Merino E, Bolivar F, Gosset G, Martinez A. Metabolic engineering of *Bacillus subtilis* for ethanol production: lactate dehydrogenase plays a key role in fermentative metabolism. *Appl Environ Microbiol*. 2007;73(16):5190–5198.

9

Biosurfactants in Nanotechnology

Recent Advances and Applications

Avinash P. Ingle[1], Shreshtha Saxena[1], Mangesh Moharil[1], Mahendra Rai[2,], and Silvio S. Da Silva[3]*

[1] Biotechnology Centre, Department of Agricultural Botany, Dr. Panjabrao Deshmukh Krishi, Vidyapeeth, Akola, Maharashtra, India
[2] Department of Biotechnology, Sant Gadge Baba Amravati University, Amravati, Maharashtra, India
[3] Engineering School of Lorena, University of Sao Paulo, Department of Biotechnology, Area I, Estrada Municipal Do Campinho, s/n, Lorena, SP, Brazil
* Corresponding author

9.1 Introduction

Biosurfactants are surfactants that are commonly synthesized from biological origin. They are usually synthesized from a variety of microbes such as bacteria, yeast, and fungi. Biosurfactants generally exhibit unique and extraordinary properties which mainly include their non-toxic and biodegradable nature, excellent surface activity, and emulsification properties. Moreover, they are found in various chemical forms e.g. glycolipids, lipopeptides, fatty acids, and neutral lipids [1]. Although biosurfactants have been reported to be very effective at low concentrations and diverse environmental conditions (e.g. pH, temperature, and salinity; better environmental compatibility, lower critical micelle concentration, higher selectivity, specific activity), long-term physicochemical stability, chemically synthesized surfactants were preferably used at the industrial level only because of their low cost of production. However, currently, the trend has been changing around the globe because of the availability of cheaper and renewable biological sources which can definitely lower production costs significantly.

Due to the above-mentioned beneficial properties biosurfactants are being used for numbers of applications in several industries like agriculture, medicine, refineries for oil recovery, food, cosmetics, detergents, etc. In particular, biosurfactants can be employed as a novel drug delivery system to improve the bioavailability of several drugs that possess low aqueous solubility. Moreover, biosurfactants produced from bacteria are reported to have potential antibacterial, antifungal, and antiviral properties confirming their role in therapeutic and biomedical and many more industrial applications [2, 3]. As a consequence, the global biosurfactant market is growing very rapidly. It has been observed that as far as the overall global market is concerned, European countries are leading the world in terms

Biosurfactants and Sustainability: From Biorefineries Production to Versatile Applications, First Edition.
Edited by Paulo Ricardo Franco Marcelino, Silvio Silverio da Silva, and Antonio Ortiz Lopez.
© 2023 John Wiley & Sons Ltd. Published 2023 by John Wiley & Sons Ltd.

of the production and consumption of biosurfactants. According to one report, the global biosurfactant market was expected to grow quite rapidly with market revenues of around $25 billion during 2018–2020 (https://www.grandviewresearch.com/press-release/global-biosurfactants-market).

Apart from the above-mentioned applications, biosurfactants have recently gained considerable interest from scientific communities due to their increasing use in the synthesis of a variety of nanomaterials. Their superior properties and renewable nature make them a suitable candidate for developing a sustainable approach to nanomaterial production under the umbrella of green nanotechnology. It is well-known that nanotechnology is an emerging science of the twenty-first century having enormous applications in diverse fields. Like in other biological approaches to nanoparticle synthesis, biosurfactants act as capping and reducing agents. Moreover, they also provide self-assembly structures for encapsulation, functionalization, or templates, and act as emulsifiers in nanoemulsions [4]. Considering all these facts, in this chapter, we have focused on the recent developments in the role of biosurfactants in nanotechnology particularly in the production of nanomaterials. Moreover, special attention has been also given to various other related aspects such as the types and properties of biosurfactants, and conventional methods for biosurfactants production.

9.2 Biosurfactants and their Types

The essential criteria for classifying biosurfactants are their molecular structure and microbiological source of generation. Lipoproteins and lipopolysaccharides are classes of biosurfactants with a high molecular weight. They are also known as bio-emulsifiers [5]. On the other hand, glycolipids, lipopeptides, and phospholipids are known as low molecular weight biosurfactants [6]. The classification of biosurfactants based on their microbiological origin and chemical composition is given below

9.2.1 Glycolipid Biosurfactants

The most studied class of low molecular weight biosurfactants is glycolipids [3]. The structure of a glycolipid is made up of a hydrophilic carbohydrate moiety linked to hydrophobic fatty acid chains of varying lengths with an ester group [6]. Rhamnolipids, trehalolipids, sophorolipids, and mannosylerythritol lipids (MELs) are examples of glycolipid biosurfactants that comprise mono- and disaccharides units coupled with long-chain aliphatic acids or hydroxy-aliphatic acids [5].

9.2.2 Rhamnolipids

Rhamnolipids are a type of glycolipid that are well-known for their excellent physicochemical characteristics [3]. One or two fatty acid chains ranging in length from 8 to 16 carbons are connected to one or two rhamnose sugar molecules to form rhamnolipids [7]. *Pseudomonas* and *Burkholderia* species are the principal producers of rhamnolipids. Other promising rhamnolipid-producing microorganisms are *Pseudomonas cepacia* [8], *Pseudomonas* sp. [9], *Lysinibacillus sphaericus* [10], *Serratia rubidaea* [11].

The biosynthesis of the sugar part containing rhamnose from D-glucose and the synthesis of the hydrophobic acid part from fatty acids is the first step in the formation of rhamnolipids [7]. Most bacteria have the enzymes required for this first step, however, the specific enzymes required for rhamnolipid biosynthesis are found mainly in *P. aeruginosa* and *Burkholderia* sp. In *P. aeruginosa*, five distinct enzymes have been linked to the formation of rhamnolipids: RhlA, RhlB, RhlC, RhlG, and RhlI [12].

9.2.3 Trehalolipids

Trehalolipids are lipids that include trehalose disaccharides linked to a fatty acid (mycolic acid), they are mostly produced by *Rhodococcus, Nocardia, Mycobacterium*, and *Corynebacterium* species [13]. Surface and interfacial tension in culture broth is reduced by 25–40 and 1–5 mNm, respectively, using trehalose lipids from *Rhodococcus erythropolis* and *Arthrobacter spp* [14].

9.2.4 Sophorolipids

Sophorolipids are a class of glycolipids synthesized mainly by yeasts e.g. *Starmerella bombicola* [13], *Candida sphaerica* [15], and *Cutaneotrichosporon mucoides* [16]. These are made up of the hydrophilic disaccharide sophorose, which is made up of two monomers connected by beta-1,2 bonds. The sophorose is then linked to C16 or C18 hydroxylated fatty acid chains, which can be acetylated or non-acetylated, via a glycosidic linkage [6].

9.2.5 Mannosylerythritol Lipids

Mannosylerythritol lipids (MEL) are one of the most promising biosurfactants. *Pseudozyma antarctica* produces MEL in large quantities from vegetable oils. MELs are identified by the presence of mannose sugar connected to fatty acid and are classified as per the elongation of the hydrophobic fatty acid chain and the degree of saturation and/or acetylation in the monosaccharide's C4 and C6 positions [17].

9.2.6 Lipopeptide Biosurfactants

Lipopeptide surfactants are another type of low-molecular-weight biosurfactant. They possess a cyclic structure composed of hydrophilic peptide sequences that are usually 7 to 10 amino acids long, with a C_{13}–C_{18} fatty acid chain as the hydrophobic component. *Bacillus* or *Pseudomonas* species are the most commonly used microorganism for lipopeptide surfactant production [18].

The amino acid cyclic sequences of lipopeptides can be used to classify them into several types. For example, lipopeptides produced by *Bacillus subtilis* strains belong to the subcategory of surfactin, iturin, and fengycin families, all of which have a well-defined general structure. Surfactin is among the most effective biosurfactants ever discovered. Surfactin comes in a variety of forms, each with its own set of amino acids and fatty acid residues [19]. *Bacillus licheniformis* produces several biosurfactants called lichenysin [20]. Lichenysin is similar to surfactin in terms of stability under high temperature, pH, and salt environments [14].

9.2.7 Phospholipid Biosurfactants

A substantial quantity of fatty acids and phospholipid surfactants in great amounts are synthesized by bacteria and yeasts when they feed on n-alkanes [13]. *Thiobacillus thiooxidans*, a sulfur-reducing bacteria, produces more of this biosurfactant [21]. "Phosphatidyl ethanolamine-rich" vesicles are formed in Acinetobacter spp. which forms optically transparent micro-emulsions of alkanes in water [14].

9.2.8 Polymeric Biosurfactants

Many bacterial species of various genera produce exocellular polymeric surfactants, which are made up of a mixture comprising biopolymers such as proteins, polysaccharides, and lipopolysaccharides [21]. Polysaccharides are covalently bonded to fatty acids via o-ester bonds.

Emulsan, liposan, alasan, lipomanan, and other polymeric biosurfactants are examples of polymeric biosurfactants [22]. Emulsan is a highly effective emulsifier for hydrocarbons in water, even at fixation levels as low as 0.001–0.01% [23]. *Candida lipolytica* produces liposan, an extracellular water-soluble emulsifier made up of 83% carbohydrates and 17% protein. Alasan and lipomannan are the other polymeric biosurfactant produced by *Acinetobacter radioresistens* and *Candida tropicalis*. Table 9.1 shows a list of different kinds of biosurfactants obtained from microbial origin. Moreover, Figure 9.1 shows the schematic representation of various classes of biosurfactants and their members.

Table 9.1 Different kinds of biosurfactants from microbial origin.

Biosurfactants	Class	Microbial source	References
Rhamnolipids	Glycolipids	*Pseudomonas aeruginosa, Pseudomonas* sp. *Burkholderia* sp., *T. aquaticus*, and *Meiothermus rubber*	Herman et al. (1997), Hörmann et al. (2010), Maier and Soberón-Chávez (2000), Rezanka et al. (2011)
Trehalose lipids	Glycolipids	*Nocardia* sp.,*Rhodococcus erithropolis,Mycobacterium* sp., *Corynebacterium* sp., *Gordonia* sp.*Torulopsis bombicola, T. apicola, Candida kuoi, Rhodotorula bogoriensis*, and *Wickerhamiella domericqiae*	Macdonald et al. (1981), Franzetti et al. (2010)
Sophorolipids	Glycolipids	*Torulopsis bombicola, T. apicola, Candida kuoi, Rhodotorula bogoriensis, Wickerhamiella domericqiae,Candida sphaerica,Starmerella bombicola, Cutaneotrichosporon mucoides*	Celligoi et al. (2020), Price et al. (2012), Van Bogaert et al. (2007), Chen et al. (2006), Deshpande and Daniels (1995)Luna et al, 2011 Liu et al, 2019 Kurtzman et al, 2010Marcelino, et al, 2019

Table 9.1 (Continued)

Biosurfactants	Class	Microbial source	References
Mannosylerythritol lipids	Glycolipids	*Arthobacter* sp., *Candida antarticaPseudozyma* sp., and *Ustilago scitaminea*	Kim et al. (2002b), Arutchelvi et al. (2008), Saravanan and Vijayakuma (2015), Morita et al. (2015)
Surfactin	Lipopeptides	*Bacillus subtilis, Bacillus licheniformis,* and *Bacillus mojavensis*	Arima et al. (1968), From et al. (2007), Chen et al. (2015), Pecci et al. (2010)
Iturin	Lipopeptides	*Bacillus subtilis* and *Bacillus amyloliquefaciens*	Dang et al. (2019), Romero et al. (2007)
Lichenysin	Lipopeptides	*Bacillus licheniformis*	Ali et al. (2019)
Viscosin	Lipopeptides	*Pseudomonas fluorescens*	Alsohim et al. (2014)
Serrawettin	Lipopeptides	*Serratia marcescens*	Sunaga et al. (2004)
Arthrofactin	Lipopeptides	*Arthrobacter* sp.	Morikawa et al. (1993)
Polymyxin	Lipopeptides	*Bacillus polymyxa*	Muthusamy et al. (2008)
Emulsan	Polymeric	*Acinetobacter calcoaceticus* RAG-1	Zosim et al. (1982)
Liposan	Polymeric	*Candida lipolytica*	Cirigliano and Carman (1985)
Biodispersan	Polymeric	*Acinetobacter calcoaceticus* A2	Shabtai, (1990)
Lipomanan	Polymeric	*Candida tropicalis*	Rosenberg and Ron (1999)
Mannoproteins	Polymeric	*Acinetobacter* sp. and *Saccharomyces cerevisiae*	Cameron et al. (1988), Jagtap et al. (2010)
Alasan	Polymeric	*Acinetobacter radioresistens* KA53	Navon-Venezia et al. (1995)
Corynomycolic acid	Fatty acids, phospholipids, and neutral lipids	*Corynebacterium lepus* and *C. diphtheriae*	Fujii et al. (1999), Brennan et al. (1970)
Spiculisporic acid	Fatty acids, phospholipids, and neutral lipids	*Penicillium spiculisporum* and *Talaromyces trachyspermus*	Moriwaki-takano, (2021), Ishigami et al. (1983)
Phosphatidylethanolamine	Fatty acids, phospholipids, and neutral lipids	*Acinetobacter* sp. and *Rhodococcus erythropolis*	Kappeli and Finnerty (1979), Kretschmer et al. (1982)
Vesicles	Particulate biosurfactants	*Acinetobacter* sp., *P. marginilis,* and *Serratia marcescens*	Kappeli and Finnerty (1979), Matsuyama et al. (1986), Saharan et al. (2012)

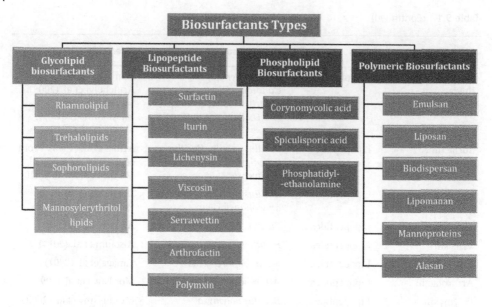

Figure 9.1 Classification and types of different microbial biosurfactants.

9.3 Properties of Biosurfactants

The following are some of the characteristics of biosurfactants. They are found in the majority of biosurfactants. Figure 9.2 shows the schematic representation of different properties exhibited by the biosurfactants.

9.3.1 Surface and Interface Activity

The condition known as surface tension is caused by systematized forces between liquid molecules. Surface tension and interfacial tension are similar in that both involve cohesive forces. However, the major forces associated with interfacial tension include adhesive forces (tension) between one substance's liquid phase and another substance's solid, liquid, or gas phase.

An excellent surfactant may reduce water's surface tension (ST) by 72 to 35 mN m^{-1} and water/interfacial hexadecane's tension (IT) from 40 to 1 mN m^{-1} [24]. Surfactin from *B. subtilis*, rhamnolipids from *P. aeruginosa*, and sophorolipids from *T. bombicola* have all been found to reduce the surface tension and interfacial tension of water and n-hexadecane, respectively. *B. subtilis* surfactin may lower the ST of water by 25 mN m^{-1} and the IT of water/hexadecane by 1 mN m^{-1} [25]. *P. aeruginosa* rhamnolipids reduced the ST of water by 26 mN m^{-1} and the IT of water/hexadecane by 1 mN m^{-1} [26].

Biosurfactants are 10–40 times more efficient than synthetic surfactants in comparison. The biosurfactant generated by *Bacillus mojavensis JF-2* has been shown to lower surfactant concentrations while reducing interfacial tension in crude oil [27].

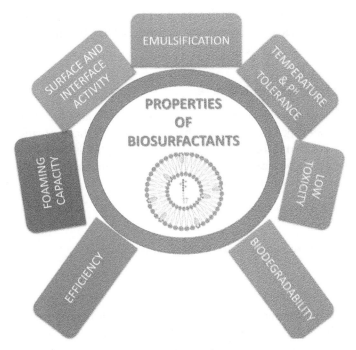

Figure 9.2 Different properties of biosurfactants.

9.3.2 Efficiency

Biologically derived surfactants are found to be highly efficient as compared to synthetic emulsifiers. Synthetic surfactants such as Tween 60 [28], SDS and polyoxyethylene [29], sorbitan monooleate, and Pluronic F-68 [30] were found to be less effective than *P. aeruginosa*-derived rhamnolipid biosurfactants.

9.3.3 Foaming Capacity

Most biosurfactants are excellent in producing and stabilizing foams. For example, surfactin's hybrid structure stimulates the evaluation of its potential to create and stabilize foams since it has both a lipidic chain, which is typically seen in small surfactant molecules, and a peptidic chain, which is the fundamental constituent of proteins. Surfactin's foaming properties are studied and compared to sodium dodecyl sulfate (SDS) and bovine serum albumin (BSA). Surfactin has better foaming characteristics than SDS and BSA, as evidenced by its capacity to generate and stabilize foam at concentrations as low as 0.05 mg mL^{-1}. Surfactin, like BSA, created a foam with a medium maximum density and stabilized the liquid in the foam [31].

9.3.4 Emulsification/Emulsion Forming and Emulsion Breaking

Emulsifiers and de-emulsifiers are two forms of biosurfactants. An emulsion is a heterogeneous system made up of an immiscible liquid dispersed into another liquid as droplets with a diameter greater than 0.1 mm. Water-in-oil (w/o) and oil-in-water (o/w) emulsions are the most common. Emulsions have limited stability, however, adding biosurfactants can make an emulsion last months or even years [32].

9.3.5 Tolerance for Temperature and pH Tolerance

Environmental factors such as pH and temperature do not affect the surface activity of many biosurfactants [32]. As per experimental data, it has been observed that even after autoclaving (121 °C/20 min) and storing for 6 months at 18 °C, a lipopeptide isolated from *B. subtilis LB5a* remained stable; the surface activity did not alter across pH values 5–11 but also with varying NaCl concentrations up to 20% it didn't change [33].

9.3.6 Low Toxicity

Biosurfactants are considered low- or non-toxic due to their natural sources.

9.3.7 Biodegradability

Biosurfactants are more biodegradable than their chemical equivalents due to their composition [6]. Biosurfactant biodegradability has been studied without the addition of external microbial biomass, while temperature, pH, and biodegradation time were varied [34]. Microorganisms in water and soil easily break down biosurfactants, making them suitable for environmental remediation and treatment of waste [32].

9.4 Conventional Methods for Biosurfactant Production

In the production of biosurfactants, specifically, those used in the pharmaceutical and food sectors, two crucial criteria should be considered. First, biosurfactants should be produced by non-pathogenic as well as safe microorganisms in terms of avoiding pathogenicity-related issues. Although the fact that the majority of biosurfactant-producing bacteria are pathogenic, their use in large-scale industrial operations would be extremely difficult. Second, to reduce the total costs of the fermentation process, biosurfactant production should include the use of low-cost substrates. To put it another way, an efficient biosurfactant production process includes an economic system that uses low-cost inputs while producing a high output yield [35]. Biosurfactant production involves the following steps (Figure 9.3):

1) preparation of media and addition of cultures
2) fermentation and maintenance
3) product recovery
4) purification

Strains and culture conditions are maintained, whenever required they are to be subcultured. The desired microorganism is then screened for biosurfactant production using suitable assays like hemolytic assay, oil-spreading method [36], drop-collapse method [37, 38],

9.4 Conventional Methods for Biosurfactant Production

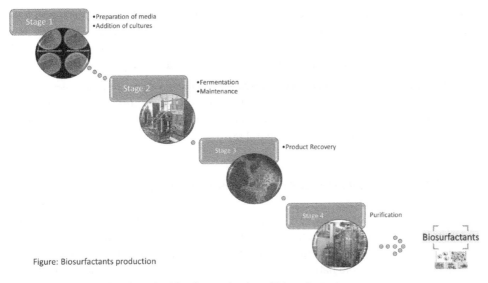

Figure 9.3 Conventional method for the production of biosurfectants.

tensiometry [39], eulsification assays [40], etc. After screening, the desired organism is selected and carried for cultivation on the specific medium of microbial species with appropriate yield-enhancing substrates. Batch fermentation techniques are used mostly for biosurfactant production. The optical density measurement of the broth is used to check for bacterial growth. For biomass analysis, a traditional method based on cell dry weight measurement is used. The biosurfactant concentrations (g l^{-1}) are mostly obtained using a calibration curve for surfactin, a commercial biosurfactant (ST (mN m^{-1}) = -8.65 x concentration (g L^{-1}) + 77, r2 = 0.973) [41]. After production biosurfactants are usually subjected to certain examination studies as follows:

The physical characteristics of biosurfactants are investigated: Different substrates were used to test the percent E24 (emulsification index) of the generated BS. Plotting the broth ST as a function of the logarithm of BS concentration reveals the CMC value of the BS solution. Pipetting cell-free BS liquid as a droplet onto parafilm as well as polystyrene surfaces revealed contact angle (CA) readings. To obtain macro shots, a digital camera is oriented at a 90° angle with the samples. The appropriate CA values are mostly calculated using Image J 1.44 software [42].

Assays for antimicrobial and anti-adhesive properties: Using the agar well diffusion method, the antimicrobial property of the BS against several pathogenic microorganisms is tested [43]. Similarly, the same microbes are used to determine anti-adhesive properties [44].

Studies on the stability of biosurfactant: To assess the pH stability of BS, freeze-dried BS is dissolved in PBS at various pH values (2.0–12.0) and the ST of each sample is calculated using the Wilhelmy-plate method at room temp (25 °C). In addition, the resistance of BS to high temperatures is assessed. The ST is determined after the BS solution has been incubated at 121 °C for 15 minutes and then cooled to room temperature [45].

Characterization of the biosurfactant structure: Bradford's technique is used to determine the protein content of BS [46]. The phenol-sulfuric acid technique of Dubois is used mostly to calculate carbohydrate content [47]. The fatty acids of BS are transformed into methyl esters (FAMEs), which are then evaluated using a gas chromatograph as well as a flame ionization detector (FID). On an FTIR instrument, Fourier transforms infrared (FTIR) spectra of lyophilized samples of BSs generated are also analyzed [35].

Analysis of statistical factors of biosurfactant: The mean values of the samples are statistically examined using SPSS software. To find any significant differences between the parameters and variables, one-way ANOVA and Duncan tests are used. In addition, paired t-tests are used to compare CA mean values [35].

9.5 Commercial Applications of Biosurfactants

Biologically produced surfactants are utilized in cosmetics, cleaning goods, and agriculture products in the biotechnology industry [48]. Biosurfactants, in addition to their ability to stabilize nanoparticles, can be employed in a variety of food, medicinal, and cosmetic goods due to their biodegradable nature and inclusion of certain functional groups. Figure 9.4 showed the schematic representation of commercial applications of biosurfactants in different fields.

9.5.1 Application of Biosurfactants in Agriculture

Surfactants are used in a variety of agricultural and agrochemical applications. These biosurfactants can be used in a variety of agricultural applications, including improving pollutant biodegradation, improving soil quality, and indirectly promoting plant growth through antimicrobial activity. They can also be used to increase plant–microbe interaction, which is beneficial to plants [49].

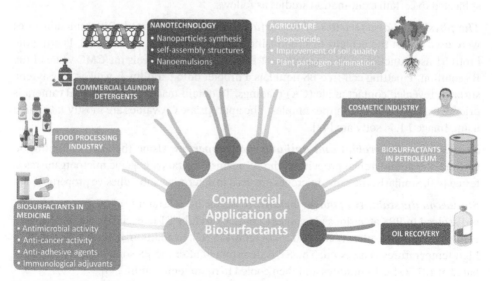

Figure 9.4 Schematic representation showing various commercial applications of biosurfactant.

9.5.1.1 Biosurfactants as Biopesticide

Broad-spectrum insecticides and pesticides are used in traditional arthropod control strategies, which often have unfavorable side effects. Furthermore, the emergence of pesticide-resistant insect populations, as well as the rising costs of new chemical pesticides, has fuelled the search for new environmentally sustainable vector control tools. Several microorganisms synthesize lipopeptide biosurfactants that have insecticidal action against fruit flies such as Drosophila melanogaster, and hence might be utilized as a biopesticide [24].

9.5.1.2 Improvement of Soil Quality

Many rhizosphere and plant-associated bacteria synthesize biosurfactants, indicating that biosurfactants can play a role in plant–microbe interactions and that they could be used in agriculture [50]. Also, biosurfactants play a role in soil remediation and better bioavailability of nutrients in the soil for better plant health [49].

9.5.1.3 Plant Pathogen Elimination

Biosurfactants derived from various microorganisms have antimicrobial activity towards plant phytopathogens and have been demonstrated to be a viable biocontrol agent for attaining sustainable agriculture. Rhamnolipid and lipopeptide are the most commonly used biosurfactants as biocontrol agents. Biosurfactants have an antagonistic impact on zoosporic plant pathogens that have developed resistance to commercial pesticides, allowing them to be used as a biocontrol agent. Biosurfactants can also stimulate plant immunity, which is being studied as an alternative to reducing disease caused by phytopathogens [51].

9.5.2 Application of Biosurfactants in Nanotechnology

Micelles and microemulsion droplets are formed, which have beneficial features for a variety of nanoscience applications. These phenomena are then investigated in a variety of settings, including nanoparticle synthesis and stabilization, oil recovery, carbon dioxide as a green solvent, and photo- and magneto-responsive surfactant systems [52].

9.5.2.1 Biosurfactant in Nanoparticles Synthesis

The use of microbial-derived compounds to aid nanomaterial synthesis is another option for contributing to a more sustainable solution. Biosurfactants have developed as environmentally friendly replacements in the manufacture of nanoscale materials [4]. Biosurfactants are essential for nanoparticle growth stability and uniform dispersion in solution. The central event in stabilization is surfactant adsorption into metal nanoparticle surfaces, which is dependent on the kind of surfactant used (ionic, non-ionic, and polymeric) and the thickness of the adsorbed layer [53, 54].

9.5.2.2 Biosurfactant Self-assembly Structures:

Biosurfactants provide self-assembled structures which ultimately assist in nanotechnology synthesis [4].

9.5.2.3 Biosurfactant in Nanoemulsions

Nanoemulsions comprise kinetically stable dispersions of two or more immiscible liquid phases stabilized by a surfactant and containing droplets smaller than 200 nanometres [55, 56]. Since biosurfactants show a variety of biological features, they can be used to substitute synthetic surfactants in nanoemulsion formulations as they can act as an active or synergic agent [4].

9.5.2.4 Biosurfactant in the Biological Activity of Nanostructure Materials

Along with the surface activity, one of the most notable characteristics of BSs is their biological activity. Antimicrobial, anticancer, anti-adhesion, anti-biofilm, immunosuppressive, immune-modulating, anti-inflammatory, and antioxidant characteristics have all been identified [57–59].

9.5.3 Applications of Biosurfactants in Commercial Laundry Detergents

Any component used in soaps and detergents should be stable at the alkaline pH level and able to endure the washing temperature as a minimum requirement. Biosurfactants with thermostable properties, such as lipopeptide biosurfactants obtained from different *B. subtilis* strains are useful in laundry applications. The thermal resilience and high alkaline pH stabilization of *B. subtilis* biosurfactants promote their use in laundry detergent formulations, which are exposed to wash performance inside a temperature range of 25–60 °C [60].

9.5.4 Application of Biosurfactants in Medicine

Biosurfactants comprise amphipathic chemicals produced by a variety of microorganisms with different biological functions which have a wide range of applications in medical science.

9.5.4.1 Antimicrobial Activity

Antimicrobial capabilities, as well as the ability to prevent adhesion and disrupt biofilm development, have all been reported for several biosurfactants. Sophorolipids may be potential molecules for use as adjuvants to many other antimicrobials in biomedical applications, inhibiting pathogen growth and/or disrupting biofilms [61].

9.5.4.2 Anti-cancer Activity

The ability of biosurfactants to affect several mammalian cell functions, and hence their ability to act as antitumor agents by interrupting some cancer growth pathways, is one of the most interesting discoveries. Because of their structural distinctiveness and diverse biophysical properties, lipopeptide, glycolipid, and other types of biosurfactants have now emerged as potential broad-spectrum agents in cancer chemotherapy/biotherapy and as safe carriers or components in drug delivery formulations [62].

9.5.4.3 Anti-adhesive Agents

Anti-adhesion treatments for bacterial infections are a substitute for antibiotics since they prevent germs from harming a host by decreasing adhesion to host tissues and cells, which is required for most infectious diseases [63]. *Pseudomonas aeruginosa* produced rhamnolipid which has anti-adhesive properties against bacterial as well as yeast strains isolated through voice prosthesis [41]. Also, it has been observed that the biosurfactants like BS-VSG4 and BSVS16 synthesized from *Bacillus* species have significant anti-adhesive properties against biofilm-forming bacteria, as well as the ability to break pre-formed bacterial biofilms [64].

9.5.4.4 Immunological Adjuvants
When combined with traditional antigens, bacterial lipopeptides are effective non-toxic, non-pyrogenic immunological adjuvants. When low-molecular-weight antigens iturin AL and herbicolin A were used, the humoral response improved [65].

9.5.4.5 Antiviral Activity
The antiviral properties of biosurfactants makes them promising alternatives for use in the treatment of a variety of viral infection-related illnesses. *Bacillus subtilis* synthesizes surfactin which is found to have antiviral activity against human immunodeficiency virus 1 (HIV-1) [66]. Also, *Rhodococcus erythropolis* produces trehalose lipid which also possesses antiviral activity against HSV and influenza virus [67].

9.5.5 Application of Biosurfactants in the Food Processing Industry

Biosurfactants are utilized in a variety of food processing applications, but they are most commonly used as food formulation ingredients and anti-adhesive agents. As a food formulation ingredient, they help to produce and stabilize emulsions by lowering interfacial tension. It is also used to prevent fat globule agglomeration, stabilize aerated systems, improve the texture and shelf life of starch-based products, change the rheological characteristics of wheat dough, and improve the consistency and texture of fat-based products [68].

9.5.6 Application of Biosurfactants in the Cosmetic Industry

Biosurfactants have been proposed to replace chemically manufactured surfactants in the cosmetic sector due to their properties like emulsification, foaming, water binding ability, spreading, and wetting properties affect viscosity and product consistency [65]. Biosurfactants may play a prebiotic role in the skin microbiota equilibrium because of their antibacterial activity against a variety of diseases. Biosurfactants may become premium skincare products due to their benefits [69]. Glycolipids are perhaps the most researched biosurfactants in personal care and cosmetics formulation [70]. Rhamnolipid-containing cosmetics have been patented and utilized as anti-wrinkle and anti-aging treatments [71].

9.5.7 Application of Biosurfactants in Petroleum

Biosurfactants and bioemulsifiers are a novel class of molecules that are among the most potent and flexible by-products that modern microbial technology has to offer in sectors like biocorrosion and biofouling, as well as enzymes and biocatalysts for petroleum upgrading. Biosurfactants are also important in petroleum extraction, transportation, upgrading, and refining, as well as petrochemical production [72].

9.5.8 Application of Biosurfactant in Microbial-enhanced Oil Recovery

Oil and oil product contamination has caused significant harm, and more emphasis is being directed to the design and delivery of novel technology to remove these contaminants. Biosurfactants have been widely employed in water and soil remediation, as well as in the harvesting, transportation, and storage phases of the oil production chain. In comparison to synthetic analogs, biodegradability, low toxicity, and higher performance under harsh conditions.

9.6 Biosurfactants in Nanotechnology (Biosurfactant Mediated Synthesis of Nanoparticles)

In material science, creating nanoparticles with great monodispersity is extremely challenging. Because of the diverse features of nanomaterials, nanoparticle synthesis has been a major research area in recent years. Furthermore, these characteristics may yield interesting results in the biological field. The size and composition of nanoparticles drive the advancement of material science research. Sol-gel, co-precipitation, laser ablation, nanoemulsion, electrospinning, and pyrolysis are some of the methods used to integrate nanomaterials [73]. The techniques mentioned above are costly, produce hazardous chemicals, and produce less stable nanoparticles [74]. High temperatures, as well as biologically harmful substances, are used in physical and chemical procedures, resulting in toxicity to human health and the environment. The most well-known approach is chemical reduction, which involves employing reducing agents to convert metal salts into metal atoms. All of these agents, however, can trigger environmental problems [75, 76].

As a result, an environmentally acceptable technique for metal nanoparticle generation is required. The use of biosurfactants to improve the biological synthesis of nanoparticles is one method. Biosurfactants and their use in nanotechnology and nanoparticle synthesis are increasingly becoming more common in green chemistry [77]. The amphiphilic nature of biosurfactants allows them to maintain interfacial tension and stabilize compositions. Biosurfactants are often used in the formulation and stabilization of nanoparticles because of their ability to act as stabilizers and capping agents (NPs). Biosurfactants operate as surface-active agents and promote the creation and stabilization of NPs by adsorbing across their surfaces. They also inhibit aggregation and stabilize NP formulations. Biosurfactants are employed as replacements for synthetic stabilizers in green synthesis as well as the stabilization of NPs due to their various properties and natural origin [78].

Several biosurfactants are investigated to see if they can be employed as a stabilizer and regulators in the manufacture of various metallic nanoparticles.

9.6.1 Glycolipids Biosurfactants Produced Nanoparticles

Due to the special features of silver nanoparticles such as antibacterial, antiviral, and antifungal nature, they have been used in a variety of disciplines. Silver nanoparticles are typically synthesized using several different, physical, and chemical processes [79]. Chemical reduction, photochemical reduction, reverse micelle-based and lamellar liquid crystal procedures, aerosol techniques, and an electrostatic spraying technique are now used to synthesize nano-sized silver nanoparticles. Since the reverse micelle mechanism has been utilized to synthesize metal nanoparticles, it has received much interest. Biosurfactant-mediated nanoparticle synthesis has now evolved as a green approach for improving nanoparticle synthesis and stability. Adsorption upon metallic nanoparticles, surface stabilization, and prevention of aggregation are some of the mechanisms of action. As a result, silver nanoparticles can be produced using biosurfactants.

9.6.2 Lipopeptides Biosurfactants Produced Nanoparticles

9.6.2.1 Gold Nanoparticles

Chemical methods are mostly used to make gold nanoparticles, which mostly pollute the environment due to the use of harmful compounds including borohydride, citrate, and acetylene. Shape-dependent characteristics are observed in gold nanoparticles [80]. Surfactants, thiols, amines, phosphines, phosphine oxides, and carboxylates are utilized as stabilizing agents in the synthesis process to control form and size. Researchers are currently emphasizing the biological production of stabilized gold nanoparticles using biosurfactants. Biosurfactants could be an effective way to boost nanoparticle synthesis in a biological environment. Microorganisms such as *Bacillus subtilis* and *Pseudomonas aeruginosa* are often employed for synthesizing gold nanoparticles with the assistance of biosurfactants. At a basic and neutral pH, gold nanoparticles are found in a spherical form and are very stable with no aggregates formed [81, 82].

9.6.2.2 Zinc Nanoparticles

Zinc nanoparticles are considered the most important nanoparticles because of their applications as a catalyst, sensor, optoelectronic, and photoelectron device [83]. Zinc nanoparticles are useful in the medical industry fields because of their high surface area and catalytic properties. Antimicrobial compounds are used in baby powder, calamine lotion, anti-dandruff shampoos, and antiseptic ointments, among other things [84]. Several methods for synthesizing zinc metal nanoparticles have been developed, including precipitation, wet chemical synthesis, solid-state pyrolytic method, and sol-gel method. However, many disadvantages are explained in the findings, such as aggregation, expanding the size of the particles, poor reproducibility, the need for high temperature and high pressure, and so on [85, 86]. Rhamnolipids produced from *Pseudomonas aeruginosa*, are biocompatible and biodegradable biosurfactants [87]. Rhamnolipids are employed as a capping and stabilizing agent for generating stable and biocompatible zinc sulfide nanoparticles [88].

9.6.2.3 Iron Nanoparticles

Iron oxide nanoparticles are becoming highly important in the development of new biological and nanotechnology applications. Because of the advantages that iron oxide nanoparticles have over other materials, they have been explored extensively in a variety of fields. Iron oxide nanoparticles are widely used because of their low cost, ability to be separated using external magnetic fields, large surface area, and excellent adsorption capacity [89–91]. Magnetic nanoparticles are commonly prepared by microemulsion, thermal decomposition, co-precipitation techniques, gamma-ray radiation, and microwave plasma synthesis [92]. The above-mentioned processes result in the creation of aggregation and the production of large particle sizes.

As a result, scientists are concentrating their efforts on developing magnetic nanoparticles via a biological process. Sophorolipids can be used to make iron nanoparticles. The acidic properties of sophorolipids made them suitable for use as surface complexing agents in the manufacture of iron oxide nanoparticles [93]. Surfactin is a form of lipopeptide that is also used to stabilize super magnetic iron oxide nanoparticles. The produced nanoparticle is generally spherical and easily dispersed in the aqueous phase [94].

9.6.2.4 Nickel Nanoparticles

Nickel nanoparticles (Ni-NPs) have attracted a lot of attention because of their remarkable properties such as high reactivity, operational simplicity, biocompatibility, microbial resistance, cost-effectiveness, abundance, anti-inflammatory activities, and environmental compatibilities. They have been used as a catalyst, an electro-catalyst, a photo-catalyst, a biosensor, and a heat exchanger [95]. Chemical reduction techniques, discharge routes, photocatalytic reduction, and other ways have all been used to make Ni-NPs. Complex processing, complex reaction conditions, high temperatures, and extended reaction periods are all shortcomings of most of these techniques [96–99]

It has been discovered that rhamnolipid biosurfactants can be employed to make nanoparticles. The findings reveal that the biosurfactants pH influences the shape of nickel nanoparticles. As a result, the nanoparticle shapes change as the pH changes without impacting the environment. The morphology altered to a spherical shape when the pH was elevated, and to a flaky shape when the pH was dropped [100].

9.7 Conclusions

Biosurfactants have several properties that can be useful for a variety of industries. By providing appropriate surface functional groups onto nanoparticles, biosurfactant as a capping agent play a critical role. It helps influence particle size, shape, agglomeration, surface energy, dispersal, and electrostatic and steric hindrance of the targeted nanoparticle. Even while chemical surfactants continue to dominate the market, green surfactants are gaining popularity. Green chemistry has been used to reduce, stabilize, and fabricate synthetic nanoparticles due to the limits of chemical capping agents. Although developments in biosurfactant manufacturing have permitted a boost in the generation of these biomolecules, there are examples of biosurfactants that have been successfully commercialized, but complete market penetration is still impeded by expensive production costs and low product titers. But more remarkable improvements (even if of a low amount) are likely to be required to make this technology economically feasible.

References

1 Vedarethinam V, Chelliah A. Current trends in the application of biosurfactant in the synthesis of nanobiosurfactant such as engineered biomolecules from various biosurfactant derived from diverse sources, nanoparticles, and nanorobots. In: Inamuddin Adetunji CO, Ahamed MI, editors. *Green Sustainable Process for Chemical and Environmental Engineering and Science*. Academic Press, UK: 2022, pp. 619–632.

2 Shaban SM, Kang J, Kim DH. Surfactants: recent advances and their applications. *Composites Commun.* 2020;22:100537.
3 Bjerk TR, Severino P, Jain S, Marques C, Silva AM, Pashirova T, Souto EB. Biosurfactants: properties and applications in drug delivery, Biotechnology and ecotoxicology. *Bioeng.* 2021;8(8):115.
4 Nitschke M, Marangon CA. Microbial surfactants in nanotechnology: recent trends and applications. *Crit Rev Biotechnol.* 2022;42(2):294–310.
5 Drakontis CE, Amin S. Biosurfactants: formulations, properties, and applications. *Curr Opin Colloid Interface Sci.* 2020;48:77–90.
6 Sarubbo LA, Silva M, da GC, Durval IJB, Bezerra KGO, Ribeiro BG, Silva IA, Twigg MS, Banat IM. Biosurfactants: production, properties, applications, trends, and general perspectives. *Biochem Eng J.* 2022;181:1–19.
7 Abdel-Mawgoud AM, Lépine F, Déziel E. Rhamnolipids: diversity of structures, microbial origins and roles. *Appl Microbiol Biotechnol.* 2010;86(5):1323–1336.
8 Silva R, de CFS, Almeida DG, Rufino RD, Luna JM, Santos VA, Sarubbo LA. Applications of biosurfactants in the petroleum industry and the remediation of oil spills. *Int J Mol Sci.* 2014;15(7):12523–12542.
9 Hassan M, Essam T, Yassin AS, Salama A. Optimization of rhamnolipid production by biodegrading bacterial isolates using plackett–burman design. *Int J Biol Macromol.* 2016;82:573–579.
10 Gaur VK, Bajaj A, Regar RK, Kamthan M, Jha RR, Srivastava JK, Manickam N. Rhamnolipid from a Lysinibacillus sphaericus strain IITR51 and its potential application for dissolution of hydrophobic pesticides. *Bioresour Technol.* 2019;272:19–25.
11 Nalini S, Parthasarathi R. Optimization of rhamnolipid biosurfactant production from Serratia rubidaea SNAU02 under solid-state fermentation and its biocontrol efficacy against Fusarium wilt of eggplant. *Ann Agrar Sci.* 2018;16(2):108–115.
12 Kiss K, Ng WT, Li Q. Production of rhamnolipids-producing enzymes of Pseudomonas in E. coli and structural characterization. *Front Chem Sci Eng.* 2017;11(1):133–138.
13 Rawat G, Dhasmana A, Kumar V. Biosurfactants: the next generation biomolecules for diverse applications. *Environ Sustain.* 2020;3(4):353–369.
14 Vijayakumar S, Saravanan V. Biosurfactants-Types, sources, and applications. *Res J Microbiol.* 2015;10(5):181–192.
15 Luna JM, Rufino RD, Albuquerque CDC, Sarubbo LA, Campos-Takaki GM. economic optimized medium for tensio-active agent production by Candida sphaerica UCP0995 and application in the removal of hydrophobic contaminant from sand. *Int J Mol Sci.* 2011;12(4):2463–2476.
16 Marcelino PRF, Peres GFD, Terán-Hilares R, Pagnocca FC, Rosa CA, Lacerda TM, dos Santos JC, da Silva SS. Biosurfactants production by yeasts using sugarcane bagasse hemicellulosic hydrolysate as new sustainable alternative for lignocellulosic biorefineries. *Ind Crops Prod.* 2019;129:212–223.
17 Ribeiro BG, Guerra JMC, Sarubbo LA. Biosurfactants: production and application prospects in the food industry. *Biotechnol Prog.* 2020;36(5):e3030.
18 Kourmentza K, Gromada X, Michael N, Degraeve C, Vanier G, Ravallec R, Coutte F, Karatzas KA, Jauregi P. antimicrobial activity of lipopeptide biosurfactants against foodborne pathogen and food spoilage microorganisms and their cytotoxicity. *Front Microbiol.* 2021;11:3398.

19 Cochrane SA, Vederas JC. Lipopeptides from Bacillus and Paenibacillus Spp.: a gold mine of antibiotic candidates. *Med Res Rev*. 2016;36(1):4–31.

20 Das AJ, Kumar R. Utilization of agro-industrial waste for biosurfactant production under submerged fermentation and its application in oil recovery from sand matrix. *Bioresour Technol*. 2018;260:233–240.

21 Rosas-Galván NS, Martínez-Morales F, Marquina-Bahena S, Tinoco-Valencia R, Serrano-Carreón L, Bertrand B, León-Rodríguez R, Guzmán-Aparicio J, Alvaréz-Berber L, Trejo-Hernández MR. Improved production, purification, and characterization of biosurfactants produced by Serratia marcescens sm3 and its isogenic smrg-5 strain. *Biotechnol Appl Biochem*. 2018;65(5):690–700.

22 Gayathiri E, Prakash P, Karmegam N, Varjani S, Awasthi MK, Ravindran B. Biosurfactants: potential and eco-friendly material for sustainable agriculture and environmental safety: a review. *Agron*. 2022;12(3):662.

23 Roy A. A review on the biosurfactants: properties, types, and its applications. *J Fundam Renewab Energy and Appl*. 2018;08(01).

24 Mulligan CN. Environmental applications for biosurfactants. *Environ Pollut*. 2005;133(2):183–198.

25 Cooper DG, Macdonald CR, Duff SJB, Kosaric N. Enhanced production of surfactin from Bacillus subtilis by continuous product removal and metal cation additions. *Appl Environ Microbiol*. 1981;42(3):408.

26 Syldatk C, Lang S, Wray V, Witte L. Chemical and physical characterization of four interfacial-active rhamnolipids from Pseudomonas spec. dsm 2874 grown on n-alkanes. *Z Naturforsch C Biosci*. 1985;40(1–2):51–60.

27 McInerney MJ, Maudgalya SK, Knapp R, Folmsbee M Development of biosurfactant-mediated oil recovery in model porous systems and computer simulations of biosurfactant-mediated oil recovery. 2004.

28 Scheibenbogen K, Zytner RG, Lee H, Trevors JT. Enhanced removal of selected hydrocarbons from soil by Pseudomonas aeruginosa ug2 biosurfactants and some chemical surfactants. *J Chem Tech Biotech*. 1994;59(1):53–59.

29 Hirata Y, Ryu M, Oda Y, Igarashi K, Nagatsuka A, Furuta T, Sugiura M. Novel characteristics of sophorolipids, yeast glycolipid biosurfactants, as biodegradable low-foaming surfactants. *J Biosci Bioeng*. 2009;108(2):142–146.

30 Pornsunthorntawee O, Wongpanit P, Chavadej S, Abe M, Rujiravanit R. Structural and physicochemical characterization of crude biosurfactant produced by Pseudomonas aeruginosa SP4 isolated from petroleum-contaminated soil. *Bioresour Technol*. 2008;99(6):1589–1595.

31 Razafindralambo H, Paquot M, Baniel A, Popineau Y, Hbid C, Jacques P, Thonart P. Foaming properties of surfactin, a lipopeptide biosurfactant from Bacillus subtilis. *J Americ Oil Chemists Socie*. 1996;73(1):149–151.

32 Santos DKF, Rufino RD, Luna JM, Santos VA, Sarubbo LA. Biosurfactants: multifunctional biomolecules of the 21st century. *Int J Mol Sci*. 2016;17(3):401.

33 Nitschke M, Pastore GM. Production and properties of a surfactant obtained from Bacillus subtilis grown on Cassava wastewater. *Bioresour Technol*. 2006;97(2):336–341.

34 Rodríguez-López L, Rincón-Fontán M, Vecino X, Moldes AB, Cruz JM. Biodegradability study of the biosurfactant contained in a crude extract from corn steep water. *J Surfactants Deterg*. 2020;23(1):79–90.

35 Ghasemi A, Moosavi-Nasab M, Setoodeh P, Mesbahi G, Yousefi G. Biosurfactant production by lactic acid bacterium Pediococcus dextrinicus shu1593 grown on different carbon sources: strain screening followed by product characterization. *Sci Rep.* 2019;9(1):1–12.
36 Youssef NH, Duncan KE, Nagle DP, Savage KN, Knapp RM, McInerney MJ. Comparison of methods to detect biosurfactant production by diverse microorganisms. *J Microbiol Methods.* 2004;56(3):339–347.
37 Kuiper I, Lagendijk EL, Pickford R, Derrick JP, Lamers GEM, Thomas-Oates JE, Lugtenberg BJJ, Bloemberg Gv. Characterization of Two Pseudomonas putida lipopeptide biosurfactants, putisolvin i and ii, which inhibit biofilm formation and break down existing biofilms. *Mol Microbiol.* 2004;51(1):97–113.
38 Tugrul T, Cansunar E. Detecting surfactant-producing microorganisms by the drop-collapse test. *World J Microbiol Biotechnol.* 2005;21(6–7):851–853.
39 Fernandes PAV, de Arruda IR, dos Santos AFAB, de Araújo AA, Maior AMS, Ximenes EA. Antimicrobial activity of surfactants produced by Bacillus subtilis R14 against multidrug-resistant bacteria. *Braz J Microbiol.* 2007;38(4):704–709.
40 Satpute SK, Banpurkar AG, Dhakephalkar PK, Banat IM, Chopade BA. Methods for investigating biosurfactants and bioemulsifiers: a review. *Crit Rev Biotechnol.* 2010;30(2):127–144.
41 Rodrigues LR, Teixeira JA, Oliveira R. Low-cost fermentative medium for biosurfactant production by probiotic bacteria. *Biochem Eng J.* 2006;32(3):135–142.
42 Gudiña EJ, Teixeira JA, Rodrigues LR. Isolation and functional characterization of a biosurfactant produced by Lactobacillus paracasei. *Colloids Surf B Biointerfaces.* 2010;76(1):298–304.
43 Madhu AN, Prapulla SG. Evaluation and functional characterization of a biosurfactant produced by Lactobacillus plantarum CFR 2194. *Appl Biochem Biotechnol.* 2014;172(4):1777–1789.
44 Heinemann C, Hylckama Vlieg JET, Janssen DB, Busscher HJ, Mei HC, Reid G. Purification and characterization of a surface-binding protein from Lactobacillus fermentum RC-14 that inhibits adhesion of Enterococcus faecalis 1131. *FEMS Microbiol Lett.* 2000;190(1):177–180.
45 Gudiña EJ, Fernandes EC, Teixeira JA, Rodrigues LR. Antimicrobial and anti-adhesive activities of cell-bound biosurfactant from Lactobacillus agilis CCUG31450. *RSC Adv.* 2015;5(110):90960–90968.
46 Bradford MM. A rapid and sensitive method for the quantitation of microgram quantities of protein utilizing the principle of protein-dye binding. *Anal Biochem.* 1976;72(1–2):248–254.
47 Dubois M, Gilles KA, Hamilton JK, Rebers PA, Smith F. Colorimetric method for determination of sugars and related substances. *Anal Chem.* 1956;28(3):350–356.
48 Nehal N, Singh P. Role of nanotechnology for improving properties of biosurfactant from newly isolated bacterial strains from Rajasthan. *Mater Today Proc.* 2022;50:2555–2561.
49 Sachdev DP, Cameotra SS. Biosurfactants in agriculture. *Appl Microbiol Biotechnol.* 2013;97(3):1005–1016.
50 Amani H, Sarrafzadeh MH, Haghighi M, Mehrnia MR. Comparative study of biosurfactant producing bacteria in MEOR applications. *J Pet Sci Eng* 2010;75(1–2):209–214.

51 Nalini S, Parthasarathi R, Inbakanadan D Biosurfactant in food and agricultural application. 2020: 75–94.
52 Eastoe J, Tabor RF. Surfactants and nanoscience. In: Berti D, Palazzo G, editors. *Colloid Foundation of Nanoscience,* **Elsevier**, UK: 2022, pp. 153–182.
53 Kvítek L, Panáček A, Soukupová J, Kolář M, Večeřová R, Prucek R, Holecová M, Zbořil R. Effect of surfactants and polymers on stability and antibacterial activity of silver nanoparticles (NPs). *J Phys Chem C*. 2008;112(15):5825–5834.
54 Chou KS, Lai YS. Effect of polyvinyl pyrrolidone molecular weights on the formation of nano-sized silver colloids. *Mater Chem Phys*. 2004;83(1):82–88.
55 Al-Bukhaiti WQ, Noman A, Wang H. Emulsions: micro and nano-emulsions and their applications in industries-a mini-review. *Int J Agri Innovat Res*. 2018;7.
56 Aswathanarayan JB, Vittal RR. Nanoemulsions and their potential applications in food industry. *Front Sustain Food Syst*. 2019;3:95.
57 Rodrigues LR, Teixeira JA. Biomedical and therapeutic applications of biosurfactants. *Adv Exp Med Biol*. 2010;672:75–87.
58 Fracchia L, Banat J, Cavallo M, et al. Potential therapeutic applications of microbial surface-active compounds. *AIMS Bioeng*. 2015;2(3):144–162.
59 Naughton PJ, Marchant R, Naughton V, Banat IM. Microbial biosurfactants: current trends and applications in agricultural and biomedical industries. *J Appl Microbiol*. 2019;127(1):12–28.
60 Mukherjee AK. Potential Application of cyclic lipopeptide biosurfactants produced by Bacillus subtilis strains in laundry detergent formulations. *Lett Appl Microbiol*. 2007;45(3):330–335.
61 Díaz De Rienzo MA, Banat IM, Dolman B, Winterburn J, Martin PJ. Sophorolipid biosurfactants: possible uses as antibacterial and antibiofilm agent. *N Biotechnol*. 2015;32(6):720–726.
62 Gudiña EJ, Rangarajan V, Sen R, Rodrigues LR. Potential therapeutic applications of biosurfactants. *Trends Pharma Sci*. 2013;34(12):667–675.
63 Cozens D, Read RC. Anti-adhesion methods as novel therapeutics for bacterial infections. *Expert Rev Anti Infect Ther* 2012;10(12):1457–1468.
64 Giri SS, Ryu EC, Sukumaran V, Park SC. Antioxidant, antibacterial, and anti-adhesive activities of biosurfactants isolated from Bacillus strains. *Microb Pathogen* 2019;132:66–72.
65 Gharaei-Fathabad E. Biosurfactants in pharmaceutical industry: a mini review. *Am J Drug Discov Dev*. 2011;1:58–69.
66 Vollenbroich D, Ozel M, Vater J, et al. Mechanism of inactivation of enveloped viruses by the biosurfactant surfactin from Bacillus subtilis. *Biologicals*. 1997;25:289–297.
67 Uchida Y, Misava S, Nakahara T, et al. Factors affecting the formation of succinoyltrehalose lipids by Rhodococcus erythropolis SD-74 grown on n-alkanes. *Agric Biol Chem*. 1989;53:765–769.
68 Krishnaswamy M, Subbuchettiar G, Ravi TK, Panchaksharam S. Biosurfactants properties, commercial production and application. *Curr Sci*. 2008;94:736–747.
69 Vecino X, Cruz JM, Moldes AB, Rodrigues LR. Biosurfactants in cosmetic formulations: trends and challenges. *Crit Rev Biotechnol*. 2017;37(7):911–923.
70 Lourith N, Kanlayavattanakul M. Natural surfactants used in cosmetics: glycolipids. *Int J Cosmet Sci* 2009;31:255–261.

71. Piljac T, Piljac G Use of rhamnolipids in wound healing, treating burn shock, atherosclerosis, organ transplants, depression, schizophrenia and cosmetics. Patent WO1999043334 1999;A1.
72. Perfumo A, Rancich I, Banat IM. Possibility and challenges for biosurfactant uses in petroleum industry. *Adv Exp Med Biol*. 2010;672:135–145.
73. Tartaj P, Morales MDPM, Veintemillas-Verdaguer S, et al. Progress in the preparation of magnetic nanoparticles for applications in biomedicine. *J Phys D: Appl Phys*. 2003;42(22):182–197.
74. Anandan S, Grieser F, Ashokkumar M. Sonochemical synthesis of Au–Ag core-shell bimetallic nanoparticles. *J Phys Chem C*. 2008;112:15102–15105.
75. Patel K, Kapoor S, Dave DP, et al. Synthesis of Pt, Pd, Pt/Ag and Pd/Ag nanoparticles by microwave – polyol method. *J Chem Sci*. 2005;117(4):311–316.
76. Chen Z, Gao L. A facile and novel way for the synthesis of nearly monodisperse silver nanoparticles. *Mater Res Bull*. 2007;42(9):1657–1661.
77. Mallik T, Banerjee D. Biosurfactants: the potential green surfactants in the 21st century. *J Adv Sci Res*. 2022;13(01):97–106.
78. Sayyed RZ. Microbial surfactants. *Microbi Surfact*. 2022;3(1).
79. Naganthran A, Verasoundarapandian G, Khalid FE, Masarudin MJ, Zulkharnain A, Nawawi NM, Karim M, Abdullah CAC, Ahmad SA. Synthesis, characterization and biomedical application of silver nanoparticles. *Materials* 2022;15(2):427.
80. Wang Z, Chen J, Yang P, et al. Biomimetic synthesis of gold nanoparticles and their aggregates using a polypeptide sequence. *Appl Organometal Chem*. 2007;21:645–651.
81. Reddy AS, Chen CY, Chen CC, et al. Synthesis of gold nanoparticles via an environmentally benign route using a biosurfactant. *J Nanosci Nanotechnol*. 2009;9(11):6693–6699.
82. Tomar RS, Banerjee S, Kaushik S, et al. Microbial synthesis of gold nanoparticles by biosurfactant producing Pseudomonas aeruginosa. *Int J Adv Life Sci*. 2015;8(4):520–527.
83. Nel A, Xia T, Mädler L, et al. Toxic potential of materials at the nanolevel. *Science* 2006;311(5761):622–627.
84. Ansari SA, Husain Q, Qayyum S, et al. Designing and surface modification of zinc oxide nanoparticles for biomedical applications. *Food Chem Toxicol*. 2011;49(9):2107–2115.
85. Bai HJ, Zhang ZM, Gong J. Biological synthesis of semiconductor zinc sulfide nanoparticles by immobilized Rhodobacter sphaeroides. *Biotechnol Lett*. 2006;28(14):1135–1139.
86. Singh BR, Dwivedi S, Al-Khedhairy AA, et al. Synthesis of stable cadmium sulfide nanoparticles using surfactin produced by Bacillus amyloliquefaciens strain KSU-109. *Colloids Surf B, Biointerfaces*. 2011;85(2):207–213.
87. Singh BN, Rawat AKS, Khan W, et al. Biosynthesis of stable antioxidant ZnO nanoparticles by Pseudomonas aeruginosa rhamnolipids. *PLoS One* 2014;9(9):106937.
88. Hazra C, Kundu D, Chaudhari A, Jana T. Biogenic synthesis, characterization, toxicity and photocatalysis of zinc sulfide nanoparticles using rhamnolipids from pseudomonas aeruginosa BS01 as capping and stabilizing agent. *J Chem Tech Biotech*. 2013;88(6):1039–1048.
89. Wang X, Zhuang J, Peng Q, et al. A general strategy for nanocrystal synthesis. *Nature* 2005;437(7055):121–124.

90 Shokuhfar A, Alibeigi S, Vaezi MR, et al. Synthesis of Fe_3O_4 nanoparticles prepared by various surfactants and studying their characterizations. *Defect Diffus Forum*. 2008; 273-276:22–27.

91 Jabbar KQ, Barzinjy AA, Hamad SM. Iron oxide nanoparticles: preparation methods, functions, adsorption and coagulation/flocculation in wastewater treatment. *Environ Nanotechnol Monit Manag*. 2022;17:100661.

92 Park J, An K, Hwang Y, et al. Ultra-large-scale syntheses of monodisperse nanocrystals. *Nat Mater*. 2004;3(12):891–895.

93 Baccile N, Noiville R, Stievano L, et al. Sophorolipids-functionalized iron oxide nanoparticles. *Phys Chem*. 2013;15(5):1606–1620.

94 Liao Z, Wang H, Wang X, et al. Biocompatible surfactin-stabilized superparamagnetic iron oxide nanoparticles as contrast agents for magnetic resonance imaging. *Colloids Surf A, Physicochem Eng Aspects*. 2010;370(1–3):1–5.

95 Ahghari MR, Soltaninejad V, Maleki A. Synthesis of nickel nanoparticles by a green and convenient method as a magnetic mirror with antibacterial activities. *Sci Rep*. 10;2020:12627.

96 Li D, Komarneni S. Microwave-assisted polyol process for synthesis of Ni nanoparticles. *J Am Ceram Soc*. 2006;89:1510–1517.

97 Roselina NN, Azizan A, Hyie KM, Jumahat A, Bakar MA. Effect of pH on formation of nickel nanostructures through chemical reduction method. *Proc Eng*. 2013;68:43–48.

98 Muntean A, Wagner M, Meyer J, Seipenbusch M. Generation of copper, nickel, and CuNi alloy nanoparticles by spark discharge. *J Nanoparticle Res*. 18;2016:229.

99 El-Khatib AM, Badawi MS, Roston GD, Moussa RM, Mohamed MM. Structural and magnetic properties of nickel nanoparticles prepared by arc discharge method using an ultrasonic nebulizer. *J Cluster Sci*. 29;2018:1321–1327.

100 Palanisamy P. Biosurfactant mediated synthesis of NiO nanorods. *Mater Lett*. 2008;62(4–5):743–746.

10

Interaction of Glycolipid Biosurfactants with Model Membranes and Proteins

Francisco J. Aranda, Antonio Ortiz, and José A. Teruel

Department of Biochemistry and Molecular Biology-A, Faculty of Veterinary, University of Murcia, 30100 Murcia, Spain

10.1 Introduction

The family of glycolipid biosurfactants contains some of the most interesting biosurfactants known so far, regarding their physicochemical, physiological and biological properties. These compounds are essentially formed by a fatty acid portion linked to a carbohydrate moiety. The nature of the fatty acid and the carbohydrate moieties is diverse, giving rise to a variety of structures. Among the various sugar portions found the most relevant are rhamnose, trehalose, sophorose, manosylerithritol, and cellobiose [1], and accordingly, rhamnolipids [2], trehalose lipids [3], and sophorolipids [4] (Figure 10.1) are the most remarkable glycolipid biosurfactants.

Rhamnolipids were first described in 1949 as a fraction obtained from *Pseudomonas aeruginosa* cultures [5]. The main species of rhamnolipids are monorhamnolipid (monoRL) and dirhamnolipid (diRL), formed by one or two rhamnoses linked to a hydroxy fatty acid. Rhamnolipids are perhaps the most studied glycolipid biosurfactants so far, because of their exceptional properties and variety of applications [2], from which their antimicrobial activities are the most remarkable [6].

Uchida and coworkers provided an early description of the accumulation of a trehalose lipid biosurfactant in cultures of a strain of *Rhodococcus erythropolis* grown on n-alkanes [7]. These biosurfactants are formed by a mycolic acid linked to the disaccharide trehalose. Their physicochemical and biological properties also have granted them a prominent place within glycolipid biosurfactants [3, 8].

In contrast to rhamnolipids and trehalose lipids, sophorolipid biosurfactants are mainly synthesized by yeasts [4], and have attracted the attention of researchers because of their excellent physicochemical properties and biological activities. The structure of sophorolipids comprises a sophorose moiety linked through a β-glycosidic bond to a long chain fatty acid. Two main advantages point sophorolipids as one of the most promising glycolipid biosurfactants: they can be produced in high yields, and the yeast strains are non-pathogenic, in contrast to most of the biosurfactant producing bacteria.

Biosurfactants and Sustainability: From Biorefineries Production to Versatile Applications, First Edition.
Edited by Paulo Ricardo Franco Marcelino, Silvio Silverio da Silva, and Antonio Ortiz Lopez.
© 2023 John Wiley & Sons Ltd. Published 2023 by John Wiley & Sons Ltd.

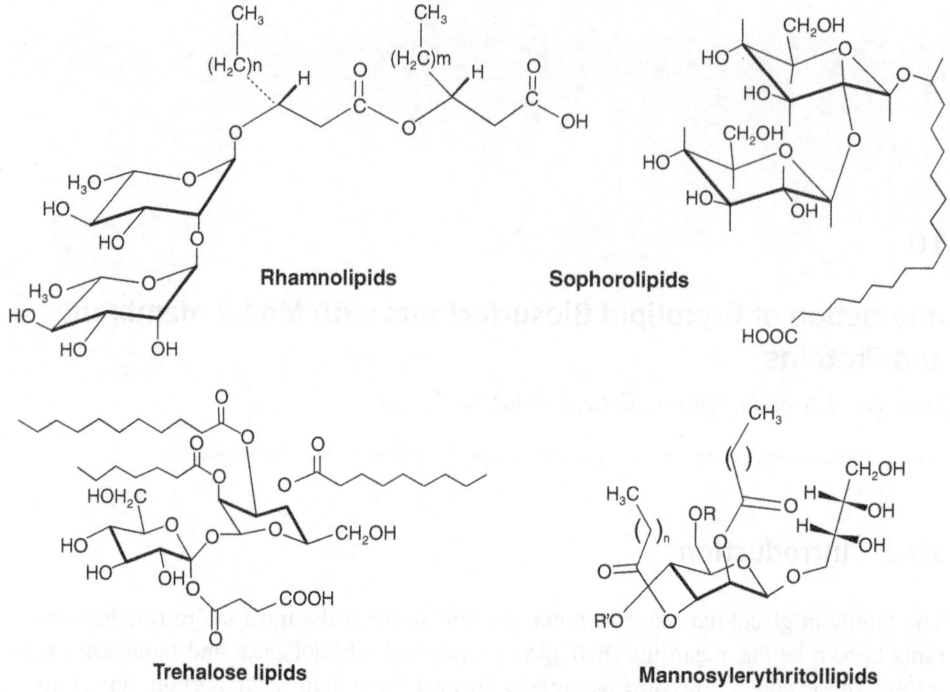

Figure 10.1 General structure of the main glycolipid biosurfactants.

10.2 Interaction of Glycolipid Biosurfactants with Model Membranes

The amphiphilic and surface-active properties of glycolipid biosurfactants allow them to interact with cellular membranes, and in this way, they may modify the membrane structure and therefore affect the function of the cell. There is an increasing body of evidence showing the potential biological functions of glycolipid biosurfactants, including antibacterial, antiviral, and anticancer activities [9, 10]. The molecular mechanisms underlying these effects are mostly incomplete, and for that reason the study of the interaction between glycolipid biosurfactants and membranes is necessary. Because of the complexity of biological membranes, many biomimetic model systems have been devised and utilized to examine biosurfactant–membrane interactions. It is relevant to emphasize that the simplification and consequently a full control of that kind of elaborate structure is crucial to understanding the interaction at the molecular level. Next, we will review the studies on the interaction between glycolipid biosurfactants and phospholipid model membranes composed by simple and mixed bilayers and monolayers, differing in phospholipid acyl chain length and headgroup, which have been carried out in order to get insights into the molecular mechanism of the biosurfactant–membrane interaction.

10.2.1 Rhamnolipids

The nature of the interaction between dirhamnolipids (diRL) and phospholipid membranes was initially investigated by Ortiz et al. using biomimetic membranes formed by different phosphatidylcholines [11]. Differential scanning calorimetry (DSC) was used to characterize the influence of a purified diRL on the thermotropic properties of saturated phosphatidylcholines bearing acyl chains with different (14, dimyristoylphosphatidylcholine, DMPC; 16, dipalmitoylphosphatidylcholine, DPPC; 18, distearoylphosphatidylcholine, DSPC). DSC is a straightforward and potent non-perturbing physical technique, which is well appropriated to observe and depict the thermotropic phase transition of membranes [12]. The thermotropic pretransition of the different phosphatidylcholines was greatly affected by the presence of a very low concentration of diRL, being already abolished at a diRL mole fraction of 0.01. Increasing concentrations of diRL progressively made the transition less cooperative as demonstrated by the increase in width of the main transition and caused a shift to lower temperatures with the appearance of a second endothermic component in the thermograms. The effect of diRL on the main phase transition was qualitatively similar for the different phosphatidylcholines, however it was larger in the case of the shorter homologue DMPC where the broadening of the transition and the separation between both endotherms were more evident. These effects were explained by the establishment of molecular interactions between the phospholipid acyl chains and the diRL molecule. This interaction was considered to be the consequence of the intercalation of the diRL molecules between the phospholipids, where the alignment of the hydrocarbon chains of diRL with the phospholipid acyl chains could disrupt the phospholipid packing, reduce the cooperativity of the transition, and shift the phase transition temperature to lower values.

In the same study, X-ray diffraction was used to obtain information about the structural properties and packing characteristics of the diRL-phosphatidylcholine system. X-Ray diffraction is an acknowledged exploratory technique which allows the examination of the overall structural properties of model membranes [13]. Samples containing diRL always gave rise to two or three reflections which related as 1:1/2:1/3 in the whole range of temperatures under study, confirming that the presence of this biosurfactant did not alter the lamellar structural organization of phosphatidylcholines. However, the interlamellar repeat distance was found to be between 5 and 13 Å (depending on phosphatidylcholine acyl chain and temperature) and was larger in the presence diRL than in the absence of the glycolipid. It was interesting to note that the presence or diRL broadened the reflections and lowered their intensities until only broad scattered bands were observed, indicating that the presence of diRL reduced the long-range order in the multilamellar system. The results also indicated that diRL was able to perturb the packing of the phosphatidylcholine acyl chains and that this perturbation did not depend on the acyl chain length. Partial phase diagrams showed good miscibility between diRL and phosphatidylcholines both in the gel and the liquid crystalline phases except for the appearance of fluid immiscibility in the case of the shorter homologue (DMPC).

In this early study, Ortiz et al. [11] also used Fourier-transform infrared spectroscopy (FTIR) to obtain data about the chemical structure and the membrane physical state of the diRL-phosphatidylcholine system. The effect of diRL on the phosphatidylcholine acyl chains was examined by monitoring the changes occurring in the CH_2 stretching vibration bands which occur between 2800 and 3000 cm^{-1}. The CH_2 stretching region of the infrared spectrum of phosphatidylcholines contains two major bands centred near 2850 and 2920 cm^{-1}, which arise from the symmetric and asymmetric methylene stretching vibrations, respectively [14]. The results suggested that the incorporation of diRL into DMPC bilayers resulted in an overall increase in hydrocarbon chain disorder in both the gel and the liquid crystalline states. This disordering effect was less marked when diRL was incorporated into DPPC and disappeared when it was incorporated into DSPC systems, indicating that phosphatidylcholines with shorter acyl chains are less able to accommodate the diRL molecules into the phospholipid palisade than do the ones with larger acyl chains, and therefore are more sensible to the presence of the glycolipid.

A more in depth study of the interaction of diRL with phosphatidylcholine membranes was carried out by Sánchez et al. [15] by using FTIR and fluorescence probe polarization. The use of ^{2}H-DPPC and ^{13}C=O-DPPC allowed the authors to simultaneously monitor the temperature-dependent changes occurring in the acyl chains and carbonyl groups of both diRL and DPPC, because the methylene and carbonyl absorption bands corresponding to the labelled phospholipid are shifted toward lower wavenumbers due to the isotope effect [16, 17]. Incorporation of increasing amounts of diRL into ^{2}H-DPPC membranes broadened the transition and shifted the transition temperature toward lower values, according to the effect on the CD_2 stretching vibration. Examination of the ^{13}C=O stretching band of ^{13}C-DPPC indicated that, both below and above the phase transition, diRL produced a shift of the band frequency toward higher values, indicating a strong dehydration of the phospholipid C=O groups, and therefore of the interfacial region of the membrane. Fluorescence polarization measurements of diphenylhexatriene (DPH) probes provided additional support to hypothesize on the location of diRL within the bilayer. Interestingly, the stretching bands of the acyl chains and the carbonyl group of diRL did not show any change upon interaction with ^{2}H-DPPC within the same temperature range. The authors concluded that the molecular interactions that diRL acyl chains establish with DPPC acyl chains would disrupt the phospholipid lipid packing resulting in the broadening of the transition and shifting to lower temperatures that was reported. The strong dehydration of the phospholipid carbonyl groups, and therefore of the interfacial region of the membrane was attributed to the ability of diRL to establish a large number of hydrogen bonds with the surrounding water and leaving fewer water molecules available to interact with the phospholipid.

Rhamnolipids can form supramolecular complexes with DPPC as was shown by Pashynska et al. using electrospray ionization mass spectrometry techniques [18]. The formation of the non-covalent complexes of rhamnolipids with membrane phospholipids was considered as a possible mechanism for the membrane action of these glycolipids.

Haba et al. studied the properties of a mixture of eight homologues of diRL and DPPC using X-ray diffraction [19], confirming the previous finding that the presence of diRL molecules maintains the DPPC bilayer in the fluid phase. It was concluded that it is the anionic group in the rhamnolipid molecule which has an influence on the elasticity of the DPPC bilayer, and that the bilayer thickness, that is, the non-aqueous portion of the lamellar phase, increases as the content of diRL increases.

Using fluorescence quenching, Mousa et al. studied the fluidity of DSPC liposomes in the presence of various concentrations of diRL [20]. Using pyrene as an external fluorescence probe and KI as a hydrophilic quencher, the authors found that the Stern–Volmer constant (K_{SV}) remarkably decreased in the presence of 10% diRL. The study was complemented using cetylpyridinium bromide as a hydrophobic quencher and curcumin as an additional external fluorescence probe. Considering that pyrene is positioned close to the headgroups of the bilayer [21] and that curcumin intercalates parallel to the lipid bilayers of the liposome [22], it was concluded that diRL increased the permeability and fluidity of DSPC liposomes. The study established that intercalation of diRL exceptionally increased partition of curcumin into the solid gel phase of DSPC liposomes, whereas this increase was moderate in the liquid crystalline phase. Using the profile of fluorescence intensity of curcumin vs. temperature to monitor the phase transition of the phospholipid, the authors found that the presence of diRL broadened and shifted the phase transition temperature to lower values; supporting the previous finding that diRL disrupts the close packing of the phospholipid membrane.

Phosphatidylethanolamines (PE) are the second major component of membrane phospholipids in animal cells, and in general, higher amounts of PEs are found in the inner membrane of Gram-negative bacteria. One feature of PE, which distinguishes it from phosphatidylcholines, is its tendency to undergo a lamellar-to-inverted hexagonal-H_{II} phase transition [23]. The ability of unsaturated PE species, like dielaidoylphosphatidylethanolamine (DEPE) to adopt such non-lamellar phases is favored by the relatively small size of the hydrophilic headgroup as compared to the larger volume of the acyl chains [24]. In this context, the interaction between diRL and DEPE liposomes has been studied by Sánchez et al. using a number of physical techniques such as differential scanning calorimetry, FTIR, small angle X-ray (SAX) diffraction, and dynamic light scattering [25]. DSC thermograms indicated that diRL incorporates into DEPE bilayers and interacts with the phospholipid, decreasing the cooperativity of the gel to liquid crystalline phase transition and giving rise to the formation of domains within the membrane with lower transition temperatures. DiRL displaced the bilayer to the inverted hexagonal H_{II} phase transition to higher values, indicating a clear stabilization of the bilayer organization. This stabilization was confirmed by SAX diffraction which showed that, at 70 °C, diRL-containing samples adopted a lamellar organization, as compared to pure DEPE which was organized in the hexagonal phase. The authors explained their findings using the dynamic shape theory [26] in which PE is described as a lipid with a dynamic cone shape, conferring negative curvature to bilayers, and thus having a great tendency to form non-lamellar phases. DiRL having a large polar headgroup and a smaller hydrophobic portion would be an inverted cone molecule which confers positive curvature to membranes, counteracting the shape of DEPE, and thus stabilizing the lamellar phase. This shape-structure concept also explained their quasielastic light scattering measurements which indicated that the presence of diRL counteracts the tendency of DEPE to form vesicular aggregates of large size, forming vesicles of smaller diameter which most probably have a lower lamellarity index.

In the same investigation, Sánchez et al. [25] used FTIR to examine the perturbations exerted by diRL on the different parts of the DEPE molecule. Incorporation of diRL into DEPE bilayers increased the population of *gauche* conformers both in the gel and the liquid crystalline phase, as indicated by the shifts toward higher values observed in the frequencies of the CH$_2$ stretching bands. Therefore, the interaction of diRL with DEPE caused an additional disordering of the phospholipid acyl chains even in the fluid phase. Interestingly, diRL also perturbed the bilayer aqueous interface as showed by the effects on the C=O stretching band. The carbonyl groups of diacylphospholipids in lipid vesicles may be in a hydrated and a dehydrated state, with their proportion depending on the physical state of the phospholipid bilayer [17, 27]. Thus, the carbonyl band in the spectrum of pure DEPE represents a summation of component bands, which has been attributed to dehydrated and hydrated C=O groups [14]. The presence of diRL increased bilayer hydration at temperatures below the gel to liquid crystalline phase transition, but reduced this parameter in the fluid bilayer. This was a surprising result that could be related to the gel phase immiscibility exhibited in the phase diagram presented by the authors. Thus, the lateral separation of a diRL/DEPE compound of fixed stoichiometry could facilitate water penetration into the polar part of the bilayer, increasing water hydrogen bonding of the C=O groups. However, in the fluid phase, where there was good miscibility of both lipids, the result is a decrease in water penetration due to a tighter packing of diRL and DEPE molecules, compatible with the observed formation of smaller vesicles with a narrower size distribution.

Among the negatively charged glycerophospholipids the most relevant is phosphatidylserine (PS). Although PS is present in small proportions, it presents a number of outstanding physicochemical and biological properties of crucial importance for cell structure and function [28]. PS location is mainly limited to the inner monolayer in normal cells; nevertheless, it presents important actions involving both the inside and the outside of cells. It can be mentioned that PS plays an important role in hemostasis [29], and that it is exposed to the extracellular environment in apoptotic cells [30].

Given the importance of phosphatidylserine as a component of biological membranes, Oliva et al. [31] recently took up the study of the phase behavior of mixtures of diRL with 1,2-dimyristoylphosphatidylserine (DMPS). As seen by DSC, the presence of increasing concentrations of diRL in the bilayer resulted in a progressive broadening of the phase transition, with the transition temperature shifted towards lower values as compared to the pure phospholipid, indicating a significant loss of the cooperativity of the phase transition. This decrease in the melting of the phospholipid is the result of the intercalation of the acyl chains of diRL between those of DMPS. This effect was similar to those described previously for phosphatidylcholine [11] and phosphatidylethanolamine [25]. The partial phase diagram of the DMPS/diRL mixture showed a gel-phase immiscibility with the presence of a eutectic point at 0.07 diRL molar fraction, indicating formation of a diRL/DMPS complex with this stoichiometry, which separated from the bulk of the phospholipid, and melted at lower temperature.

The steady-state fluorescence polarization of the hydrophobic probe DPH is frequently used to study the molecular order of the acyl chains of phospholipids [32] in such a way that depolarization of DPH fluorescence allows changes in membrane fluidity to be studied [33]. Increasing diRL concentrations gave rise to a slight decrease of DPH fluorescence polarization at temperatures below the phase transition and, although this small change may reflect some influence of diRL on DMPS acyl chains order in the gel phase, the addition of diRL showed no effect in the liquid crystalline state. This lack of alteration of the acyl chain order in the already disordered liquid crystalline phase was in agreement with what was already described for DPPC systems [11].

In order to get insight into the influence of diRL on DMPS acyl chains and polar headgroup regions, the authors used FTIR. The presence of diRL gave rise to a progressive upward shift of maximum frequency of the methylene absorption band, indicating an increase in the *gauche* rotamers and hence an increase of the disorder of the phospholipid acyl chains. The analysis of the asymmetric stretching of the carboxylate group of pure DMPS showed a band centred near 1620 cm^{-1} which was also sensitive to the phase transition, with a shift in frequency indicative of a carboxylate group of DMPS in a fully hydrated state in both phases [34]. Addition of diRL resulted in an important shift of the frequency toward higher values, both below and above the phase transition, which provided a solid support for the idea that diRL strongly affected the phospholipid headgroup region [31]. Oliva et al. also investigated the effect of diRL on the polar region of DMPS bilayers, examining the carbonyl ester bond stretching band. Addition of the biosurfactant clearly affected the shape of this band at all temperatures, but in a more marked way above the phase transition, which was also shifted to lower temperatures. The bands were subjected to band fitting using a Gaussian–Lorentzian function rendering two component bands at 1727 and 1742 cm^{-1} (corresponding to hydrated and dehydrated states respectively), and whose relative areas were determined. The presence of diRL in the membrane produced an increase of the hydrated component both in the gel and in the liquid crystalline phase. This effect was opposite to that described before for mixtures with phosphatidylcholine, where a dehydrating effect was observed [15], and presented more similarities to that described for diRL/DEPE systems, where hydration was increased in the gel phase, and slightly decreased in the fluid phase [25]. The authors concluded that the change in the location of the carboxylate group of DMPS induced by diRL, would change the conformation of the phospholipid moving the region of the carbonyl groups toward a more hydrated local environment, and thus facilitating hydrogen bonding with water molecules [31].

In their study, Oliva et al. [31] also described the interactions and behavior of diRL molecules in a DMPS bilayer using molecular dynamics simulations. The analysis of size distributions of diRL molecular clusters showed that diRL aggregation was higher in the gel phase. The mass distribution of diRL molecules along the normal to the bilayer (z-axis) in a fluid membrane located the molecule at the centre of the acyl chain palisade of the DMPS monolayer, between the DMPS carbonyl group and the centre of the bilayer, were it increased hydrogen bond formation between water and the carbonyl group of DMPS and the contents of the acyl chains *gauche* conformers.

The interaction between rhamnolipids bearing only one rhamnose moiety (monoRL) and simple model membranes has been also of interest for investigations. Abbasi et al. [35] studied the interaction of monoRL with phosphatidylcholine membranes. In general, upon increasing the glycolipid concentration, the main gel to liquid crystalline phase transition was progressively broadened, and the onset temperature shifted to lower temperatures, however DMPC thermograms were more complicated than for the other longer homologues, with broad features at high temperatures, due to the shorter chain length. Partial phase diagrams for DPPC and DSPC indicated near-ideal behavior. However, the DMPC diagram indicated fluid phase immiscibility which led to domain formation. FTIR showed that interaction of monoRL with the phospholipid acyl chains did not result in a large additional disordering of the acyl chain region of the fluid bilayer. Analysis of the C=O stretching band of DPPC indicated a dehydration effect of monoRL on the interfacial region of phosphatidylcholine bilayers, which was less important that that previously reported for diRL [15].

The study of the interaction between monoRL and DEPE systems was carried out by Abbasi et al. [36]. As seen by DSC, incorporation of increasing proportions of monoRL into DEPE shifted the onset temperature of the gel to fluid phase transition to lower values, whereas the completion temperature was less affected. At low monoRL concentrations, the transition became very asymmetric, with a wide lower melting portion, and a sharper higher temperature transition. The effect of monoRL on the lamellar to H_{II} phase transition was similar, shifting the onset and completion temperatures toward lower values. Increasing the concentration of monoRL into DEPE resulted in a progressive decrease of the enthalpy change of both the gel to liquid crystalline and the lamellar to hexagonal H_{II} phase transitions. The examination of the structural organization of DEPE was analysed by X-ray diffraction showing that pure DEPE in the gel phase displayed a first-order reflection with an interlamellar repeat distance of 64.6 Å, which decreased to 53.7 Å in the fluid state, due to a decrease in the effective acyl chain length, or in the thickness of the hydration layer between bilayers. Only in the gel phase, incorporation of monoRL into DEPE gave rise to the appearance of a second reflection at 59 Å which became more intense upon increasing the concentration, which would correspond to the coexistence of both gel and fluid phases. Pure DEPE at 70 °C showed four reflections, which related as 1: $1/\sqrt{3}$: $1/\sqrt{4}$: $1/\sqrt{7}$, indicating a pure hexagonal H_{II} phase [37]. At 60 °C, pure DEPE showed a single reflection at 52.2 Å (lamellar liquid crystalline), but upon addition of monoRL the characteristic reflections of the hexagonal H_{II} phase were also present, indicating coexistence of both phases, i.e. monoRL was inducing the hexagonal H_{II} phase at temperatures at which pure DEPE was lamellar. This non-bilayer stabilizing effect was attributed to the cone shaped molecule of monoRL which, by adding to the cone shape of DEPE, conferred negative curvature, and acted as a lamellar destabilizer. The opposite effect described for diRL [25] being compatible with the pronounced inverted cone shape of diRL, with a bulky polar disaccharide moiety, as compared to the single rhamnose of monoRL. The phase diagram for the phospholipid component indicated a near-ideal behavior, with better miscibility of monoRL into DEPE in the fluid phase than in the gel phase.

FTIR allowed Abbasi et al. [36] to investigate the molecular interactions of monoRL with the various functional groups of the DEPE molecule, characteristics of the polar headgroup or the acyl chain regions. The observation of the CH_2 symmetric stretching mode at around 2850 cm^{-1}, pure DEPE showed two increases in wavenumber, the first one corresponding to the gel to liquid crystalline acyl chain melting transition, and the second to the additional conformational disorder introduced by the lamellar to hexagonal H_{II} transition. The increase in the frequency of this band induced by monoRL, both below and above the phase transition temperature, was associated with an overall increase in the conformational disorder of the hydrocarbon chains of DEPE. When monitoring the contours of the C=O stretching bands of DEPE, the authors found that they were much wider and only resolvable into at least three component bands centred around 1742, 1728, and 1714 cm^{-1}. The additional band at 1714 cm^{-1} is indicative of another population of hydrogen-bonded C=O groups, which were not present in hydrated phosphatidylcholine bilayers [38]. The overall effect of monoRL was a shift of the maximum frequency toward higher wavenumbers indicating a strong dehydration of the interface both in the gel and the fluid phase, as confirmed by fitting of the component bands.

The thermodynamics of the interaction of diRL with phospholipid membranes was studied by Aranda et al. [39], using isothermal titration calorimetry (ITC) experiments. ITC is generally applied for measuring the thermodynamic behavior of the binding interactions between molecules and ligands, heat either generated or absorbed can be monitored when two substances associate or interact [40]. In this case, a lipid dispersion was titrated to a biosurfactant solution at a concentration below its critical micelle concentration (cmc) and the partition constant, the membrane partition enthalpy, and the heat of dilution were obtained. The partition constants and the heats of dilution of diRL into palmitoyl-oleoyl-phosphatidylcholine (POPC) vesicles at pH 7.4 and 4.0 were very similar, which suggested that the membrane binding mechanism of negatively charged and neutral diRL were not different. Incorporation of cholesterol into POPC resulted in a drastic reduction of the partition constant, and this was interpreted as a consequence of the general stabilization of the membrane caused by cholesterol, which resulted in a tighter lipid packing and made the binding of the surfactant more difficult. Notably interesting was the finding that the presence of lysophosphatidylcholine and palmitoyl-oleoyl-phosphatidylethanolamine (POPE) into the POPC vesicles displayed opposite effects: partition of diRL into POPC membranes that contained POPE was more favourable, whereas the addition of lysophosphatidylcholine resulted in a large reduction of the partition constant. These data were analysed in terms of the dynamic cone shape of PE conferring negative curvature to bilayers, and the inverted-cone shape of lysophosphatidylcholine conferring positive curvature. In this way, the authors concluded that diRL, having a large polar headgroup and a smaller hydrophobic portion, has an inverted-cone shape by itself, which confers positive curvature to membranes, and it was in same way complementary to that of POPE, which facilitates diRL membrane insertion when present in a bilayer, and is similar to the shape of lysophosphatidylcholine, which interfered with diRL membrane insertion.

The study of the kinetic and structural aspects of phospholipid membrane permeabilization induced by diRL in model membrane systems was accomplished by Sánchez et al. [41], by monitoring the release of carboxyfluorescein (CF) entrapped inside phospholipid unilamellar vesicles of different composition. DiRL was able to induce leakage of CF entrapped into POPC unilamellar vesicles, and the observed results indicated that diRL-induced membrane permeabilization occurred at concentrations below the cmc, without a concomitant decrease in turbidity, i.e. without membrane solubilization. In comparison to pure POPC vesicles, incorporation of lysophosphatidylcholine gave rise to an increase of both the initial rate as well as the extent of CF leakage at a given diRL concentration. On the other hand, other lipids, namely POPE and cholesterol, reduced the rate and extent of diRL-induced CF leakage. The authors explained these results according to the shape structure concept, in which the presence of sublytic amounts of lysophosphatidylcholine facilitates diRL-induced permeabilization, due to the summation of their effects on positive membrane curvature, whereas the negative curvature promotion of POPE would also explain the protective effect against membrane permeabilization caused by this phospholipid.

The ability of dIRL to stabilize bilayer organization in PE systems has been explored in order to increase the stability and performance of PE based pH-sensitive liposomes.

Sánchez et al. [42] studied the potential of diRL for liposome formation and stabilization using dynamic light scattering analysis of mixtures with DOPE, on the basis that, in general, stable lipid mixtures produce multilamellar vesicles (MLV) with an average size in the range of nanometers, whereas lipids in the H_{II} phase form large aggregates with very heterogeneous size distributions. The authors found that incorporation of diRL into DOPE resulted in the formation of structures characteristic of that of MLV, while acidification of the system to pH 5.0 resulted in the formation of highly heterogeneous large size structures which led to disintegration of the vesicular structure and formation of stacked bilayers. DOPE/diRL mixtures were capable of forming non-leaky liposomes which were stable upon storage for long time periods. In this study, an assay consisting of monitoring the release of calcein entrapped into calcein-containing large unilamellar vesicles (LUV) at various pH values was used. The authors established that diRL/DOPE vesicles were essentially stable up to a pH of 6.5, but as the pH was lowered below this value the rate of leakage increased rapidly. This break point coincided rather well with the pKa value of 5.6 for the carboxyl group of diRL previously determined [43], and thus showed that headgroup protonation of the amphiphilic diRL caused by a pH decrease resulted in liposomal destabilization and leakage. The authors demonstrated that membrane fusion preceded leakage, indicating that protonation of the carboxylic group of diRL at pH 5.6 resulted in vesicle interaction leading to lipid mixing. An interesting finding of this study was that these DOPE/diRL pH-sensitive liposomes were incorporated into cultured cells through the endocytic pathway, delivering its contents into the cytoplasm, which established a potential use of these liposomes for the delivery of foreign substances into living cells [42].

Nasir et al. [44] synthetized hybrid glycolipids with a rhamnose headgroup and a single fatty acid chain differing only at the level of the terminal group (–CH_3, alk-RL; –COOH, ac-RL), and studied their interaction with biomimetic membranes. The authors used systems made of pure palmitoyl-linoleoylphosphatidylcholine (PLPC), pure stigmasterol or a PLPC/stigmasterol mixture aiming to model plant plasma membrane lipids. Langmuir monolayer experiments and computational simulation showed that alk-RL was more favorably inserted into lipid membranes and formed more stable molecular assemblies with phospholipids and sterols than ac-RL. When the interaction with membrane lipids was assessed by FTIR, the authors found that the synthetic rhamnolipids did not affect the C–H stretching bands from different vibrational modes of the hydrocarbon chains but they did affect the phosphate band. Interestingly, the opposite was found when stigmasterol was present in the system, where a clear fluidizing effect on the alkyl chains was evidenced. In both cases a significant shift to higher wavenumbers was detected when the carbonyl ester stretching band was examined, the latter being in agreement with the previously reported carbonyl dehydration effect of rhamnolipids [15] on phosphatidylcholine systems.

A study on the interaction of natural rhamnolipids, both diRL and monoRL, with plant and fungi membrane models taking into account the presence of unsaturation, charges, and different lipid families, was carried out by Monnier et al. [45] by using FTIR and deuterium nuclear magnetic resonance (^{2}H-NMR) techniques. PLPC was used as a simplified model of the plant membrane and the FTIR analysis of the terminal methyl group and the methylene groups of the acyl chains showed that these bands were unaffected by the presence of rhamnolipids. However, the maximum absorbance of the carbonyl and phosphate bands were shifted to higher wavenumbers by both monoRL or diRL, which indicated a less hydrogen-bonded state of these groups. These results were consistent with previous work on saturated PC [11, 15] and suggested that rhamnolipids exert the same behavior with saturated and unsaturated lipids of similar chain length. The authors performed a molecular dynamics simulation of the behavior of a mixture monoRL/diRL with POPC/PLPC system, and determined that the rhamnolipids were located near the lipid phosphate group adjacent to the phospholipid glycerol backbone. Solid-state NMR spectroscopy is the method of choice to study the modifications induced by an external molecule on membrane model dynamics and the use of deuterium nucleus informs on the hydrophobic core dynamic using chain deuterated lipids [46]. In this work, the system PLPC/POPC-^{2}H$_{31}$ was studied in the absence and the presence of rhamnolipids, and in both cases a symmetric spectrum characteristic of lamellar fluid phase [47] was obtained with no significant differences in the first-order parameter, which suggested that RLs did not strongly disturb the dynamic of this phospholipid membrane model. In order to mimic more closely the plant plasma membrane, the authors increased the complexity of the model membrane by using a system composed of soy PC/POPE-^{2}H$_{31}$/soy phosphatidylinositol (PI)/soy phosphatidylglycerol (PG)/β-sitosterol/soy glucosylceramide. Even in this complex system, no significant differences between spectra were found, which indicated that rhamnolipids did not disrupt the lipid dynamics. However, when the author tried to mimic the fungi membrane by using a system formed by POPC-^{2}H$_{31}$/palmitoyl-oleoylphosphatidylglycerol (POPG)/ergosterol, a clear decrease of the spectral width was found which corresponded to an increase in the dynamic of the membrane hydrophobic core. This fluidity increase appeared to be correlated to the presence of the specific structure of ergosterol and highlighted the impact of sterol nature on the membrane destabilization induced by rhamnolipids.

To obtain further information on the interaction of rhamnolipids with lipid membranes, Herzog et al. [48] studied the interaction between a purified monoRL and heterogeneous anionic model membranes systems, employing different spectroscopy and microscopy techniques. In order to visualize lipid domains, the authors used the fluorescent probe Laurdan [49] which emits at a specific wavelength depending on the lipid phase [50]. Thus, Laurdan blues in ordered lipid phases and greens in disordered phases, with emission maxima at 440 and 490 nm, respectively. Moreover, this probe is distributed equally between ordered and disordered phases and gives access to a mathematical parameter used to quantify membrane global order, GP [51]. The presence of monoRL in DPPC systems produced a decrease of the gel to fluid phase transition temperature and a broadening of the transition, which indicated a disordering effect of the gel-phase upon incorporation of the biosurfactant, in agreement with the previous work by Abbasi et al. [35]. The results obtained from Laurdan fluorescence in the DOPE/DPPC system showed no significant differences in the presence of monoRL, but when the system was composed by a heterogeneous five-component anionic raft membrane model (DPPC:DPPG:DOPC:DOPG:cholesterol), the liquid ordered/liquid disordered two-phase transition region was slightly broadened and shifted to higher temperatures, indicating significant changes of the lateral organization of the membrane and an increase of the overall lipid order parameter in the fluid phase.

Herzog et al. [48] also carried out atomic force microscopy (AFM) measurements to study the interaction of monoRL and the anionic five-component raft mixture model. The addition of monoRL to the system produced an immediate doubling of the bilayer thickness difference between the two liquid phases and a reversal of the normal phase proportion, which suggested about a 1 nm thinning of the bilayer by partitioning of the monoRL in the fluid phase. The latter being in accordance with the increase of the lamellar d-spacing of a DPPC-bilayer induced by monoRL using X-ray diffraction reported by Abbasi et al. [35]. The authors employed giant unilamellar vesicles (GUV) of the five-component anionic raft system labeled with different fluorescence probes partitioning in distinctive phases to visualize morphological changes using fluorescence microscopy techniques. Their results demonstrated that the incorporated monoRL not only changed the phase lateral organization of phase-separated heterogeneous membranes, but also that monoRL partitioned into all-fluid one-component lipid bilayers, and caused leakage and changes in size and morphology of the lipid vesicles over time.

Following the latter approach, Herzog et al. [52] extended their study to examine the interaction of purified monoRL, diRL and the precursor of the rhamnolipids, the 3-(3-hydroxydecanoyloxy) decanoic acid with heterogeneous anionic model membrane systems. Using AFM, confocal fluorescence microscopy, DSC, and Laurdan fluorescence spectroscopy, the authors found that the partitioning of the three compounds into phospholipid bilayers changed the phase behavior, fluidity, lateral lipid organization, and morphology of the phospholipid membranes dramatically depending on the headgroup structure of the rhamnolipid, and affecting its packing and hydrogen bonding capacity. The incorporation into a GUV made of a heterogeneous anionic raft membrane system revealed budding of domains and fission of daughter vesicles and small aggregates for all three rhamnolipids, with major destabilization of the lipid vesicles upon insertion of monoRL, and also formation of huge GUV upon the incorporation of diRL.

Recently, Come et al. [53] used analysis of optical microscopy data from GUV dispersed in aqueous solutions containing an available rhamnolipid mixture. GUV was made up of a single lipid POPC and a ternary system containing DOPC/sphingomyelin/ cholesterol, which mimicked lipid raft platforms. Their results demonstrated that rhamnolipids have a low partition into the lipid bilayer with respect to the total molecules in solution. The authors assumed that rhamnolipids insert in the outer leaflet with low propensity to flip-flop. In the case of POPC GUV, the insertion of rhamnolipid molecules in the outer leaflet impaired changes in spontaneous membrane curvature with incubation time. When the model membranes contained phase coexistence, the interaction was higher for the liquid disordered phase, and that could alter the membrane spontaneous curvature which, coupled to the change in the line tension associated to the domains boundary, conducted to liquid ordered domain protrusion.

10.2.2 Trehalose *Lipids*

One of the most reported trehalose lipid (TL) is trehalose 6,6'-dimycolate (TDM), which is an a-branched chain mycolic acid esterified to the C_6 position of each glucose. This TL is the basic component of the cell wall glycolipids in *Mycobacteria* and *Corynebacteria*, commonly named cord factor, and constitutes an important factor in the virulence of these bacteria [54]. TDM has been extensively studied from a medical point of view due to the fact that it plays a central role in pathogenesis during infection and also showed a number of different biological activities [54].

The interaction between TDM and model membranes was originally studied by Durand et al. [55] who used monolayers at the air water interface to examine the interaction between TDM (from synthetic and natural origins) and phosphatidylcholines (DPPC and egg yolk lecithin). TDM was found to interact strongly with phosphatidylcholines resulting in a decrease of the apparent molecular area of the phospholipid in both the liquid expanded and gel phases. This film condensation was postulated to arise from changes in the hydration of the phospholipid polar head at the interface, originating from some hydrogen linkage between the phosphate and hydroxyl functions of the two lipids, detrimental to the phosphate–water hydrogen bonding required for polar head hydration.

Imasato et al. [56] studied the effect of TDM on water permeability and electrical capacitance of lecithin bilayer membranes, using black lipid membranes prepared with Mueller–Rudin brush techniques [57]. It was found that TDM induced a decrease of water osmotic permeability suggesting that this glycolipid led to an increase in the degree of packing of the constituent lipid molecules. A condensing effect of TDM was also apparent from a decrease in membrane electrical capacitance measurement, which is suggested to reflect an increase in membrane thickness.

Using the fluorescence assay of the ANTS-DPX [58], Spargo et al. [59] showed that the presence of TDM was extraordinarily effective at inhibiting calcium induced fusion between large unilamellar vesicles composed of phosphatidic acid and PE, suggesting that TDM could affect fusion either by increasing the hydration force which is known to be an important primary barrier to fusion, or acting as a steric barrier to fusion. Later, the same authors studied the interaction of TDM with other phospholipids [60] and found that TDM increased molecular area when measured by isothermal compression of a monolayer film, and increased the overall hydration of bilayers of DPPC. Fluorescence anisotropy and FTIR techniques showed that the presence of calcium increased molecular area and headgroup hydration in phosphatidylserine liposomes

The surface properties of TDM and its interactions with phosphatidylinositol were investigated by Almog and Mannella [61] using the monolayer technique. The observed change in surface properties suggested a molecular rearrangement of long and short chains of the TDM molecules, which in the presence of phosphatidylinositol increases the lateral packing density and rigidification at concentrations close to the relative composition of TDM in the mycobacterial cell wall.

Harland et al. [62] devised and experimental model that mimicked the structure of mycobacterial envelopes in which an immobile hydrophobic layer supported a TDM-rich, two-dimensionally fluid leaflet. It was found that TDM monolayers, in contrast to phospholipid membranes, could be dehydrated and rehydrated without loss of integrity, this protection from dehydration extended to TDM-phospholipid mixtures. This observation was the first reported instance of dehydration resistance provided by a membrane glycolipid, and the desiccation protection to membranes was later confirmed using a synthetic trehalose glycolipid [63].

TDM, when mixed and hydrated with cardiolipin, produced liposomes. Rath et al. [64] carried out ^{2}H-NMR experiments that revealed that the acyl chain composition and degree of unsaturation of both lipids were adapted to produce fluid bilayers at the bacterial growth temperature, and that TDM moderately influenced the cardiolipin membrane fluidity in the mixture.

The other group of important TL is composed of the different structures that have been elucidated particularly in *Rhodococcus* genus. These TL have gained increased interest for their potential applications in a number of fields due to their ability to lower interfacial tension and increase pseudosolubility of hydrophobic compounds [54].

In order to gain insight into the localization of TL into the phospholipid bilayer and its effect on membrane structure, the interaction of TL with model membranes was studied by Aranda et al. [65] using phosphatidylcholines of different chain length. DSC measurements showed that the presence of increasing TL concentrations abolished the pretransition and the main gel to liquid crystalline phase transition was broadened and shifted to lower temperatures. Interestingly, a second peak above the main transition becomes apparent, which suggested that mixed phases were present. The constructed partial phase diagrams revealed that fluid phase immiscibility was taking place in DMPC and DPPC membranes. In order to visualize these TL domains, the authors examined the topology of supported bilayers prepared with POPC by means of AFM, and found that two different domains with irregular size and shape were evident with a height difference of 1 nm, indicating that the TL tended to form domains within the POPC matrix. Using FTIR, it was found that incorporation of TL into phosphatidylcholine membranes produced a small shift of the methylene antisymmetric stretching band toward higher wavenumbers, indicating a weak increase in fluidity, and an increase of the proportion of the dehydrated component in the carbonyl stretching band of the phospholipids. SAX diffraction measurements showed that in the samples containing TL the interlamellar repeat distance was larger than in those of pure phospholipids.

The interaction between TL and PE membranes of different chain length and saturation was studied by Ortiz et al. [66]. Using differential scanning calorimetry, SAX and WAX diffraction and FTIR, the authors found that TL affected the gel to liquid crystalline phase transition of PEs, broadening and shifting the transition to lower temperatures. TL did not modify the macroscopic bilayer organization of these saturated phospholipids and presented good miscibility both in the gel and the liquid crystalline phases. Infrared experiments evidenced an increase of the hydrocarbon chain conformational disorder and an important dehydrating effect of the interfacial region of the saturated PEs. Interestingly, TL when incorporated into DEPE systems greatly promoted the formation of the inverted hexagonal H_{II} phase.

The same authors continued the study of the interaction between TL and model membranes and analyzed the effect of TL on bilayers formed by an anionic phospholipid. DSC, X-ray diffraction and FTIR techniques were used to gain insight into the interaction between TL and DMPS [67]. DSC and X-ray diffraction showed that TL broadened and shifted the phospholipid gel to liquid crystalline phase transition to lower temperatures, and did not modify the macroscopic bilayer organization and presented good miscibility both in the gel and the liquid crystalline phases. Infrared experiments showed that TL increased the fluidity of the phosphatidylserine acyl chains, changed the local environment of the polar headgroup, and decreased the hydration of the interfacial region of the bilayer. TL was also able to affect the thermotropic transition of DMPS in the presence of calcium.

Combining different biophysical techniques, Ortiz et al. examined the effect of TL on 1,2-dimyristoylphosphatidylglycerol (DMPG) membranes [68]. DSC data supported that TL was able to incorporate into DMPG membranes and to intercalate between the phospholipid molecules, where it could reduce the cooperativity and lower the transition temperature of the gel to liquid crystalline phase transition. The partial phase diagram showed good miscibility between trehalose lipid and DMPG, both in the gel and liquid crystalline phases. X-ray diffraction measurements indicated that TL did not affect the macroscopic bilayer organization of DMPG, but the presence of the biosurfactant produced a small decrease of the bilayer thickness. Infrared experiments revealed that the biosurfactant increased the fluidity of the phospholipid acyl chains and decreased the hydration of the interfacial region of the membrane. Finally, fluorescence polarization of membrane probes provided evidence that trehalose lipid disordered the DMPG membrane in the gel phase while producing a small ordering effect in the liquid crystalline phase.

DMPG has received particular attention because of its unusual phase properties, it appears to form single bilayers when dissolved in water at low lipid and low salt concentration [69], exhibiting a very unusual thermal profile, with a broad transition with a width greater than 15 °C. The interaction between TL and DMPG was extended to include the condition of a low ionic strength [70]. The authors found that there were extensive interactions between TL and DMPG involving the perturbation of the thermotropic intermediate phase of the phospholipid, the destabilization and shifting of the DMPG gel to liquid crystalline phase transition to lower temperatures, the perturbation of the sample transparency, and the modification of the order of the phospholipid palisade in the gel phase. They also reported an increase of fluidity of the phosphatidylglycerol acyl chains and dehydration of the interfacial region of the bilayer. These changes would increase the monolayer negative spontaneous curvature of the phospholipid explaining the destabilizing effect on the intermediate state exerted by this biosurfactant.

The interactions of TL with phospholipid vesicles, leading to membrane permeabilization, were studied by Zaragoza et al. [71] by means of calorimetric and fluorescence and absorption spectroscopic techniques in search for a molecular model. The cmc of TL was determined, by surface tension measurements, to be 300 μM. Binding of trehalose lipid to POPC membranes was studied by means of ITC. The partition constant, in conjunction with the cmc, indicates that TL behaved as a weak detergent, which preferred membrane incorporation over micellization. Addition of TL to POPC large unilamellar vesicles resulted in a size-selective leakage of entrapped solutes to the external medium. The experimental evidence supported the requirement of a stage of flip-flop prior to membrane permeabilization, and the rate of flip-flop was measured using fluorescent probes assays. The lipid composition of the target membrane was found to modulate the leakage process to a great extent. It was proposed that TL incorporated into phosphatidylcholine membranes and segregated within lateral domains which might constitute membrane defects or "pores", through which the leakage of small solutes could take place.

10.2.3 Other Glycolipids

There has been great interest in mannosylerythritol lipids (MEL) due to their pharmaceutical applications and versatile biochemical functions [72]. Imura et al. [73] using trapping efficiency for calcein and turbidity measurements together with confocal laser scanning and freeze fracture microscopies, were the first to evidence the formation of thermodynamically stable vesicles from MEL-A and DLPC mixtures. The authors suggested that the mechanism of formation of such vesicles underlay in the asymmetric distribution of MEL-A and DLPC in the two vesicle monolayers due to their different spontaneous curvature.

Madihalli et al. [74], in their investigation on the physicochemical properties of MEL-A, used DSC measurements to study the interaction between MEL-A and DPPC and found that upon addition of MEL-A there was a complete disappearance of the pretransition and a broadening in the main transition region.

Recently, Fan et al. [75] formulated stable vesicles by assembling egg yolk lecithin and MEL-A, evaluating their structural characterization, stability and encapsulation properties, and showed that the predominant average diameters varied ranging from 200 nm to 700 nm with MEL-A/PC ratio changed. The authors obtained binding information of MEL-A to PC, by using ITC measurements, and concluded that the formation of the vesicles was likely driven by the asymmetric distribution of MEL-A and PC which caused the lipid curvature changes associated with the partitioning of MEL-A into the layer of PC rather than by the strong binding interaction between both molecules.

The interest in the use of sophorolipids (SL) in commercial products have produced and increase in the number of publications in the area, indicating that SL have a very vast application in research and industries [76]. In order to get insight into the mechanisms underlying their membrane interactions, Singh et al. [77] used a single liposome assay to observe directly and quantify the kinetics of interaction of SL micelles with model membrane systems. The authors employed quantitative single particle microscopy [78], and arrays of surface tethered liposomes as model cell membranes to observe directly several successive docking events of SL assemblies on individual nanoscale liposomes. Their results revealed several repetitive docking events on individual liposomes and quantified how pH and membrane charges affected the docking of SL micelles on model membranes, providing direct evidence that acidic pH (6.5) and charges (5%), like the one found in cancerous cells, is a dominant feature underlying their interaction with membranes.

Marcelino et al. [79] carried out a biophysical study to unveil the molecular details of the interaction of an acidic SL with a model phospholipid membrane made of DPPC. Using DSC it was found that SL altered the phase behavior of DPPC at low molar fractions, producing fluid phase immiscibility with the result of formation of biosurfactant-enriched domains within the phospholipid bilayer. FTIR showed that SL interacted with DPPC increasing ordering of the phospholipid acyl chain palisade and hydration of the lipid–water interface. X-ray diffraction experiments showed that SL did not modify bilayer thickness in the biologically relevant fluid phase. SL was found to induce contents leakage in POPC unilamellar liposomes, at sublytic concentrations below the cmc. The authors concluded that this SL-induced membrane permeabilization at concentrations below the onset for membrane solubilization could be the result of the formation of laterally segregated domains, which might contribute to providing a molecular basis for the reported antimicrobial actions of SL.

10.3 Interaction of Glycolipid Biosurfactants with Proteins

Ligand–protein interactions are responsible for a majority of the biologically relevant cellular processes. Within this context, those studies aimed to unveil the molecular details of the interaction of biosurfactants and proteins are of relevance, given the wide application of surfactant/protein systems both in laboratory and industrial processes, including food, cosmetic, and pharmaceutical. However, as it is shown below, in the particular case of glycolipid biosurfactants the number of these studies is scarce. We will briefly summarize the most relevant results shown so far.

10.3.1 Rhamnolipids

Rhamnolipids probably constitute the most relevant group of glycolipid biosurfactants known so far, and accordingly the number of papers devoted to study their interactions with proteins is greater. Sánchez et al. studied the interactions of a *Pseudomonas aeruginosa* diRL with model bovine serum albumin [80]. It was shown that up to two molecules of diRL bound to albumin with high affinity. This interaction shifted the unfolding temperature toward higher values, i.e. increased protein stability, but the secondary structure of the protein was essentially unaffected. These results thus showed that diRL could be useful tools for protein studies, and in particular for membrane protein purification procedures.

The interaction of a mixture of monoRL and diRLs with α-lactalbumin and myoglobin was studied through the application of spectroscopic and calorimetric techniques [81] as well as X-ray scattering [82]. It was found that the biosurfactant mixture denatured both proteins, either below or above the cmc [81], and formed smaller complexes than with SDS [82]. However, these same authors showed that that the outer membrane protein A can be folded and stabilized by anionic biosurfactant rhamnolipid at concentrations above its cmc [83]. Given the importance of surfactants in relation to the applications of many enzymes in various industries, Madsen and co-workers studied the interaction of rhamnolipids with industrial enzymes including cellulase, phospholipase, and α-amylase [84]. It was found the rhamnolipids did not affect the tertiary or secondary structure of these enzymes, did not strongly affect thermal stability and bound at low stoichiometry, very similar results as those described before for the interaction with serum albumin [80]. These results indicated that rhamnolipid biosurfactants did alter the structure of these industrially relevant enzymes, making it a feasible alternative to synthetic surfactants for industrial applications.

In another line of evidence it has been shown that rhamnolipids produced by pathogenic *Pseudomonas aeruginosa* may weaken host defenses by facilitating degradation of key host lysozyme through its binding to the enzyme [85]. This interaction rhamnolipid-lysozyme does not denature the enzyme by itself, but exposes sites in the protein that can then suffer proteolytic attack.

10.3.2 Trehalose Lipids

Few works have addressed the interaction of bacterial trehalose lipids with proteins. It has been shown that addition of *Rhodococcus* sp. trehalose lipid to bovine serum albumin protects the protein toward thermal denaturation, mainly by avoiding formation of β-aggregates, as shown by FTIR [86]. However, this same biosurfactant facilitated the thermal unfolding of cytochrome C, increasing the proportion of β-aggregates [86].

Rhodococcus erythropolis 51T7 trehalose lipid was also shown to interact with porcine pancreatic phopholipase A_2 and to inhibit its catalytic activity [87]. Upon incubation with trehalose lipid the proportion of unordered and aggregate structures increased considerably at the expense of α-helix, and β-sheets structures, indicating protein unfolding [87], probably responsible for the effect on enzyme activity.

10.3.3 Mannosylerythritol Lipids

The interaction of assembled monolayers of mannosylerythritol lipids (MEL) with various immunoglobulins has been studied through surface plasmon resonance and atomic force microscopy [88]. Immunoglobulins bound to MEL lipids monolayers in a bivalent mode, showing that MEL monolayers would be useful as ligand systems for these immunogobulines. Later, the molecular details of the interaction of MEL with β-glucosidase were depicted [89]. Using a variety of techniques, such as differential scanning calorimetry, circular dichroism spectroscopy, isothermal titration calorimetry, and docking simulation, these authors showed a concentration dependent effect of MEL on glucosidase activity: below the cmc enzyme activity was increased, and decreased at concentrations of the biosurfactants above the cmc. The unfolding temperature was shifted to higher values upon interaction with the biosurfactant, concomitantly to an increase in α-helix, β-turn, or random coil contents, at the expense of β-sheet structures. Nevertheless, it was shown that binding of MEL to glucosidase was driven by weak hydrophobic interactions.

10.4 Conclusions

During the last decades researchers have used a variety of experimental techniques to study the interaction of glycolipid biosurfactants and model membranes and proteins. The use of DSC provided information regarding the effect of the biosurfactants on the gel to liquid crystalline phase transition as well as on the lamellar to hexagonal H_{II} phase transition. ITC was used to determine the thermodynamics of the interaction between biosurfactants and the bilayer. X-ray diffraction supplied data regarding the thickness of the bilayer and the phase organization adopted by the system, whereas FTIR was employed to gain insight into the perturbation exerted by biosurfactant on the order of the phospholipid acyl chains and also on the hydration of the interfacial region of the membrane. The use of fluorescence spectroscopy techniques allowed information on the fluidity and permeability properties of the membrane in the presence of biosurfactants to be obtained. Langmuir monolayers, black lipid membranes, ^{2}H-NMR, AFM, and different spectroscopy and microscopy techniques also contributed to gaining information on the interaction between biosurfactants and bilayers of complex composition. The results obtained from these studies have been substantial in order to get a general picture of the molecular interactions between glycolipid biosurfactants and the membrane. On the other hand, a number of studies have provided information on the interaction between glycolipid biosurfactants and proteins. Surface plasmon resonance, AFM, circular dichroism spectroscopy, DSC, ITC, FTIR, and X-ray scattering techniques have been used to obtain information about binding of biosurfactants to proteins and its effects on enzyme activity, denaturation, and secondary and tertiary structure of the proteins. Taken together, we believe that all the work dedicated to the study of the interaction between glycolipid biosurfactants and model membranes and proteins have been useful and valuable, and will contribute to the understanding of the molecular mechanisms underlying the diverse biological functions exhibited by glycolipid biosurfactants.

References

1 Kitamoto D, Isoda H, Nakahara T. Functions and potential applications of glycolipid biosurfactants — from energy-saving materials to gene delivery carriers. *J Biosci Bioeng.* 2002;94(3):187–201. https://doi.org/10.1263/jbb.94.187

2 Thakur P, Saini NK, Thakur VK, Gupta VK, Saini RV, Saini AK. Rhamnolipid the glycolipid biosurfactant: emerging trends and promising strategies in the field of biotechnology and biomedicine. *Microb Cell Fact.* 2021;20(1):1–16. https://doi.org/10.1186/s12934-020-01497-9

3 Kuyukina MS, Ivshina IB, Baeva TA, Kochina OA, Gein SV, Chereshnev VA. Trehalolipid biosurfactants from nonpathogenic rhodococcus actinobacteria with diverse immunomodulatory activities. *N Biotechnol.* 2015;32:559–568. https://doi.org/10.1016/j.nbt.2015.03.006

4 Ma X, Meng L, Zhang H, Zhou L, Yue J, Zhu H, et al. Sophorolipid biosynthesis and production from diverse hydrophilic and hydrophobic carbon substrates. *Appl Microbiol Biotechnol.* 2020;104(1):77–100. https://doi.org/10.1007/s00253-019-10247-w

5 Jarvis FG, Johnson MJ. A glyco-lipide produced by pseudomonas aeruginosa. *J Am Chem Soc.* 1949;71(12):4124–4126. https://doi.org/10.1021/ja01180a073

6 Crouzet J, Arguelles-Arias A, Dhondt-Cordelier S, Cordelier S, Pršić J, Hoff G, et al. Biosurfactants in plant protection against diseases: rhamnolipids and lipopeptides case study. *Front Bioeng Biotechnol.* 2020;8:1–11. https://doi.org/10.3389/fbioe.2020.01014

7 Uchida Y, Tsuchiya R, Chino M, Hirano J, Tabuchi T. Extracellular accumulation of mono- and di-succinoyl trehalose lipids by a strain of rhodococcus erythropolis grown on n-alkanes. *Agric Biol Chem.* 1989;53(3):757–763. https://doi.org/10.1080/00021369.1989.10869385

8 Kuyukina MS, Kochina OA, Gein SV, Ivshina IB, Chereshnev VA. Mechanisms of immunomodulatory and membranotropic activity of trehalolipid biosurfactants (a review). *Appl Biochem Microbiol.* 2020 1;56(3):245–255. https://doi.org/10.1134/S0003683820030072

9 Rodrigues L, Banat IM, Teixeira J, Oliveira R. Biosurfactants: potential applications in medicine. *J Antimicrob Chemother.* 2006;57(4):609–618. https://doi.org/10.1093/jac/dkl024

10 Semkova S, Antov G, Iliev I, Tsoneva I, Lefterov P, Christova N, et al. Rhamnolipid biosurfactants-Possible natural anticancer agents and autophagy inhibitors. *Separations.* 2021;8(7):92. https://doi.org/10.3390/separations8070092

11 Ortiz A, Teruel JA, Espuny MJ, Marqués A, Manresa A, Aranda FJ. Effects of dirhamnolipid on the structural properties of phosphatidylcholine membranes. *Int J Pharm.* 2006;325:1–2. https://doi.org/10.1016/j.ijpharm.2006.06.028

12. Lewis RNAH, Mannock DA, McElhaney RN. Differential scanning calorimetry in the study of lipid phase transitions in model and biological membranes. *Methods Mol Biol.* 2007;400:171–195. https://doi.org/10.1007/978-1-59745-519-0_12
13. Semeraro EF, Marx L, Frewein MPK, Pabst G. Increasing complexity in small-angle X-ray and neutron scattering experiments: from biological membrane mimics to live cells. *Soft Matter.* 2021;17(2):222–232. https://doi.org/10.1039/c9sm02352f
14. Mantsch HH, McElhaney R. Phospholipid phase transitions in model and biological membranes as studied by infrared spectroscopy. *Chem Phys Lipids.* 1991;57(2–3):213–226. https://doi.org/10.1016/0009-3084(91)90077-O
15. Sánchez M, Aranda FJ, Teruel JA, Ortiz A. Interaction of a bacterial dirhamnolipid with phosphatidylcholine membranes: a biophysical study. *Chem Phys Lipids.* 2009;161(1):51–55. https://doi.org/10.1016/j.chemphyslip.2009.06.145
16. Casal HL, Mantsch HH. Polymorphic phase behaviour of phospholipid membranes studied by infrared spectroscopy. *BBA - Rev Biomembr.* 1984;779(4):381–401. https://doi.org/10.1016/0304-4157(84)90017-0
17. Lewis RN, McElhaney RN, Pohle W, Mantsch HH. Components of the carbonyl stretching band in the infrared spectra of hydrated 1,2-diacylglycerolipid bilayers: a reevaluation. *Biophys J.* 1994;67(6):2367–2375. https://dx.doi.org/10.1016%2FS0006-3495(94)80723-4
18. Pashynska VA. Mass spectrometric study of rhamnolipid biosurfactants and their interactions with cell membrane phospholipids. *Biopolym Cell.* 2009;25(6):504–508. http://dx.doi.org/10.7124/bc.0007FE
19. Haba E, Pinazo A, Pons R, Pérez L, Manresa A. Complex rhamnolipid mixture characterization and its influence on DPPC bilayer organization. *Biochim Biophys Acta - Biomembr.* 2014;1838(3):776–783. http://dx.doi.org/10.1016/j.bbamem.2013.11.004
20. Moussa Z, Chebl M, Patra D. Interaction of curcumin with 1,2-dioctadecanoyl-sn-glycero-3-phosphocholine liposomes: intercalation of rhamnolipids enhances membrane fluidity, permeability and stability of drug molecule. *Colloids Surf B Biointerfaces.* 2017;149:30–37. http://dx.doi.org/10.1016/j.colsurfb.2016.10.002
21. Hoff B, Strandberg E, Ulrich AS, Tieleman DP, Posten C. 2H-NMR study and molecular dynamics simulation of the location, alignment, and mobility of pyrene in POPC bilayers. *Biophys J.* 2005;88(3):1818–1827. http://dx.doi.org/10.1529/biophysj.104.052399
22. El Khoury E, Patra D. Length of hydrocarbon chain influences location of curcumin in liposomes: curcumin as a molecular probe to study ethanol induced interdigitation of liposomes. *J Photochem Photobiol B Biol.* 2016;158:49–4. http://dx.doi.org/10.1016/j.jphotobiol.2016.02.022
23. Siegel DP, Tenchov BG. Influence of the lamellar phase unbinding energy on the relative stability of lamellar and inverted cubic phases. *Biophys J.* 2008;94(10):3987–3995. http://dx.doi.org/10.1529/biophysj.107.118034
24. Brown PM, Steers J, Hui SW, Yeagle PL, Silvius JR. Role of head group structure in the phase behavior of amino phospholipids. 2. Lamellar and nonlamellar phases of unsaturated phosphatidylethanolamine analogues. *Biochemistry.* 1986;25(15):4259–4267. https://pubs.acs.org/doi/abs/10.1021/bi00363a013
25. Sánchez M, Teruel JA, Espuny MJ, Marqués A, Aranda FJ, Manresa Á, et al. Modulation of the physical properties of dielaidoylphosphatidylethanolamine membranes by a dirhamnolipid biosurfactant produced by *Pseudomonas aeruginosa*. *Chem Phys Lipids.* 2006;142(1–2):118–127. https://doi.org/10.1016/j.chemphyslip.2006.04.001

26 Cullis PR, Hope MJ, Tilcock CPS. Lipid polymorphism and the roles of lipids in membranes. *Chem Phys Lipids*. 1986;40(2–4):127–144. https://doi.org/10.1016/0304-4157(79)90012-1
27 Blume A., Hübner W, Messner G. Fourier transform infrared spectroscopy of ^{13}C=O-labeled phospholipids hydrogen bonding to carbonyl groups. *Biochemistry*. 1988;27(21):8239–8249. https://doi.org/10.1021/bi00421a038
28 Leventis PA, Grinstein S. The distribution and function of phosphatidylserine in cellular membranes. *Annu Rev Biophys*. 2010;39(1):407–427. https://doi.org/10.1146/annurev.biophys.093008.131234
29 Wiliiamson P., Bevers EM., Smeets EF, Comfurius P, Schlegel RA, Zwaal RFA. Continuous analysis of the mechanism of activated transbilayer lipid movement in platelets. *Biochemistry*. 1995;34(33):10448–10455. https://doi.org/10.1021/bi00033a017
30 Huster D, Müller P, Arnold K, Herrmann A. Dynamics of lipid chain attached fluorophore 7-nitrobenz-2-oxa-1,3-diazol-4-yl (NBD) in negatively charged membranes determined by NMR spectroscopy. *Eur Biophys J*. 2003;32(1):47–54. https://doi.org/10.1007/s00249-002-0264-9
31 Oliva A, Teruel JA, Aranda FJ, Ortiz A. Effect of a dirhamnolipid biosurfactant on the structure and phase behaviour of dimyristoylphosphatidylserine model membranes. *Colloids Surf B Biointerfaces*. 2020;185:110576. https://doi.org/10.1016/j.colsurfb.2019.110576
32 Toptygin D, Brand L. Determination of DPH order parameters in unoriented vesicles. *J Fluoresc*. 1995;5(1):39–50. https://doi.org/10.1007/bf00718781
33 Lentz BR. Use of fluorescent probes to monitor molecular order and motions within liposome bilayers. *Chem Phys Lipids*. 1993;64(1–3):99–116. https://doi.org/10.1016/0009-3084(93)90060-g
34 Casal HL, Martin A, Mantsch HH, Paltauf F, Hauser H. Infrared studies of fully hydrated unsaturated phosphatidylserine bilayers. Effect of litium and calcium. *Biochemistry*. 1987;26(23):7395–7401. https://pubs.acs.org/doi/abs/10.1021/bi00388a033
35 Abbasi H, Noghabi KA, Ortiz A. Interaction of a bacterial monorhamnolipid secreted by Pseudomonas aeruginosa MA01 with phosphatidylcholine model membranes. *Chem Phys Lipids*. 2012;165(7):745–752. http://dx.doi.org/10.1016/j.chemphyslip.2012.09.001
36 Abbasi H, Aranda FJ, Noghabi KA, Ortiz A. A bacterial monorhamnolipid alters the biophysical properties of phosphatidylethanolamine model membranes. *Biochim Biophys Acta - Biomembr*. 2013;1828(9):2083–2090. https://doi.org/10.1016/j.bbamem.2013.04.024
37 Luzzati V. X-ray diffraction studies of lipid-water systems. In: Chapman D, editors. *Biological Membranes*. Academic Press, New York; 1968. pp. 71–123.
38 Lewis RN, McElhaney RN. Calorimetric and spectroscopic studies of the polymorphic phase behavior of a homologous series of n-saturated 1,2-diacyl phosphatidylethanolamines. *Biophys J*. 1993;64(4):1081–1096. https://dx.doi.org/10.1016%2FS0006-3495(93)81474-7
39 Aranda FJ, Espuny MJ, Marqués A, Teruel JA, Manresa Á, Ortiz A. Thermodynamics of the interaction of a dirhamnolipid biosurfactant secreted by Pseudomonas aeruginosa with phospholipid membranes. *Langmuir*. 2007;23(5):2700–2705. https://doi.org/10.1021/la061464z
40 Heerklotz H, Seelig J. Titration calorimetry of surfactant-membrane partitioning and membrane solubilization. *Biochim Biophys Acta - Biomembr*. 2000;1508(1–2):69–85. https://doi.org/10.1016/S0304-4157(00)00009-5
41 Sánchez M, Aranda FJ, Teruel JA, Espuny MJ, Marqués A, Manresa A, et al. Permeabilization of biological and artificial membranes by a bacterial dirhamnolipid produced by *Pseudomonas aeruginosa*. *J Colloid Interface Sci*. 2010;341(2):240–247. https://doi.org/10.1016/j.jcis.2009.09.042

42 Sánchez M, Aranda FJ, Teruel JA, Ortiz A. New pH-sensitive liposomes containing phosphatidylethanolamine and a bacterial dirhamnolipid. *Chem Phys Lipids*. 2011;164(1):16–23. https://doi.org/10.1016/j.chemphyslip.2010.09.008

43 Sánchez M, Aranda FJ, Espuny MJ, Marqués A, Teruel JA, Manresa A, et al. Aggregation behaviour of a dirhamnolipid biosurfactant secreted by Pseudomonas aeruginosa in aqueous media. *J Colloid Interface Sci*. 2007;307(1):246–253. https://doi.org/10.1016/j.jcis.2006.11.041

44 Nasir MN, Lins L, Crowet JM, Ongena M, Dorey S, Dhondt-Cordelier S, et al. Differential interaction of synthetic glycolipids with biomimetic plasma membrane lipids correlates with the plant biological response. *Langmuir*. 2017;33(38):9979–9987. https://doi.org/10.1021/acs.langmuir.7b01264

45 Monnier N, Furlan AL, Buchoux S, Deleu M, Dauchez M, Rippa S, et al. Exploring the dual interaction of natural rhamnolipids with plant and fungal biomimetic plasma membranes through biophysical studies. *Int J Mol Sci*. 2019;20(5):1009 https://doi.org/10.3390/ijms20051009.

46 Davis JH. The description of membrane lipid conformation, order and dynamics by ^{2}H-NMR. *BBA - Rev Biomembr*. 1983;737(1):117–171 https://doi.org/10.1016/0304-4157(83)90015-1.

47 Davis JH, Maraviglia B, Weeks G, Godin DV. Bilayer rigidity of the erythrocyte membrane 2H-NMR of a perdeuterated palmitic acid probe. *BBA - Biomembr*. 1979;550(2):362–366. https://doi.org/10.1016/0005-2736(79)90222-0

48 Herzog M, Tiso T, Blank LM, Winter R. Interaction of rhamnolipids with model biomembranes of varying complexity. *Biochim Biophys Acta - Biomembr*. 2020;1862(11):183431. https://doi.org/10.1016/j.bbamem.2020.183431

49 Furlan AL, Laurin Y, Botcazon C, Rodríguez-Moraga N, Rippa S, Deleu M, et al. Contributions and limitations of biophysical approaches to study of the interactions between amphiphilic molecules and the plant plasma membrane. *Plants*. 2020;9(5):648. https://dx.doi.org/10.3390%2Fplants9050648

50 Sezgin E, Schwille P. Fluorescence techniques to study lipid dynamics. *Cold Spring Harb Perspect Biol*. 2011;3:009803. https://doi.org/10.1101/cshperspect.a009803

51 Bagatolli LA. To see or not to see: lateral organization of biological membranes and fluorescence microscopy. *Biochim Biophys Acta - Biomembr*. 2006;1758(10):1541–1556. https://doi.org/10.1016/j.bbamem.2006.05.019

52 Herzog M, Li L, Blesken CC, Welsing G, Tiso T, Blank LM, et al. Impact of the number of rhamnose moieties of rhamnolipids on the structure, lateral organization and morphology of model biomembranes. *Soft Matter*. 2021;17(11):3191–3206. https://doi.org/10.1039/D0SM01934H

53 Come B, Donato M, Potenza LF, Mariani P, Itri R, Spinozzi F. The intriguing role of rhamnolipids on plasma membrane remodelling: from lipid rafts to membrane budding. *J Colloid Interface Sci*. 2021;582:669–677. https://doi.org/10.1016/j.jcis.2020.08.027

54 Franzetti A, Gandolfi I, Bestetti G, Smyth TJP, Banat IM. Production and applications of trehalose lipid biosurfactants. *Eur J Lipid Sci Technol*. 2010;112(6):617–627. https://doi.org/10.1002/ejlt.200900162

55 Durand E, Welby M, Laneelle G, Tocanne J-F. Phase behaviour of cord factor and related bacterial glycolipid toxins: a monolayer study. *Eur J Biochem*. 1979;93(1):103–112. https://doi.org/10.1111/j.1432-1033.1979.tb12799.x

56 Imasato H, Procópio J, Tabak M, Ioneda T. Effect of low mole fraction of trehalose dicorynomycolate from Corynebacterium diphtheriae on water permeability and electrical capacitance of lipid bilayer membranes. *Chem Phys Lipids*. 1990;52(3–4):259–262. https://doi.org/10.1016/0009-3084(90)90122-8

57 Mueller R, Rudin D, Tien H, Wescott M. Reconstitution of cell membrane structure in vitro and its transformation into a excitable system. *Nature*. 1962;194:979–980. https://doi.org/10.1038/194979a0

58 Düzgüneş N. Fluorescence assays for liposome fusion. *Methods Enzymol*. 2003;372:260–274. https://doi.org/10.1016/S0076-6879(03)72015-1

59 Spargo B, Crowe L, Ioneda T, Beaman B, Crowe JH. Cord factor (alpha, alpha-trehalose 6,6'-dimycolate) inhibits fusion between phospholipid vesicles. *Proc Natl Acad Sci USA*. 1991;88(3):737–740. https://doi.org/10.1073/pnas.88.3.7

60 Crowe L, Spargo B, Ioneda T, Beaman B, Crowe J. Interaction of cord factor (α,α'-trehalose-6,6'-dimycolate) with phospholipids. *Biochim Biophys Acta*. 1994;1194:53–60. https://doi.org/10.1016/0005-2736(94)

61 Almog R, Mannella CA. Molecular packing of cord factor and its interaction with phosphatidylinositol in mixed monolayers. *Biophys J*. 1996;71(6):3311–3319. http://dx.doi.org/10.1016/S0006-3495(96)79523-1

62 Harland CW, Rabuka D, Bertozzi CR, Parthasarathy R. The mycobacterium tuberculosis virulence factor trehalose dimycolate imparts desiccation resistance to model mycobacterial membranes. *Biophys J*. 2008;94(12):4718–4724. https://doi.org/10.1529/biophysj.107.125542

63 Harland CW, Botyanszki Z, Rabuka D, Bertozzi CR, Parthasarathy R. Synthetic trehalose glycolipids confer desiccation resistance to supported lipid monolayers. *Langmuir*. 2009;25(9):5193–5198. https://doi.org/10.1021/la804007a

64 Rath P, Saurel O, Czplicki G, Tropis M, Daffé M, Ghazi A, et al. Cord factor (trehalose 6,6'-dimycolate) forms fully stable and non-permeable lipid bilayers required for a functional outer membrane. *Biochim Biophys Acta*. 2013;1828:2173–2181. https://doi.org/10.1016/j.bbamem.2013.04.021

65 Aranda FJ, Teruel JA, Espuny MJ, Marqués A, Manresa Á, Palacios-Lidón E, et al. Domain formation by a Rhodococcus sp. biosurfactant trehalose lipid incorporated into phosphatidylcholine membranes. *Biochim Biophys Acta - Biomembr*. 2007;1768(10):2596–2604. https://doi.org/10.1016/j.bbamem.2007.06.016

66 Ortiz A, Teruel JA, Espuny MJ, Marqués A, Manresa Á, Aranda FJ. Interactions of a Rhodococcus sp. biosurfactant trehalose lipid with phosphatidylethanolamine membranes. *Biochim Biophys Acta - Biomembr*. 2008;1778(12):2806–2813. http://dx.doi.org/10.1016/j.bbamem.2008.07.016

67 Ortiz A, Teruel JA, Espuny MJ, Marqués A, Manresa A, Aranda FJ. Interactions of a bacterial biosurfactant trehalose lipid with phosphatidylserine membranes. *Chem Phys Lipids*. 2009;158(1):46–53. https://doi.org/10.1016/j.chemphyslip.2008.11.001

68 Ortiz A, Teruel JA, Manresa A, Espuny MJ, Marqués A, Aranda FJ. Effects of a bacterial trehalose lipid on phosphatidylglycerol membranes. *Biochim Biophys Acta - Biomembr*. 2011;1808(8):2067–2072. https://doi.org/10.1016/j.bbamem.2011.05.003

69 Gershfeld NL, Stevens WF, Nossal RJ. Equilibrium studies of phospholipid bilayer assembly: coexistence of surface bilayers and unilamellar vesicles. *Faraday Discuss Chem Soc.* 1986;81:19–28. https://doi.org/10.1039/dc9868100019

70 Teruel JA, Ortiz A, Aranda FJ. Interactions of a bacterial trehalose lipid with phosphatidylglycerol membranes at low ionic strength. *Chem Phys Lipids.* 2014;181:34–39. https://doi.org/10.1016/j.chemphyslip.2014.03.005

71 Zaragoza A, Aranda FJ, Espuny MJ, Teruel JA, Marqués A, Manresa A, et al. Mechanism of membrane permeabilization by a bacterial trehalose lipid biosurfactant produced by *Rhodococcus sp. Langmuir.* 2009;25(14):7892–7898. https://doi.org/10.1021/la900480q

72 Arutchelvi JI, Bhaduri S, Uppara PV, Doble M. Mannosylerythritol lipids: a review. *J Ind Microbiol Biotechnol.* 2008;35(12):1559–1570. https://doi.org/10.1007/s10295-008-0460-4

73 Imura T, Yanagishita H, Ohira J, Sakai H, Abe M, Kitamoto D. Thermodynamically stable vesicle formation from glycolipid biosurfactant sponge phase. *Colloids Surf B Biointerfaces.* 2005;43(2):115–121. https://doi.org/10.1016/j.colsurfb.2005.03.015

74 Madihalli C, Sudhakar H, Doble M. Production and investigation of the physico-chemical properties of MEL-A from glycerol and coconut water. *World J Microbiol Biotechnol.* 2020;36(6):88. https://doi.org/10.1007/s11274-020-02857-8

75 Fan L, Chen Q, Mairiyangu Y, Wang Y, Liu X. Stable vesicle self-assembled from phospholipid and mannosylerythritol lipid and its application in encapsulating anthocyanins. *Food Chem.* 2021;344:128649. https://doi.org/10.1016/j.foodchem.2020.128649

76 Van Bogaert INA, Saerens K, De Muynck C, Develter D, Soetaert W, Vandamme EJ. Microbial production and application of sophorolipids. *Appl Microbiol Biotechnol.* 2007;76(1):23–34. https://doi.org/10.1007/s00253-007-0988-7

77 Singh PK, Bohr SSR, Hatzakis NS. Direct observation of sophorolipid micelle docking in model membranes and cells by single particle studies reveals optimal fusion conditions. *Biomolecules.* 2020;10(9):1291. https://doi.org/10.3390/biom10091291

78 Ruthardt N, Lamb DC, Bräuchle C. Single-particle tracking as a quantitative microscopy-based approach to unravel cell entry mechanisms of viruses and pharmaceutical nanoparticles. *Mol Ther.* 2011;19(7):1199–1211. http://dx.doi.org/10.1038/mt.2011.102

79 Marcelino PRF, Ortiz J, da Silva SS, Ortiz A Interaction of an acidic sophorolipid biosurfactant with phosphatidylcholine model membranes. *Colloids Surf B Biointerfaces.* 2021;207:112029. https://doi.org/10.1016/j.colsurfb.2021.112029

80 Sánchez M, Aranda FJ, Espuny MJ, Marqués A, Teruel JA, Manresa Á, et al. Thermodynamic and structural changes associated with the interaction of a dirhamnolipid biosurfactant with bovine serum albumin. *Langmuir.* 2008;24(13):6487–6495. https://doi.org/10.1021/la800636s

81 Andersen KK, Otzen DE. Denaturation of α-lactalbumin and myoglobin by the anionic biosurfactant rhamnolipid. *Biochim Biophys Acta - Proteins Proteom.* 2014;1844(12):2338–2345. http://dx.doi.org/10.1016/j.bbapap.2014.10.005

82 Mortensen HG, Madsen JK, Andersen KK, Vosegaard T, Deen GR, Otzen DE, et al. Myoglobin and α-lactalbumin form smaller complexes with the biosurfactant rhamnolipid than with SDS. *Biophys J.* 2017;113(12):2621–2633. https://doi.org/10.1016/j.bpj.2017.10.024

83 Andersen KK, Otzen DE. Folding of outer membrane protein A in the anionic biosurfactant rhamnolipid. *FEBS Lett.* 2014;588(10):1955–1960. http://dx.doi.org/10.1016/j.febslet.2014.04.004

84 Madsen JK, Pihl R, Møller AH, Madsen AT, Otzen DE, Andersen KK. The anionic biosurfactant rhamnolipid does not denature industrial enzymes. *Front Microbiol.* 2015;6:292. https://dx.doi.org/10.3389%2Ffmicb.2015.00292

85 Andersen KK, Vad BS, Scavenius C, Enghild JJ, Otzen DE. Human lysozyme peptidase resistance is perturbed by the anionic glycolipid biosurfactant rhamnolipid produced by the opportunistic pathogen Pseudomonas aeruginosa. *Biochemistry.* 2017;56(1):260–270. https://doi.org/10.1021/acs.biochem.6b01009

86 Zaragoza A, Teruel JA, Aranda FJ, Marqués A, Espuny MJ, Manresa Á, et al. Interaction of a Rhodococcus sp. trehalose lipid biosurfactant with model proteins: thermodynamic and structural changes. *Langmuir.* 2012 Jan;28(2):1381–1390. https://doi.org/10.1021/la203879t

87 Zaragoza A, Teruel JA, Aranda FJ, Ortiz A. Interaction of a trehalose lipid biosurfactant produced by Rhodococcus erythropolis 51T7 with a secretory phospholipase A2. *J Colloid Interface Sci.* 2013;408(1):132–137. http://dx.doi.org/10.1016/j.jcis.2013.06.073

88 Ito S, Imura T, Fukuoka T, Morita T, Sakai H, Abe M, et al. Kinetic studies on the interactions between glycolipid biosurfactant assembled monolayers and various classes of immunoglobulins using surface plasmon resonance. *Colloids Surf B Biointerfaces.* 2007;58(2):165–171. https://doi.org/10.1016/j.colsurfb.2007.03.003

89 Fan L, Xie P, Wang Y, Huang Z, Zhou J Biosurfactant-protein interaction: influences of mannosylerythritol lipids-A on β-glucosidase. *J Agric Food Chem.* 2018;66(1):238–246. https://doi.org/10.1021/acs.jafc.7b04469

11

Biosurfactants

Properties and Current Therapeutic Applications

Cristiani Baldo, Maria Ines Rezende, and Fabiana Guillen Moreira Gasparin

Department of Biochemistry and Biotechnology, Centre of Exact Science, Londrina State University, Londrina, Brazil

11.1 Production of Microbial Biosurfactants

Biosurfactants are tensioactive amphiphilic biomolecules produced by bacteria, fungi, and yeasts. Structurally, biosurfactants are composed of hydrophilic (polar) and hydrophobic (non-polar) parts (Figure 11.1). The hydrophobic portion is a long chain of fatty acid, hydroxy fatty acid, or α-alkyl β-hydroxy fatty acid. The hydrophilic part can be a carbohydrate, amino acid, cyclic peptide, phosphate, acid carboxylic acid, or alcohol. Based on their chemical nature, biosurfactants are classified as glycolipids, lipopolysaccharides, lipopeptides and lipoproteins, phospholipids, fatty acids. Biosurfactants have properties very similar to petrochemical surfactants, but present several advantages such as biodegradability, biocompatibility, low toxicity, high specificity, chemical, and functional diversity, being superior especially for their ecological and sustainable characteristic [1].

Considering the high mortality rates caused by diseases such as cancer, bacterial and virus infections, and inflammatory syndromes, the search for new drugs by the pharmaceutical industry is crucial. In this context, microbial biosurfactants exhibit a myriad of applications in the medical and health industry sectors. They present a variety of biological actions as emulsifying, anti-adhesive, anti-oxidant, anti-microbial, anti-tumor, anti-inflammatory activities. In addition, they can be used as drug delivery, thrombolytic, platelet aggregation, and anti-hypertensive agents [1].

The process of production and purification of biosurfactants is complex and involves different methodologies (Figure 11.1).

One of the main factors that influences the production of biosurfactants is the hydrophilic and hydrophobic source in the culture medium. In addition, other factors such as nitrogen source, mineral salts, pH, temperature, and agitation must be optimized during the fermentation process [2]. The scale-up of the fermentation process is crucial to obtain large amounts of the biomolecule (Figure 11.1). Considering the application of these molecules in medical/health sciences, the downstream steps of the fermentative process may

Biosurfactants and Sustainability: From Biorefineries Production to Versatile Applications, First Edition.
Edited by Paulo Ricardo Franco Marcelino, Silvio Silverio da Silva, and Antonio Ortiz Lopez.
© 2023 John Wiley & Sons Ltd. Published 2023 by John Wiley & Sons Ltd.

be to obtain a high degree of purity. Economically and environmentally friendly methods to improve the recovery yield and purity of biosurfactants are required. Extraction and separation is generally performed using organic solvents such as methanol and ethyl acetate. The purification of these molecules requires several steps as shown in the Table 11.1. Biochemical characterization of biosurfactants is also usually executed by high-performance liquid chromatography (HPLC), Fourier transformed infrared (FT-IR) and nuclear magnetic resonance (NMR) [3]. The industrial biosurfactant production is still faced with the high cost of microbial cultivation and biosurfactant recovery.

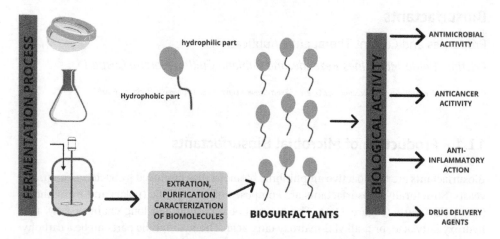

Figure 11.1 Production and biochemical characterization of biosurfactants.

Table 11.1 Biosurfactants purification methods.

Principle of purification method	Methods	Biosurfactants class
Separation	Distillation	Rhamnolipid [4]
	solid–liquid extraction,	
	sweep floc coagulation,	
	liquid–liquid extraction	
Precipitation	Acid precipitation	Lipopeptide [5]
	Ammonium sulphate precipitation	Rhamnolipid [6]
		Glicolipid [4]
Extraction	Ethyl-acetate extraction	Sophorolipid [7]
	Mixture of chloroform and ethanol extraction (2:1)	Lipopeptide [4]
		Rhamnolipid [6]
	n-Hexano	Lipopeptide [4]
		Lipopeptide [4]
Fractionation	Foam fractionation	Sophorolipid [8]

Table 11.1 (Continued)

Principle of purification method	Methods	Biosurfactants class
Crystallization	Spontaneous crystallization	Sophorolipid [9]
	Crystallization	Sophorolipid [10]
	Lyophilization	Rhamnolipid [11]
Ultrafiltration	Ultrafiltration	Rhamnolipid [12]
		Surfactin [13]
Chromatography	Silica gel chromatography	Lipopetide [14]
	Ion exchange chromatography	Lipopeptide [15]
	Size exclusion chromatography	Lipopeptide [16]
	Fast Protein Liquid Chromatography	

11.2 Anti-tumoral Activity of Biosurfactants

Cancer is the second most common cause of death globally, after coronary diseases. Published data for 2020 recorded an incidence of approximately 19 million cancer cases worldwide, with 10 million deaths [17]. Cancer is characterized by the uncontrolled division of cells, which escape the control mechanisms of the cell cycle and apoptosis, invading other tissues, promoting angiogenesis and metastases [18, 19]. Also, cancer cells secrete a series of angiogenic factors responsible for the formation of new blood vessels that supply the tumor's needs [20]. Considering the high mortality and morbidity rates in cancer patients, the search for new molecules to inhibit tumor development is essential. In this context, biosurfactants have been extensively investigated aim at the development of future drugs for cancer therapy (Table 11.2).

A variety of mechanisms have been proposed to describe the anti-cancer activity of biosurfactants such as apoptosis, cell cycle evolution delay, natural killer T cells stimulation, inhibition of signaling pathways, and angiogenesis reduction [21]. Moreover, biosurfactants are able to disrupt the cell membranes by lysis and increasing the permeability of cellular membrane [22].

Sophorolipids are important microbial biosurfactants, that are of great interest in pharmaceutical areas as potential anti-cancer medicines [23]. Several studies have described the anti-cancer activities of sophrolipids against several types of cancer cells [23–27]. Sophorolipids showed anti-tumor activity against human pancreatic carcinoma cells [24]. A natural mixture of sophorolipids or select derivatives (ethyl ester, methyl ester, ethyl ester monoacetate, ethyl ester diacetate, acidic sophorolipid, lactonic sophorolipid diacetate) induced cytotoxicity in pancreatic carcinoma cells at all concentrations tested. However, methyl ester derivative mediated much greater levels of cytotoxicity (63%) compared with other derivative sophorolipids. Importantly, the treatment of peripheral blood mononuclear cells with sophorolipids did not affect their viability demonstrating the selectivity of these molecules against cancer cells [24]. In another study, the cytotoxic effects of sophorolipids produced from *Wickerhamiella domercqiae* were evaluated on cancer cells of H7402 (liver cancer line), A549 (lung cancer line), HL60 and K562 (leukemia lines). The results indicated a decrease of cell viability in dose-dependent manner (≤ 62.5 µg mL^{-1}), suggesting that the sophorolipids produced by *W. domercqiae* present anti-tumor activity [28].

Sophorolipid molecules differing in degree of sophorose acetylation, unsaturation of hydroxyl fatty acid, and lactonization were investigated as possible anti-cancer drugs [27]. The results showed that the inhibition of diacetylated lactonic sophorolipid on esophageal cancer cells was stronger than observed to monoacetylated lactonic sophorolipid. In addition, the unsaturation degree of hydroxyl fatty acid of molecules interferes cytotoxicity to esophageal cancer cells. The sophorolipid with one double bond in the fatty acid part had the strongest cytotoxic effect on two esophageal cancer cells [27].

The cytotoxic effects of lactonic sophorolipids (C18:3) on breast cancer cells were also studied [29]. The results showed that lactonic sophorolipids induced the inhibition of MDA-MB-231 cell migration without affecting the cell viability. The increase of intracellular reactive oxygen species was also observed. Interestingly, the cytotoxic effect of lactonic form was substantially higher than that observed for acidic sophorolipids [29]. In another study, the diacetylated lactonic sophorolipids from *Starmerella bombicola* were also tested against human cervical cancer cells (HeLa). The treatment of HeLa cells with sophorolipids induced apoptosis and cell cycle interruption at the G0 phase and partly at the G2 phase. Additionally, the expression of CHOP and Bip/GRP78 was also detected induced. C/EBP homologous protein (CHOP) expression is a direct consequence of endoplasmic reticulum stress and mediated the apoptosis pathway [30]. The increase in the concentration of cytosolic calcium would cause the accumulation of unfolded proteins in endoplasmic reticulum, leading to the induction of GRP78/Bip chaperone [31]. The activation of caspase-3 and caspase-12 were also detected. Then, the authors concluded that the apoptosis of HeLa cells was possibly triggered through the endoplasmic reticulum signaling pathway. In addition, sophorolipids also are able to induce the morphological alterations in cancer cells. Joshi-Navare and co-workers showed the effect of sophorolipids which induced several morphological changes such as the formation of long thread-like extensions, cell alignment, cell elongation, and bundle formation in a dose-dependent manner on the glioma cell line (LN-229) [26].

The anti-cancer effect of sophorolipids was also evaluated by an *in vivo* approach. Doses of 5, 50, and 500 mg kg^{-1} of lactonic sophorolipids resulted in 29.90, 41.24, and 52.06 % of inhibition without induced significant toxicity to mice, respectively [23]. *In vivo* apoptosis of tumor sections from mice treated with 50 mg kg^{-1} of sophorolipids were assessed by TUNEL assay. The results showed lactonic sophorolipids induced apoptosis of tumor *in vivo*, but the signaling pathway of apoptosis-induction *in vivo* still needs to be clarified [23].

Rhamnolipids are microbial amphipathic biosurfactants. They are produced mainly by *Pseudomonas aeruginosa* and have shown a very potent effect on human and animal cancer cells [32]. Anti-tumor activity of a mono-rhamnolipid and di-rhamnolipid against Hl-60 (acute myeloid Leukemia cancer cells), BV-173 (chronic myeloid leukemia cancer cells), SKW-3 (T-cell lymphocytic leukemia cells), and JMSU-1 (malignant ascitic urinary bladder carcinoma cells) were studied [33]. The results showed that mono-rhamnolipid was more effective than di-rhamnolipid to reduce the cancer cell viability. Mono-rhamnolipid present a great toxic effect on BV-173 and SKW-3 cells, but a lower toxicity effect on HL-60 and JMSU-1 cells was observed. It was also detected that mono-rhamnolipid induced morphology alterations in leukemic cells such as plasma membrane blebbing, condensation of chromatin, and apoptosis of BV-173 cells [33].

Anti-tumor activity of fractionated mono- and di-rhamnolipids produced by *P. aeruginosa* MR01 against human breast cancer cells MCF-7 was also evaluated [34]. The results showed that both mono- and di-rhamnolipids induced a decrease of viability of MCF-7 cells in a dose-dependent manner after 48 h of treatment. Several morphological alterations were also detected. Cells treated with rhamnolipids present apoptotic characteristics as round, shrunken shapes detached from the surface of the well plates. The expression of the p53 gene were also observed. In another study, di-rhamnolipids from *P. aeroginosa* MR01 also inhibited the growth of the HeLa cancer cells [35].

Surfactin, produced by different species of the *Bacillus*, is one of the most studied biosurfactant. Several studies showed the anti-cancer activity of surfactin against different cancers cells lines. According to the literature, surfactin is able to suppress the tumor growth arresting the cell cycle, inducing apoptosis, and also impairing the metastasis [21, 36–38]. The effect of surfactin on the human colon carcinoma cell line LoVo was studied and the authors described a strong inhibitory activity of sufactin by inducing apoptosis and cell cycle arrest [36].

In 2010, Cao and co-workers studied the anti-cancer activity of surfactin from *Bacillus natto* TK-1 in human breast cancer MCF-7 cell lines. In this study, surfactin induced reactive oxygen species (ROS) generation and triggered the ROS/JNK-mediated mitochondrial/caspase pathway of MCF-7 cell lines [38]. Lee and co-workers also showed that surfactin inhibited the growth of MCF7 human breast cancer cells in a dose-dependent manner [39]. Liu et al. evaluated the effect of lipopeptides by *Bacillus subtilis* HSO121 on Bcap-37 breast cancer cell lines and demonstrated that these compounds induced apoptosis in a dose-dependent manner [37]. Duarte et al. showed the a surfactin from *B. subtilis* was able to inhibit T47D and MDAMB-231 cells (breast cancer cells) in a time- and dose dependent manner. Additionally, the authors showed that the 24 h exposure to surfactin (0.05 g l^{-1}) resulted in the inhibition of cell proliferation with the interruption of the cell cycle in the G1 phase [21].

Iturins are cyclic lypopeptides produced by bacterial Bacillus sp including *Bacillus subtilis*, *Bacillus amyloliquefaciens*, *Bacillus licheniformis*, *Bacillus thuringiensis*, and *Bacillus methyltrophicus*. The iturins family is mainly represented by Iturin A, C, D, and E. The anti-tumoral activity of iturin against the human hepatoma cell line HepG2, breast cancer MDA-MB-231, and MDA-MB-468 epithelial colorectal adenocarcinoma (Caco-2) cells have already been described [40–43]. In summary, numerous studies have shown the anti-cancer properties of biosurfactants against different cancer cell lines, showing the great potential of these compounds for cancer treatment (Table 11.2).

Table 11.2 Anti-tumor activity of biosurfactants against cancer cells lines.

Biossurfactant	Source	Tumoral cell line	References
Surfactin	B. subtilis	T47D and MDAMB-231 cells (breast cancer cells)	[21]
Sophorolipids	W. domercqiae	H7402 human liver	[25]
Sophorolipid	W. domercqiae	On esophageal cancer cells	[27]
Sophorolipid	Starmerella bombicola	Breast cancer cells	[29]
Sophorolipid	S. bombicola	Human cervical cancer cells (HeLa)	[30]
Rhamnolipid	Pseudomonas aeruginosa	Hl-60 (acute myeloid leukemia cancer cells), BV-173 (chronic myeloid leukemia cancer cells), SKW-3 (T-cell lymphocytic leukemia cells), and JMSU-1 (malignant ascitic urinary bladder carcinoma cells)	[33]
Rhamnolipids	P. aeruginosa	Human breast cancer cells MCF-7	[34]
Rhamnolipids	P. aeruginosa	Human breast cancer cells MCF-7	[35]
surfactin	Bacillus subtilis	Human colon carcinoma cell line LoVo	[36]
Surfactin	B. subtilis	Breast cancer cell lines	[37]
Surfactin	Bacillus natto	Breast cancer MCF-7 cell lines	[38]
surfactin	B. subtilis	MCF7 human breast cancer	[39]
Iturin A on	B. subtilis	HepG2 cell	[40]
Iturin A	B. subtilis	Human hepatoma cell line HepG2	[41]
Iturin A	B. megaterium	Breast cancer MDA-MB-231 and MDA-MB-468 cells w	[42]
Iturin A	B. subtillis	Epithelial colorectal adenocarcinoma (Caco-2)	[43]

11.3 Anti-inflammatory Activity of Biosurfactants

Inflammation is a defense response that occurs after several stimuli as damage caused by microorganisms, physical agents (radiation, trauma, burns), chemicals (toxins, caustic substances), and tissue necrosis. The acute inflammatory reaction is characterized by a series of events such as increase of vascular permeability, fluid exudation (edema), pain, migration, and accumulation of leukocytes [44]. Macrophages are activated in response to the stimuli and release several pro-inflammatory mediators, such as interleukin-1 beta (IL-1β), tumor necrosis factor-alpha (TNF-α), nitric oxide (NO), prostaglandin E2 (PGE2), and reactive oxygen (ROS), among others [45]. Chronic inflammation is commonly observed in different pathologies as rheumatoid arthritis and osteoarthritis, due to the production of pro-inflammatory mediators [46]. The anti-inflammatory drugs current available for the treatment of chronic inflammatory diseases, when effective, usually induce different side effects. Therefore, the search for new anti-inflammatory drugs is extremely important.

11.3 Anti-inflammatory Activity of Biosurfactants

The anti-inflammatory effect of surfactin has been well-studied in macrophage cells. The mechanism of the anti-inflammatory effect of surfactin on macrophages involves several mechanisms as the TLR4 (toll-like receptor 4) and nuclear factor-κB (NF-κB) modulation, cytosolic phospholipase A_2 (PLA_2) interaction, lipoteichoic acid (LTA)-induced [47]. The PLA_2 are responsible for acid arachidonic acid release. Arachidonic acid is a substrate biosynthesis of prostaglandins, prostacyclins and thromboxanes which functions in the maintenance of the inflammatory process [48].

The anti-inflammatory effect of glycolipid biosurfactant produced by *Rhodococcus ruber* IEGM 231 was also reported. The results showed that the glycolipid was able to stimulate the production of TNF-α, IL-1β, and IL-6 in adherent human peripheral blood monocyte culture. Moreover, the glycolipid show no cytotoxicity effect against human lymphocytes and could be used as a potential immunomodulating and anti-tumor agent [49]. The anti-inflammatory effect of surfactin from *B. subtilis* was investigated in macrophages stimulates with lipopolysaccharide [50]. In this study, surfactin significantly reduced the increase in the IFN-γ, IL-6, iNOS, and nitric oxide expression. Surfactin also induced the downregulation of macrophages LPS-induced TLR4 protein expression suggesting that the surfactin-mediated signal pathway was involved with TLR4. The anti-neuroinflammatory properties of surfactin in lipoteichoic acid (LTA)-stimulated BV-2 microglial cells was investigated. The results showed that surfactin was able to induce the inhibition of TNF-α, IL-1β, IL-6, monocyte chemoattractant protein-1, PGE2, NO, and reactive oxygen species. Surfactin also suppressed the expression of matrix metalloproteinase-9 (MMP-9), inducible NO synthase (iNOS), and cyclooxygenase-2 (COX-2) [51].

The effect of surfactin on the antigen-presenting property of macrophages was also studied. The results indicated that surfactin significantly inhibits the lipopolysaccharide-induced expression of CD40, CD54, CD80, and MHC-II. Macrophages treated with surfactin also presented the compromised phagocytosis and the expression of IL-12 reduced. Surfactin also inhibited the activation of CD4 + T cells and impaired the NF-kappaB translocation. According to the authors, these results bring novel insights into the immunopharmacological role of surfactin in autoimmune disease and transplantation [52].

Some studies exploring the anti-inflammatory properties of sophorolipids were also performed. Hagler and co-workers studied the effect of sophorolipids from *Candida bombicola* in U266 cells (IgE producing myeloma cells). The results indicated that sophorolipids were able to decrease immunoglobulin E in a dose dependent manner. In addition, sophorolipid was able to decrease the mRNA expression of TLR-2, STAT3, and IL-6 at 24 h of incubation [53]. The anti-inflammatory effect of sophorolipid was also evaluated by *in vivo* models, and a decrease of pulmonary inflammation in a mouse asthma model was observed after treatment [54]. Hardin et al. showed that administration of sophorolipids after induction of intra-abdominal sepsis improved the survival in the rat model. Rats treated with 5 mg kg^{-1} of sophorolipids enhanced the survival with cecal ligation and puncture-induced septic shock [55]. The effects of sophorolipid treatment on cytokine production was also evaluated by *in vitro* and *in vivo* models of experimental sepsis [56]. The results showed that sophorolipid treatment significantly decreased the expression of pro-inflammatory cytokines (TNF-α IL-1a IL-1b) in both models. In addition, sophorolipids mediated the decrease of sepsis related mortality [56]. Sophorolipids also were capable of impairing the fatal effects of septic shock in rats with cecal ligation and puncture [57]. The anti-inflammatory action of mannosylerythritol lipids produced by *Pseudomonas Antarctica* also inhibited the inflammatory mediators, inhibiting the secretion of inflammatory mediators from mast cells [58]. The authors showed that mannosylerythritol inhibited the secretion of leukotriene C4 and TNF-α.

Recently, the anti-inflammatory action of biosurfactants against the inflammatory response induced by SARS-CoV-2 was hypothesized. The infection caused by SARS-CoV-2 involved the invasion of a human host cell through the specific receptors. The immune system immediately responded by activating immune cells to respond against the virus and recruiting the antigen-presenting cells [59]. Wang and co-workers reported that COVID-19-infected pneumonia patients with acute respiratory distress syndrome exhibited high levels of cytokines [60]. In this sense, considering the vast role of biosurfactants as anti-inflammatory and immunosuppressive agents, they could be used in combination with other drugs to relieve inflammatory responses caused due to SARS-CoV-2 infection [48]. Taken together, these results showed that microbial biosurfactants showed an anti-inflammatory action and are potential candidates for inflammatory disease treatment.

11.4 Anti-microbial Activity of Biosurfactant

The anti-microbial property of biosurfactants was first described by Javis and Jonson in 1949 who demonstrated the anti-bacterial activity of glycolipids from *Pseudomonas aeruginosa* against *Mycobacterium tuberculosis* [61]. Since then, many researchers have described the excellent anti-microbial properties of different biosurfactant as glycolipids, lipopeptides, phospholipids, fatty acid, and polymeric structures [4, 62]. The anti-microbial activity of biosurfactants against bacteria, fungus, and viruses are described in Table 11.3.

Table 11.3 Mechanism of action of the biosurfactants with anti-bacterial, anti-fungal and anti-viral application.

Therapeutic application	Biosurfactant/mechanism	References
Anti-bacterial	Surfactin induce the formation of ion channels in lipid bi-layer membranes and anti-adhesive properties that inhibit biofilm production and the adhesion of bacteria; preference of the lipopeptide to interact with prokaryotic membranes and so modify its molecular order	[63, 64, 65]
	Fatty acid action by diverse mechanisms: cell lysis; suppression of enzyme activity; deprivation of nutrient uptake; production of toxic peroxidation and auto oxidation products; disruption of electron transport chain and interruption of oxidative phosphorylation.	
Anti-fungal	Destabilization of cellular membrane causing cytoplasmic extrusions and eventually resulting in the rupture of cells	[66]
Anti-viral	Depends on the hydrophobicity of the fatty acid moiety, on the charge of the peptide moiety as well as on the virus species;	[67, 68]
	Insertion into the lipid bilayer viral envelope causing reduction the membrane fusion rate.	

11.4.1 Biosurfactants as Anti-bacterial Agents

Infections caused by bacteria are of great concern to public health worldwide, as they often result in the hospitalization of patients, which increases healthcare costs, and significantly increases morbidity and mortality rates. In addition, the emergence of many multidrug-resistant bacterial strains limit the treatment and cure of patients [69]. The increase on anti-microbial resistance to traditional antibiotics has attracted the attention of many researchers to search for novel anti-microbial compounds against common antibiotic-resistant pathogens. Biosurfactants have been evaluated as better alternatives to replace the usual antibiotics (Table 11.4).

Sharma and Sharan showed the anti-microbial and anti-adhesive properties of biosurfactants produced by *Lactobacilus helveticus* MRTL91 against *Escherichia coli, Pseudomonas aeruginosa, Salmonella typhi, Shigella flexneri, Staphylococcus aureus, Staphylococcus epidermidis, Listeria monocytogenes, Listeria innocua*, and *Bacillus cereus* [70]. In another study, the anti-bacterial activity of lipopeptides produced by *Bacillus subtilis* VSG4 and *Bacillus licheniformis* VS16 against *Staphylococcus aureus, Bacillus cereus, Escherichia coli, Salmonella typhimurium*, and *Vibrio parahaemolyticus*, was also investigated [71]. According to the authors, the biosurfactants present anti-bacterial activity against pathogenic and non-pathogenic microorganisms, suggesting that these biosurfactants could be exploited further for possible use as potential anti-microbials in the biomedical industry [71].

The anti-bacterial activity of surfactin has also been described. According to the literature, surfactin is able to induce the formation of ion channels in lipid bi-layer membranes and inhibits the biofilm production and the adhesion of bacteria [63]. The interaction of surfactin with eukaryotic and prokaryotic membranes was investigated using palmitoyl-oleoyl-phosphatidylglycerol (POPG) and 1-palmitoyl-oleoyl-glycero-phosphocholine (POPC) lipid bilayer models. The results indicated that the surfactin interacts preferentially with POPG membranes leading to its destabilization, suggesting the preference of surfactin to interact with prokaryotic membranes [64].

Table 11.4 Anti-bacterial activity against human pathogens of biosurfactants.

Biosurfactant class	Source	Pathogens	References
Lipopeptide	*Bacillus subtilis* VSG4; *B. licheniformis* VS16 *B. subtilis* KLP2015	*Staphylococcus aureus, Bacillus cereus, Escherichia coli, Salmonella typhimurium*, and *Vibrio parahaemolyticus* *Klebsiella pnemoniae; Salmonella typhimurium* NCTC 74, *Staphylococcus aureus* ATCC 6538; *Escherichia coli* NCTC 10418	[71, 73]
Glicolipid	*Pseudomonas plecoglossicida* BP03	*Staphylococcus aureus, Bacillus subtilis; Aeromonas hydrophila.*	[77]
Fatty acid	----	*Porphyromona gingivalis* KCTC,	[81]
Polymers (type bioemulsan)	*Acinetobacter baumanii* AC5	*E. coli; Salmonella* sp, *S. aureus; P. aeruginosa,*	[86]

Multidrug-resistant bacteria tolerate traditional antibiotics using strategies such as mutations, efflux pump expression, or up-regulation of defense-associated enzymes. Antimicrobial photodynamic therapy (APDT) has been developed as an alternative to conventional antibiotic treatments due to several advantages such as great efficacy, safety, and ease of implementation. The APDT mechanism allows the accumulation of hydrophilic photosensitizer in the cytoplasmic membrane or intracellular target and induces irreversible damage to bacteria by cytotoxic singlet oxygen. However, hydrophilic photosensitizers cannot efficiently accumulate due to the high diffusion resistance by the hydrophobic cytoplasmic membrane. Interestingly, surfactin that presents a unique interaction with cell membrane was able to complex to methylene blue photosensitizers enabling the accumulation and inactivation of bacteria [72].

In addition, the anti-bacterial activity of surfactin produced by *B. subtilis* KLP2015 against *Klebsiella pnemoniae, Salmonella typhimurium* NCTC 74, *Staphylococcus aureus* ATCC 6538, and *Escherichia coli* NCTC 10418 was also evaluated. The researchers observed that pathogenic bacterial biofilms were reduced from 58.1 to 10.23%, depending on the bacteria studied. The greatest activity was observed on *S. aureus*, and the lowest effect was detected for *S. typhimurium* biofilms [73]. *Staphylococcus aureus* are responsible for causing a variety of skin diseases including acne, boils, pustules, folliculitis, and carbuncles, among others. Commonly beauty and personal care products incorporate some anti-bacterial preservatives. Nevertheless, synthetic surfactants can induce skin irritation and allergic reactions by interaction with proteins such as keratin (cytoskeletal proteins) or collagen and elastin (extracellular matrix proteins). Thus, natural compounds with anti-bacterial activity against *S. aureus* are very interesting for the pharmaceutical industry [74].

The anti-bacterial activity of rhamnolipids had also been described. Sabarinathan, et al. described the anti-bacterial action of rhamnolipids from *Pseudomonas plecoglossicida* against *Staphylococcus aureus, Bacillus subtilis* and *Aeromonas hydrophila*. According to authors, the highest anti-bacterial action was detected to *B. subtilis* [75]. In another study, it was found that rhamnolipid from *Pseudomonas plecoglossicida* BP03 exhibited an anti-biofilm effect against *S. aureus*. The best results were observed using 100 µg mL^{-1} of rhamnolipids. The authors also identified a reduction on exopolysaccharide production of *S. aureus*, a biopolymer involved in the biofilm formation. The synergistic effect of rhamnolipid with caprylic acid to inhibit biofilms of *P. aeruginosa* and *S. aureus* was also described [76].

The effectiveness of iturin in inhibiting bacterial growth has also been evaluated. According Wan et al. the anti-bacterial effect of iturin depends on the number of carbon atoms on the fatty acid side chain, which may be attributed to the increased interaction with bacterial biofilms [77].

Since 1972 it has been found that fatty acids exhibit efficacy against bacteria and yeast. Lauric acid was found to be the most effective indicating that hydrophobicity plays a role in microefficacy [78]. The anti-bacterial activity of fatty acids against human pathogens has already been reported [79, 80]. The mechanisms of the anti-bacterial action of fatty acids involve cell-to-cell signaling interference, interruption of bacterial adhesion and biofilm formation, disruption of electron transport and oxidative phosphorylation, and membrane lysis [65].

The efficacy of saturated and unsaturated fatty acids against oral pathogens has been studied [81]. The authors identified the strong anti-microbial effect of the ε-3 and the ε-6 poly unsaturated fatty acids against *Porphyromona gingivalis* KCTC, a periodontitis etiological agent [82]. Huang et al. studied the anti-bacterial activity of omega-6, -7, -9 (n-6, n-7, n-9) fatty acids against various oral microorganisms. The fatty acids were tested and their esters showed anti-microbial activity against the periodontopathogens, *A. actinomycetemcomitans* and *P. gingivalis* [83]. Although in these studies commercial sources of fatty acid were used, the microorganisms were able to produce large quantities of fatty acids during growth on specific substrates [62]. Free fatty acids can disrupt cell-to-cell signaling and can also stop bacterial adhesion and biofilm formation. Saturated fatty acids which have carbon numbers greater than five can stop the swarming behavior of the urinary tract pathogens. Saturated fatty acids with carbon number up to five can inhibit *Proteus mirabilis* grow [65].

Anti-microbial activity of polymeric surfactants (alasan, liposan, lipomannan, emulsan) have also been described [84]. Emulsan, a biosurfactant produced by *Acinetobacter calcoaceticus*, is a polyanionic amphipathic heteropolysaccharide bioemulsifier [46]. These compounds exhibited anti-bacterial activity [85] and can function at higher temperatures. Hyder evaluated the production of bioemulsifier by *Acinetobacter* sp. isolates. The results showed that bioemulsifier reduced the growth of *E. coli*, *Salmonella* sp, *S. aureus*, and *P. aeruginosa*, with the greatest activity on the growth of *S. aureus*, followed by *P. aeruginosa*, and *E. coli* and *Salmonella* sp., respectively [86].

11.4.2 Biosurfactants as Anti-viral Agents

Viruses are related to apoptosis induction cycle arrest and survival signaling numerous human diseases affecting the global health and economy. Pandemics caused by viruses can cause the death of thousands of people such as the Coronavirus disease in 2019 (COVID-19) that has already caused 5 million deaths worldwide according to the World Health Organization [87]. Specific anti-viral treatments including vaccines are not sufficient to control the emergence and reemergence of viral diseases. Consequently, the search for new anti-viral compounds is still necessary. In this sense, several studies have shown the great potential of biosurfactants as anti-viral drug candidates.

Surfactins have been studied against different virus families such as vesicular the stomatitis virus (VSV, rhabdoviridae), suid herpes virus type 1 (SHV-1, herpesviridae, a model virus for human herpes viruses), and Semliki Forest virus (SFV, Togaviridae, a model virus for the hepatitis G virus). According to the study, the anti-viral effect of surfactin depends on the hydrophobicity of the fatty acid moiety, the charge of the peptide moiety as well as on the virus species. Indeed, a surfactin containing a fatty acid chain moiety of 15 carbon and one negative charge presented the highest anti-viral activity [67].

Lipopeptides have been also tested against the influenza virus. Wu et al. showed that the super-short membrane-active lipopeptides were able to inhibit the viral replication of the influenza virus with IC50 value of 7.30 ± 1.57 and 8.48 ± 0.74 mg L^{-1} against A/Puerto Rico/8/34 strain, and 6.14 ± 1.45 and 7.22 ± 0.67 mg L^{-1} against A/Aichi/2/68 strain, respectively. According to the authors, the lipopeptide inhibits the virus entry in the host cell by interacting with the HA2 subunit of hemagglutinin (HA) [88].

Surfactins also exhibit a broad-spectrum of anti-viral activity. Yuan et al. [68] showed that surfactins possess anti-viral activity against a variety of enveloped viruses, including the herpes simplex virus (HSV-1, HSV-2), vesicular stomatitis virus (VSV), simian immunodeficiency virus (SIV), and Newcastle disease virus (NDV). Due to the cytotoxicity of surfactin the synthetic surfactins have been in development in order to generate a nontoxic molecule [68]. The protection against enveloped viral infections in mammal cells with biosurfactants has been studied. Yuan, et al. demonstrated that oral administration of surfactin protects piglets from porcine epidemic diarrhea virus infection. According to the authors, the insertion of surfactin into the viral envelope lipids reduces the probability of viral fusion. This study indicates that surfactin has great potential against enveloped viruses in the digestive tract [89]. In another study, Vollenbroich et al. also observed that surfactin caused the inactivation of enveloped viruses, especially herpes- and retroviruses [90]. Anti-viral activity of lipopeptides (pumilicidin A and B) from *Bacillus pumilus* against HSV-1 had already been observed [91]. Itokawa, et al. showed that the anti-viral activity of surfactin produced by *Micromonospora* sp. CPCC 202787 against HIV [92]. Jonson et al. studying the role for bacteria in shaping coronavirus infection observed that surfactin inhibited this virus. The authors proposed that the peptidoglycan-associated surfactin is a potent viricidal compound that disrupts virus integrity with broad activity against enveloped viruses [93]. The results showed the inhibitory effect of surfactin against a broad range of enveloped viruses, including influenza, Ebola, Zika, Nipah, chikungunya, Una, Mayaro, Dugbe, and Crimean-Congo hemorrhagic fever [93].

11.4.3 Biosurfactants as Anti-fungal Agents

An increasing number of fungal infection are also reported worldwide. Cutaneous and subcutaneous mycoses are an important source of morbidity, especially in immunocompromised patients [94]. Dermatophytosis is responsible for 20–25% of fungal infections worldwide [95] The most common fungal skin infections are caused by yeasts (such as *Candida albicans* or *Malassezia furfur*) *Epidermophyton, Microsporum* and *Trichophyton*. Dermatophytes [96].

The activity of biosurfactants against fungal infections might be attributed to the destabilization of cellular membranes causing cytoplasmic extrusions and eventually resulting in the rupture of cells [66]. *Rhodotorula babjevae* YS3, isolated from an agricultural field in Assam, Northeast India, exhibited promising anti-fungal activity against a very broad group of pathogenic fungi including *Trichophyton rubrum*, a human pathogen fungal infection found in superficial mycoses on glabrous skin and nails [66,97,98].

Candida albicans is a pathogen able to colonize the gastrointestinal tract, vagina, mouth, mucosa, and skin, which causes systemic infections due contaminated implants, urethral catheters, nasolaryngeal tubes, or stents [99]. *Serratia marcescens* showed anti-microbial activity towards *C. albicans* due the production of a type of glycolipid biosurfactant composed of glucose and palmitic acid. In this study, the anti-adhesive properties were tested and the attachment of *C. albicans* to polystyrene microtiter plate surfaces were significantly inhibited up to 76% using 50 mg ml^{-1} of biosurfactant. In addition, preformed biofilms of the yeast were disrupted up to 60% and 88% at concentrations of 50 and 100 mg ml^{-1}

[100]. A bioemulsifier produced by *Serratia marcescens* S10 (isolated from the gut of the American cockroach) inhibited the growth of *Candida albicans, Aspergillus niger* and *Geotricum* spp [101].

Lipopeptide biosurfactant surfactin-C15 and its complexes with divalent counterions were evaluated against *Candida albicans*. The SF-C15 isoform was originally obtained from *Bacillus subtilis* #309. The results showed the inhibitory effect on biofilm formation and preformed biofilms. Moreover, both isoform and its metal(II) complexes reduced the mRNA expression of hypha-specific genes. Therefore, these results suggested the application of lipopeptide biosurfactants against *C. albicans* related infections [99].

B. subtilis sub sp *subtilis* strains A52 isolated from a marine sediment produced two antimicrobial peptides (AMPs). The authors showed that the peptides inhibited the growth of *Candida* and filamentous fungi. In addition, the emulgel formulation of surfactin-like lipopeptide showed anti-fungal activity and did not show toxicity effects in mice. Furthermore, surfactin-like lipopeptides exhibited synergistic action with fluconazole against *Candida*, suggesting the potential therapeutic use of these molecules in topical applications [102]. Nelson et al. also investigated the potential of vaginal lactic acid bacteria (LAB) as anti-adhesive agents against *Candida* spp. In this study, it was verified that the LAB produce biosurfactants identified as surfactin, iturin, and lichenysin that showed strong anti-adherence activity. According to the authors, these results indicated the possibility of utilizing biosurfactants to prevent infections by pathogenic *Candida* spp [103]. Indeed, iturin has been already used as an anti-fungal for the treatment of human and animal mycoses due to its low toxicity and the lack of allergic effects [77, 104]. Taken together, these studies indicate that biosurfactants function as potential anti-fungal substances against fungal infections.

11.5 Other Therapeutic Applications of Biosurfactants

Beyond the applications mentioned above, biosurfactants also have other pharmacological activity as thrombolytic agents, inhibition of platelet aggregation, and anti-hypertensive agents, among others. Using a rat pulmonary embolism model, Kikuchi and Hasumi reported that surfactin was able to increase the plasma clot lysis when it was injected together with prourokinase. Surfactin shows many advantages over other traditional thrombolytic agents inducing fewer side effects permitting its utilization for long-term use [105]. In another study, it was found that surfactin was able to prevent platelet aggregation leading to the inhibition of fibrin clot formation, and enhance the fibrinolysis [106]. The anti-hypertensive activity of biosurfactants had also been described. The lipopeptides produced by *Bacillus mojavensis* showed angiotensin-converting enzyme (ACE) inhibitory activity and could be a promising anti-hypertensive agent [107].

In addition, biosurfactants can function as drug delivery agents. Inoh and co-workers reported that a mannosylerythritol lipid dramatically increases the gene transfection by cationic liposomes, via membrane fusion [108]. In another study, the role of surfactin as an ingredient of a microemulsion systems was demonstrated. In these systems, microbial surfactants may be used to replace the synthetic compounds [22]. Zhang and co-workers described the ability of surfactin to deliver insulin with a higher bioavailability when compared to the oral insulin administration, showing its potential for insulin delivery in blood glucose control [109]. In addition, lipopeptides produced by bacteria could be used as immunological adjuvants when mixed with conventional antigens [110].

11.6 Concluding Remarks

Natural products are a powerful source for discovering new compounds that can be used to develop effective anti-tumor, anti-inflammatory, and anti-microbial drugs. Microbial biosurfactants are biomolecules obtained mainly from bacteria, yeast, and fungi. They present a biocompatible nature that exhibits emulsifying, anti-adhesive, anti-oxidant, anti-microbial, anti-tumor, and anti-inflammatory activities. Therefore, biosurfactants have a wide range of applications in household detergents, cosmetics, agriculture, and in pharmaceutical industries. However, the mechanism of action and specific targets of biosurfactants responsible for their biological effects need to be further investigated. Furthermore, the development of new strategies for improving the large-scale synthesis and to reduce production costs, is also essential. Despite that, biosurfactants from safe microorganisms can be used in green technologies with little impact on the environment and great biological properties.

References

1 Markande AR, Patel D, Varjani S. A review on biosurfactants: properties, applications and current developments. *Bioresour Technol.* 2021;30:124963.
2 Ribeiro BG, Guerra JMC, Sarubbo LA. Biosurfactants: production and application prospects in the food industry. *Biotechnol Prog.* 2020;36(5):e3030.
3 Ribeiro IA, Rosário Bronze M, Castro MF, Ribeiro MHL. Optimization and correlation of HPLC–ELSD and HPLC–MS/MS methods for identification and characterization of sophorolipids. *J Chromatogr B.* 2012;899:72–80.
4 Venkataraman S, Rajendran DS, Kumar PS, Vo DVN, Vaidyanathan VK. Extraction, purification and applications of biosurfactants based on microbial-derived glycolipids and lipopeptides: a review. *Environ Chem Lett.* 2022;20:949–970.
5 Chen HL, Juang RS. Recovery and separation of surfactin from pretreated fermentation broths by physical and chemical extraction. *Biochem Eng J.* 2008;38(1):39–46.
6 Joy S, Khare SK, Sharma S. Synergistic extraction using sweep-floc coagulation and acidification of rhamnolipid produced from industrial lignocellulosic hydrolysate in a bioreactor using sequential (fill-and-draw) approach. *Process Biochem.* 2020;90:233–240.
7 Hubert J, Ple K, Hamzaoui M, Nuissier G, Hadef I, Reynaud R, Guilleret A, Renault JH. New perspectives for microbial glycolipid fractionation and purification processes. *C R Chim.* 2012;15(1):18–28.
8 Dolman BM, Wang F, Winterburn JB. Integrated production and separation of biosurfactants. *Process Biochem.* 2019;83:1–8.
9 Yang X, Zhu L, Xue C, Chen Y, Qu L, Lu W. Recovery of purified lactonic sophorolipids by spontaneous crystallization during the fermentation of sugarcane molasses with Candida albicans O-13-1. *Enzyme Microb Technol.* 2012;51(6–7):348–353.

10 Bajaj V, Tilay A, Annapure U. Enhanced production of bioactive sophorolipids by Starmerella bombicola NRRL Y-17069 by design of experiment approach with successive purification and characterization. *J Oleo Sci.* 2012;61(7):377–386.

11 Li Z, Zhang Y, Lin J, Wang W, Li S. High-yield di-rhamnolipid production by Pseudomonas aeruginosa YM4 and its potential application in MEOR. *Molecules.* 2019;24(7):1433.

12 Satpute SK, Banpurkar AG, Dhakephalkar PK, Banat IM, Chopade BA. Methods for investigating biosurfactants and bioemulsifiers: a review. *Crit Rev Biotechnol.* 2010;30(2):127–144.

13 Coutte F, Lecouturier D, Leclère V, Béchet M, Jacques P, Dhulster P. New integrated bioprocess for the continuous production, extraction and purification of lipopeptides produced by Bacillus subtilis in membrane bioreactor. *Process Biochem.* 2013;48(1):25–32. https://doi.org/10.1016/j.procbio.2012.10.005.

14 Faria AF, Stefani D, Vaz BG, Silva IS, Garcia JS, Eberlin MN, et al. Purification and structural characterization of fengycin homologues produced by Bacillus subtilis LSFM-05 grown on raw glycerol. *J Ind Microbiol Biotechnol.* 2011;38:863–871.

15 Meena KR, Dhiman R, Singh K, Kumar S, Sharma A, Kanwar SS, et al. Purification and identification of a surfactin biosurfactant and engine oil degradation by *Bacillus velezensis* KLP2016. *Microb Cell Fact.* 2021;20(26):https://doi.org/10.1186/s12934-021-01519-0.

16 Vigneshwaran C, Vasantharaj K, Krishnanand N, Sivasubramanian V. Production optimization, purification and characterization of lipopeptide biosurfactant obtained from Brevibacillus sp. AVN13. *J Environ Chem Eng.* 2021;9(1):104867. https://doi.org/10.1016/j.jece.2020.104867.

17 Ferlay J, Ervik M, Lam F, Colombet M, Mery L, Piñeros M, et al. Global cancer observatory: cancer today. Lyon: International Agency for Research on Cancer. 2020 https://gco.iarc.fr/today accessed november 2021.

18 Hanahan D, Weinberg RA. Hallmarks of cancer: the next generation. *Cell.* 2011;144(5):646–674.

19 Popper HH. Progression and metastasis of lung cancer. *Cancer Metastasis Rev.* 2016;35:75–91.

20 Niethammer AG, Xiang R, Becker JC, Wodrich H, Pertl U, Karsten G, Eliceiri BP, Reisfeld RA. A DNA vaccine against VEGF receptor 2 prevents effective angiogenesis and inhibits tumor growth. *Nature Med.* 2002;8(12):1369–1375.

21 Duarte C, Gudiña EJ, Lima CF, Rodrigues LR. *Effects of biosurfactants on the viability and proliferation of human breast cancer cells AMB Express.* 2014;15(4):40.

22 Gudiña EJ, Rangarajan V, Sen R, Rodrigues LR. Potential therapeutic applications of biosurfactants. *Trends Pharmacol Sci.* 2013;34(12):667–675.

23 Li H, Guo W, Ma XJ, Li JS, Song X. In vitro and in vivo anticancer activity of sophorolipids to human cervical cancer. *Appl Biochem Biotechnol.* 2017;181(4):1372–1387.

24 Fu SL, Wallner SR, Bowne WB, Hagler MD, Zenilman ME, Gross R, Bluth MH. Sophorolipids and their derivatives are lethal against human pancreatic cancer cells. *J Surg Res.* 2008;148:77–82.

25 Chen J, Song X, Zhang H, Qu YB, Miao JY. Production, structure elucidation and anticancer properties of sophorolipid from *Wickerhamiella domercqiae*. *Enzyme Microb Technol.* 2006;39:501–506.

26 Joshi-Navare K, Shiras A, Prabhune A. Differentiation-inducing ability of sophorolipids of oleic and linoleic acids using a glioma cell line. *Biotechnol J.* 2011;6:509–512.

27 Shao L, Song X, Ma X, Li H, Qu Y. Bioactivities of sophorolipid with different structures against human esophageal cancer cells. *J Surg Res.* 2012;173:286–291.
28 Chen J, Song X, Zhang H, Qu YB, Miao JY. Sophorolipid produced from the new yeast strain *Wickerhamiella domercqiae* induces apoptosis in H7402 human liver cancer cells. *Appl Microbiol Biotechnol.* 2006;72:52–59.
29 Ribeiro IA, Faustino CM, Guerreiro PS, Frade RF, Bronze MR, Castro MF, Ribeiro MH. Development of novel sophorolipids with improved cytotoxic activity toward MDA-MB-231 breast cancer cells. *J Mol Recognit.* 2015;28:155–165.
30 Gorman AM, Healy SJ, Jager R, Samali A. Stress management at the ER: regulators of ER stress-induced apoptosis. *Pharmacol Ther.* 2012;134:306–316.
31 Li J, Lee AS. Stress induction of GRP78/BiP and its role in cancer. *Curr Mol Med.* 2006;6:45–54.
32 Thakur P, Saini NK, Thakur VK, Gupta VK, Saini RV, Saini AK. Rhamnolipid the glycolipid biosurfactant: emerging trends and promising strategies in the field of biotechnology and biomedicine. *Microb Cell Fact.* 2021;20(1):1.
33 Christova N, Tuleva B, Kril A, Georgieva M, Konstantinov S, Terziyski I, et al. Chemical structure and in vitro antitumor activity of rhamnolipids from Pseudomonas aeruginosa BN10. *Appl Biochem Biotechnol.* 2013;170:676–689.
34 Rahimi K, Lotfabad TB, Jabeen F, Ganji SM. Cytotoxic efects of mono-and di-rhamnolipids from Pseudomonas aeruginosa MR01 on MCF-7 human breast cancer cells. *Colloid Surf B.* 2019;181:943–952.
35 Lotfabad TB, Abassi H, Ahmadkhaniha R, Roostaazad R, Masoomi F, Zahiri HS, et al. Structural characterization of a rhamnolipid-type biosurfactant produced by *Pseudomonas aeruginosa* MR01: enhancement of di-rhamnolipid proportion using gamma irradiation. *Colloid Surf B.* 2010;81:397–405.
36 Kim S-Y, Kim JY, Kim S-H, Bae HJ, Yi H, Yoon SH, Koo BS, Kwon M, Cho JY, Lee C-E, Hong S. Surfactin from Bacillus subtilis displays anti-proliferative effect via apoptosis induction, cell cycle arrest and survival signaling suppression. *FEBS Lett.* 2007;581:865–871.
37 Liu X, Tao X, Zou A, Yang S, Zhang L, Mu B. Effect of the microbial lipopeptide on tumor cell lines: apoptosis induced by disturbing the fatty acid composition of the cell membrane. *Protein Cell.* 2010;1:584–594.
38 Cao XH, Wang AH, Wang CL, Mao DZ, Lu MF, Cui YQ, Jiao RZ. Surfactin induces apoptosis in human breast cancer MCF-7 cells through a ROS/JNK mediated mitochondrial/caspase pathway. *Chem Biol Interact.* 2010;183:357–362.
39 Lee JH, Nam SH, Seo WT, Yun HD, Hong SY, Kim MK, et al. The production of surfactin during the fermentation of cheonggukjang by potential probiotic Bacillus subtilis CSY191 and the resultant growth suppression of MCF-7 human breast cancer cells. *Food Chem.* 2012;131:1347–1354.
40 Zhao H, Yan L, Guo L, Sun H, Huang Q, Shao D, et al. Effects of Bacillus subtilis iturin A on HepG2 cells in vitro and vivo. *AMB Express.* 2021;11(1):67. doi.org/10.1186/s13568-021-01226-4.
41 Zhao H, Zhao X, Lei S, Zhang Y, Shao D, Jiang C, et al. Efect of cell culture models on the evaluation of anticancer activity and mechanism analysis of the potential bioactive compound, iturin A, produced by Bacillus subtilis. *Food Funct.* 2019a;10(3):1478–1489.

42 Dey G, Bharti R, Ojha PK, Pal I, Rajesh Y, Banerjee I, et al. Therapeutic implication of "Iturin A" for targeting MD-2/TLR4 complex to overcome angiogenesis and invasion. *Cell Signal*. 2017;35:24–36.

43 Zhao H, Xu X, Lei S, Shao D, Jiang C, Shi J, et al. Iturin A-like lipopeptides from Bacillus subtilis trigger apoptosis, paraptosis, and autophagy in Caco-2 cells. *J Cell Physiol* 2019b;234(5):6414–6427.

44 Zhu X, Liu Y, Liu S, Diao F, Xu R, Ni X. Lipopolysaccharide primes macrophages to increase nitric oxide production in response to *Staphylococcus aureus*. *Immunol Lett*. 2007;112:75–81.

45 Andreasen AS, Krabbe KS, Krogh-Madsen R, Taudorf S, Pedersen BK, Møller K. Human endotoxemia as a model of systemic inflammation. *Curr Med Chem*. 2008;15:1697–1705.

46 Ferguson LR, Laing WA. Chronic inflammation, mutation and human disease. *Mutat Res*. 2010;690:3–11.

47 Zhao H, Shao D, Jiang C, Shi J, Li Q, Huang Q, et al. Biological activity of lipopeptides from Bacillus. *Appl Microbiol Biotechnol*. 2017; 101(15):5951–5960.

48 Subramaniam MD, Venkatesan D, Iyer M, Subbarayan S, Govindasami V, Roy A, et al. Biosurfactants and anti-inflammatory activity: a potential new approach towards COVID-19. *Curr Opin Environ Sci Health*. 2020;17:72–81.

49 Gein S, Kuyukina M, Ivshina I, Baeva T, Chereshnev V. In vitro cytokine stimulation assay for glycolipid biosurfactant from *Rhodococcus ruber*: role of monocyte adhesion. *Cytotechnology*. 2011;63:559–566.

50 Zhang Y, Liu C, Dong B, Ma X, Hou L, Cao X, et al. Anti-inflammatory activity and mechanism of surfactin in lipopolysaccharide-activated macrophages. *Inflamm*. 2015;38:756–764.

51 Park SY, Kim J-H, Lee SJ, Kim Y. Involvement of PKA and HO-1 signaling in anti-inflammatory effects of surfactin in BV-2 microglial cells. *Toxicol Appl Pharmacol*. 2013;268:68–78.

52 Park SY, Kim Y Surfactin inhibits immunostimulatory function of macrophages through blocking NK-kappaB, MAPK and Akt pathway. *Int Immunopharmacol*. 2009;9(7–8):886–893.

53 Hagler M, Smith-Norowitz T, Chice S, Wallner S, Viterbo D, Mueller C, et al. Sophorolipids decrease IgE production in U266 cells by downregulation of BSAP (Pax5), TLR-2, STAT3 and IL-6. *J Allergy Clin Immunol*. 2007;119:S263.

54 Vakil H, Sethi S, Fu S, Stanek A, Wallner S, Gross R, et al. Sophorolipids decrease pulmonary inflammation in a mouse asthma model. *Nature*. 2010;90:392A–392A.

55 Hardin R, Pierre J, Schulze R, Mueller CM, Fu SL, Wallner SR, et al. Sophorolipids improve sepsis survival: effects of dosing and derivatives. *J Surg Res*. 2007;142:314–319.

56 Mueller CM, Lin Y, Viterbo D, Pierre J, Murray SA, Shah V, et al. Sophorolipid treatment decreases inflammatory cytokine expression in an in vitro model of experimental sepsis. 2006:A204.

57 Bluth M, Smith-Norowitz T, Hagler M, Beckford R, Chice S, Shah V, et al. Sophorolipids decrease IgE production in U266 cells *J Allergy Clin Immunol*. 2006;117:S202.

58 Morita Y, Tadokoro S, Sasai M, Kitamoto D, Hirashima N. Biosurfactant mannosyl-erythritol lipid inhibits secretion of inflammatory mediators from RBL-2H3 cells. *BBA - Gen Subjects*. 2011;1810:1302–1308.

59 Yang M, Cell pyroptosis, a potential pathogenic mechanism of 2019-nCoV infection (January 29, 2020). Available at SSRN: https://ssrn.com/abstract=3527420 or http://dx.doi.org/10.2139/ssrn.3527420.

60 Wang W, Liu X, Wu S, Chen S, Li Y, Nong L, et al. Definition and risks of cytokine release syndrome in 11 critically Ill COVID-19 patients with pneumonia: analysis of disease characteristics *J Infect Dis* 2020; 222(9):1444–1451.

61 Jarvis FG, Johnson MJ. A Glyco-lipide produced by *Pseudomonas Aeruginosa*. *J Am Chem Soc*. 1949;71(12):4124–4126.

62 Desai JD, Banat IM. Microbial production of surfactants and their commercial potential. *Microbiol Mol Biol Rev*. 1997;61:47–64.

63 Chen WC, Juang RS, Wei YH. Applications of a lipopeptide biosurfactant, surfactin, produced by microorganisms. *Biochem Eng J*. 2015;103:158–169.

64 Asadi A, Abdolmaleki A, Azizi-Shalbaf S, Gurushankar K. Molecular dynamics study of surfactin interaction with lipid bilayer membranes. *Gene, Cell and Tissue*. 2021;8(2):e112646. doi: 10.5812/gct.112646.

65 Mukherjee D, Rooj B, Manda U. Antibacterial biosurfactants. In: Inamuddin, Ahamed MI, Prasad R, editors. *Microbial Biosurfactants Preparation, Properties and Applications*. Springer, Singapore; 2021. pp. 217–291.

66 Sen S, Borah SN, Bora A, Deka S. Production, characterization, and antifungal activity of a biosurfactant produced by *Rhodotorula babjevae* YS3. *Microb Cell Factories*. 2017;16(1):95. https://doi.org/10.1186/s12934-017-0711-z.

67 Kracht M, Rokos H, Ozel M, Kowall M, Pauli G, Vater J. Antiviral and haemolytic activities of surfactin isoforms and their methyl ester derivatives. *J Antibiot*. 1999;52:613–619.

68 Yuan L, Zhang S, Peng J, Li Y, Yang Q. Synthetic surfactin analogues have improved anti-PEDV properties. *PLoS ONE*. 2019;14(4):e0215227. https://doi.org/10.1371/journal.pone.0215227.

69 Chang -H-H, Cohen T, Grad YH, Hanage WP, O'Brien TF, Lipsitch M. Origin and proliferation of multiple-drug resistance in bacterial pathogens. *Microbiol Mol Biol Rev*. 2015;79(1):101–116. doi: 10.1128/MMBR.00039-14.

70 Sharma D Saharan BS Functional characterization of biomedical potential of biosurfactant produced by *Lactobacillus helveticus*. *Biotechnol Rep*. 2016;11:27–35.

71 Giri SS, Ryu E, Sukumaran V, Park SC. Atividades antioxidantes, antibacterianas e anti-adesivas de biossurfactantes isolados de cepas de Bacillus. *Microb Pathog*. 2019;132:66–72.

72 Zhao J, Xu L, Zhang H, Zhuo Y, Weng Y, Li S, et al. Surfactin-methylene blue complex under LED illumination for antibacterial photodynamic therapy: enhanced methylene blue transcellular accumulation assisted by surfactin. *Colloids and Surf B: Biointerfaces*. 2021;207:111974. doi.org/10.1016/j.colsurfb.2021.111974.

73 Meena KR, Sharma A, Kanwar SS. Antitumoral and antimicrobial activity of Surfactin extracted from Bacillus subtilis KLP2015. *Int J Pept Res Ther*. 2020;26:423–433.

74 Vecino X, Rodríguez-López L, Ferreira D, Cruz JM, Moldes AB. Rodrigues LR Bioactivity of glycolipopeptide cell-bound biosurfactants against skin pathogens. *Int J Biol Macromol*. 2018;09:971–979.

75 Sabarinathan D, Vanaraj S, Sathiskumar S, Chandrika SP, Sivarasan G, Arumugam SS, et al. Caracterização e aplicação de ramnolipídeo de *Pseudomonas plecoglossicida* BP03 *Lett Appl Microbiol*. 2020;72:251–262.
76 Chong H, Li Q. Microbial production of rhamnolipids: opportunities, challenges and strategies. *Microb Cell Fact*. 2017;16(1):137.
77 Wan C, Fan X, Lou Z, Wang H, Olatunde A, Rengasamy KRR. Iturin: cyclic lipopeptide with multifunction biological potential. *Crit Rev Food Sci Nutr*. 2021. doi: 10.1080/10408398.2021.1922355.
78 Falk NA. Surfactants as antimicrobials: a brief overview of microbial interfacial chemistry and surfactant antimicrobial activity. *J Surfactant Deterg*. 2019;22:1119–1127.
79 Newman DJ, Cragg GM, Snader KM. Natural products as sources of new drugs over the period 1981-2002. *J Nat Prod*. 2003;66:1022–1037.
80 Choi JS, Ha YM, Joo CU, Cho KK, Kim SJ, Choi IS. Inhibition of oral pathogens and collagenase activity by seaweed extracts. *J Environ Biol*. 2012;33:115–121.
81 Choi JS, Park NH, Hwang SY, Sohn JH, Kwak I, Cho KK, et al. The antibacterial activity of various saturated and unsaturated fatty acids against several oral pathogens *J Environ Biol*. 2013; 34(4):673–676.
82 Carvalho C, Cabral CT. Role of *Porphyromonas gingivalis* in periodontal disease. *Rev Port Estomato Cir Maxilofac*. 2007;48:167–171.
83 Huang CB, George B, Ebersole JL. Antimicrobial activity of n-6, n-7 and n-9 fatty acids and their esters for oral microorganisms. *Arch Oral Biol*. 2010;55(8):555–560.
84 Sharma J, Sundar D, Srivastava P. Biosurfactants: potential agents for controlling cellular communication, motility, and antagonism. *Front Mol Biosci*. 2021;8:727070. doi: 10.3389/fmolb.2021.727070.
85 Dhagat S, Jujjavarapu SE. Isolation of a novel thermophilic bacterium capable of producing high-yield bioemulsifier and its kinetic modelling aspects along with proposed metabolic pathway. *Braz J Microbiol*. 2020;51(1):135–143.
86 Hyder HN. Production, characterization and antimicrobial activity of a bioemulsifier produced by *Acinetobacter baumanii* AC5 utilizing edible oils. *Iraq J Biotechnol*. 2015;14:55–70.
87 World Health Organization. Coronavirus (COVID-19) dashboard. 2021, December 3). Available on: https://covid19.who.int/measures.
88 Wu W, Wang J, Lin D, Chen L, Xie X, Shen X, et al. Super short membrane-active lipopeptides inhibiting the entry of influenza A virus. *Biochim Biophys Acta*. 2015 Oct;1848(10 Pt A):2344–2350. doi: 10.1016/j.bbamem.2015.06.015.
89 Yuan L, Zhang S, Wang Y, Li Y, Wang X, Yang Q. Surfactin inhibits membrane fusion during invasion of epithelial cells by enveloped viruses. *J Virol*. 2018;92:e00809–18. https://doi.org/10.1128/JVI.00809-18.
90 Vollenbroich D, Ozel M, Vater J, Kamp RM, Pauli G. Mechanism of inactivation of enveloped viruses by the biosurfactant surfactin from *Bacillus subtilis*. *Biologicals*. 1997;25:289–297.
91 Naruse N, Tenmyo O, Kobaru S, Kamei H, Miyaki T, Konishi M, et al. Pumilacidin, a complex of new antiviral antibiotics: production, isolation, chemical properties, structure and biological activity *J Antibiot*. 1990;43:267–280.

92 Itokawa H, Miyashita T, Morita H, Takeya K, Hirano T, Homma M, et al. Structural and conformational studies of [Ile7] and [Leu7] surfactins from *Bacillus subtilis Chem Pharm Bull.* 1994;42:604–607.

93 Johnson BA, Hage A, Kalveram B, Mears M, Plante JA, Rodriguez SE, et al. Peptidoglycan-associated cyclic lipopeptide disrupts viral infectivity *J Virol.* 2019;93(22):e01282-19. doi: 10.1128/JVI.01282-19.

94 Guegan S, Lanternierd F, Rouzaudf C, Duping N, Lortholaryd O. Fungal skin and soft tissue infections. *Curr Opin Infect Dis.* 2016;29:124–130.

95 Gupta AK, Versteeg SG, Shear NH. Onychomycosis in the 21st century: an update on diagnosis, epidemiology, and treatment. *J Cutan Med Surg.* 2017;21:525–539.

96 Tortorano AM, Richardson M, Roilides E, et al. European society of clinical microbiology and infectious diseases fungal infection study group. European confederation of medical mycology. ESCMID and ECMM joint guidelines on diagnosis and management of hyalohyphomycosis: *fusarium spp., Scedosporium spp.* and others. *Clin Microbiol Infect.* 2014;203:27–46.

97 Zhan P, Dukik K, Li D, Sun J, Stielow JB, Gerrits van den Ende B, et al. Phylogeny of dermatophytes with genomic character evaluation of clinically distinct *Trichophyton rubrum* and *T. violaceum. Stud Mycol.* 2018;89:153–175.

98 Peixoto I, Maquine G, Francesconi VA, Francesconi F. Dermatophytosis caused by *Tricophyton rubrum* as an opportunistic infection in patients with Cushing disease. *An Bras Dermatol.* 2010;85(6):888–890.

99 Janek T, Drzymała K, Dobrowolski A. *In vitro* efficacy of the lipopeptide biosurfactant surfactin-C15 and its complexes with divalent counterions to inhibit *Candida albicans* biofilm and hyphal formation. *Biofouling.* 2020;36(2):210–221.

100 Dusane DH, Pawar VS, Nancharaiah YV, Venugopalan VP, Kumar AR, Zinjarde SS. Anti-biofilm potential of a glycolipid surfactant produced by a tropical marine strain of *Serratia marcescens. Biofouling.* 2011;27(6):645–654.

101 Ahmed EF, Hassan SS. Antimicrobial activity of a bioemulsifier produced by Serratia marcescens S10. *ANJS.* 2013;16(1):147–155.

102 Sharma D, Singh SS, Baindara P, Sharma S, Khatri N, Grover V, et al. Surfactin like broad spectrum antimicrobial lipopeptide Co-produced with Sublancin from *Bacillus subtilis* strain A52: dual reservoir of bioactives *Front Microbiol.* 2020;11:1167. doi: 10.3389/fmicb.2020.01167.

103 Nelson J, El-Gendy AO, Mansy MS, Ramadan MA, Aziz RK. The biosurfactants iturin, lichenysin and surfactin, from vaginally isolated lactobacilli, prevent biofilm formation by pathogenic *Candida. FEMS Microbiol Lett.* 2020;367(15):fnaa126. doi: 10.1093/femsle/fnaa126.

104 Dang Y, Zhao F, Liu X, Fan X, Huang R, Gao W, et al. Enhanced production of antifungal lipopeptide iturin A by *Bacillus amyloliquefaciens* LL3 through metabolic engineering and culture conditions optimization *Microb Cell Fact.* 2019;18(1):68. doi: 10.1186/s12934-019-1121-1.

105 Kikuchi T, Hasumi K. Enhancement of reciprocal activation of prourokinase and plasminogen by the bacterial lipopeptide surfactins and iturin Cs. *J Antibiot.* 2003;56(1):34–37.

106 Lim JH, Park BK, Kim MS, Hwang MH, Rhee MH, Park SC, et al. The anti-thrombotic activity of surfactins. *J Vet Sci* 2005; 6(4):353–355.
107 Ghazala I, Bouassida M, Krichen F, Manuel Benito J, EllouzChaabouni S, Haddar A. Anionic lipopeptides from *Bacillus mojavensis* I4 as efective antihypertensive agents: production, characterization, and identifcation. *Eng Life Sci.* 2005;17:1244–1253.
108 Inoh Y, Kitamoto D, Hirashima N, Nakanishi M. Biosurfactant MEL-A dramatically increases gene transfection via membrane fusion. *J Control Release.* 2004 10;94(2–3):423–431. doi: 10.1016/j.jconrel.2003.10.020. PMID: 14744492.
109 Zhang L, Xing X, Ding J, Zhao X, Qi G. Surfactin variants for intra-intestinal delivery of insulin. *Eur J Pharm Biopharm.* 2005;115:218–228.
110 Sandeep L, Rajasree S. Biosurfactant: pharmaceutical perspective. *J Anal Pharm Res.* 2017. 10.15406/japlr.2017.04.00105.

12

Fungal Biosurfactants

Applications in Agriculture and Environmental Bioremediation Processes

Láuren Machado Drumond de Souza[1], Débora Luiza Costa Barreto[1], Lívia da Costa Coelho[1], Elisa Amorim Amâncio Teixeira[1], Vívian Nicolau Gonçalves[1], Júlia de Paula Muzetti Ribeiro[1], Natana Gontijo Rabelo[1], Stephanie Evelinde Oliveira Alves[1], Mayanne Karla da Silva[1], Laura Beatriz Miranda Martins[1], Charles Lowell Cantrell[2], Stephen Oscar Duke[3], and Luiz Henrique Rosa[1],*

[1] *Departamento de Microbiologia, Universidade Federal de Minas Gerais, Brasil*
[2] *Natural Products Utilization Research Unit, Agricultural Research Service, United States Department of Agriculture, Oxford, Mississippi, USA*
[3] *National Center for Natural Products Research, School of Pharmacy, Oxford, Mississippi, USA*
* *Corresponding author*

12.1 Biosurfactants as Agrochemicals

Agriculture represents one of the most important economic sectors in the world, which contributes to food production and food security, as well as producton of fiber, biofuels, and feedstock for industrial processes [1]. In order to increase crop productivity, agrochemical products such as herbicides, fungicides, and insecticides are widely used to control pests; however, current agricultural methods follow unsustainable practices that result in a huge amount of pesticides contaminating air, soil, and water [2]. Many studies have reported the negative impacts caused by these synthetic compounds on the environment and on human health [3]. Recent biotechnological advances in the agricultural and medical sectors are being employed to discover sustainable, natural, and biodegradable compounds which are effective against different pests while having less impact on the environment [4].

Often used in crops, agricultural adjuvants such as surfactants are substances that facilitate the application of products in order to increase their efficiency. Such substances are added to the formulation of pesticides prior to their application partly due their capabilities to favorably modify the surface properties of liquids [5]. Most active ingredients present in pesticide formulations are insoluble or weakly soluble in water, making the addition of surfactants to the solution essential, as when sprayed, the pesticide will better adhere to the surface of its target and spread adequately, optimizing the biological effectiveness of the product [6]. According to Purwasena et al. [7], around 0.2 million tons per year of surfactants are used to protect crops in pesticide formulations. Among the surfactants used in agriculture, there are those called "green surfactants" or biosurfactants (Figure 12.1), which often can be derived from microorganisms such as bacteria and fungi. Additionally, biosurfactants have many advantages when compared to synthetic surfactants, being less toxic, more biodegradable and highly effective, making them ideal for use in conventional farming applications [8].

12.1.1 Biosurfactants as Herbicide Adjuvants

In agriculture, herbicides are extensively used to manage weeds. Herbicides represent about 60% of all pesticides used worldwide. For most herbicides to exert their toxic effects on the plant, uniform coverage of the treated parts is necessary, an objective facilitated by an adjuvant [9]. Adjuvants can also enhance absorption of the herbicide and increase rainfastness [10]. Currently, there are many compounds available on the market, including

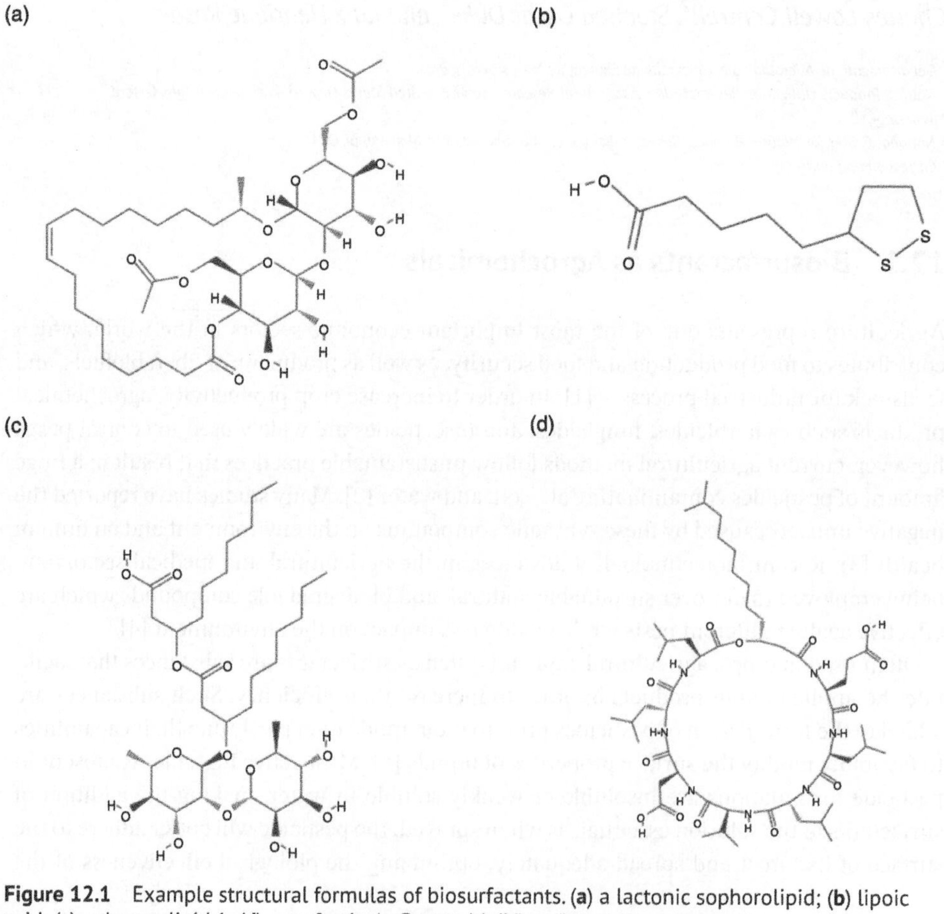

Figure 12.1 Example structural formulas of biosurfactants. (a) a lactonic sophorolipid; (b) lipoic acid; (c) a rhamnolipid 1; (d) a surfactin A. *Source:* MolView (https://molview.org).

surfactants, which are used to promote greater coverage of the leaves and increase the absorption of herbicides that also reduce the needed dose of the herbicide applied by more than 50% as compared to that used without the adjuvant [5].

Among the biosurfactants with potential for use as an adjuvant in agriculture are those produced by fungi. Vaughn et al. [11] reported sophorolipids produced by the yeast *Starmerella bombicola* as biosurfactants for post-emergence herbicides. In their study, lactone sophorolipids produced by *S. bombicola* (Sb) and non-lactone sophorolipids produced by *Candida kuoi* (Ck) were compared with a synthetic surfactant polyethoxylated tallow amine (POEA) that is used with some post-emergence commercial herbicides. *C. kuoi* sophorolipids formed longer-lasting and more stable emulsions with lemongrass oil (LGO) than Sb or T-15 (T-15 was chosen as the POEA for testing over T-5 and T-10, as it exhibited the best emulsifying abilities of the three synthetic surfactants), which is extremely important for the practical use of these compounds with lipophilic herbicides. Phytotoxicity (measured by reduced fresh and dry weight and visual damage three days after spraying) for *Senna obtusifolia* by the Ck/LGO and Sb/LGO mixtures was similar to a POEA/LGO mixture, while the visual damage to maize (*Zea mays* L.) was increased by the addition of all surfactants. When applied with the very water-soluble herbicide phosphinothricin, Ck and Sb enhanced reductions in *S. obtusifolia* fresh and dry weights and herbicide damage compared to phosphinothricin applied without surfactants. For corn, T-15 and Ck applied with PT resulted in the greatest reductions in fresh and dry weights and in HDR values. These results indicate that sophorolipids are excellent as natural surfactants for use with post-emergence herbicides.

12.1.2 Biosurfactants and Antifungal Activity

Some biosurfactants produced by fungi have antimicrobial activity against plant pathogens [8]. Most of these molecules produced by fungi have high biodegradability and low toxicity, which are important characteristics for large-scale, commercial use for plant protection in the context of sustained agriculture. In addition to antimicrobial activity, stimulation of local or systemic plant immunity can be involved in the efficiency of these molecules as biopesticides [12]. At least some of the biosurfactant antimicrobial activities might be attributed to the destabilization of cellular membranes, eventually resulting in their disfunction and rupture [13]. Different microorganisms, which produce biosurfactant molecules with antimicrobial activities, have been investigated to stem the worldwide evolution and spread of microbial antibiotic resistance [14], and may also be used against phytopathogens that have evolved resistance to antimicrobial pesticides. Biosurfactants produced by some yeast species have been assayed against some fungal plant pathogens (Table 12.1). An example is the sophorolipid biosurfactant produced by *Starmerella bombicola* that was assayed against *Phytophthora capsici*, *Phytophthora nicotianae*, *Phytophthora infestans*, *Pythium aphanidermatum* and *Pythium ultimum*, which at 2 mg mL^{-1} caused a 42% reduction of the damping off disease in pot trials [15]. Rhamnolipid, at the same concentration, reduced damping off by 33%. Another example is sophorolipids from *Rhodotorula babjevae* YS3 that showed promising activity against *Colletotrichum gloeosporioides*, a pathogen which causes post-harvest decay of apples, indicating that this yeast has a potential application as a food preservative; *R. babjevae* sophorolipid inhibited *C. gloeosporioides* (MIC = 62 μg mL^{-1}), it

Table 12.1 Biosurfactants produced by yeasts which were tested against fungal plants pathogens.

Biossurfactant class	Source	Plant pathogenic fungi	References
Sophorolipid	*Starmerella bombicola*	*Phytophthora capsici*	Yoo et al. [15]
Rhamnolipid		*Phytophthora nicotianae*	
		Phytophthora infestans	
		Pythium aphanidermatum	
		Pythium ultimum	
Sophorolipid	*Rhodotorula babjevae*	*Colletotrichum gloeosporioides*	Sen et al. [16]
		Fusarium verticilliodes	
		Fusarium oxysporum f. sp. *pisi*	
		Trichophyton rubrum	

also had a MIC of 125 µg mL^{-1} against *Fusarium verticilliodes* and *Fusarium oxysporum* f. sp. *pisi*. Unfortunately, it was only active at very high conentrations on two other fungi (MIC values of 1 mg mL^{-1} for *Trichophyton rubrum* and > 2 mg mL^{-1} for *Corynespora cassiicola*) [16].

12.1.3 Biosurfactants as Insecticidal Adjuvants

Biosurfactants have been used as adjuvants with insecticides, acting as emulsifying, dispersing, spreading, and wetting agents [17] and are considered to have less human toxicicity, to be more environmentally safe, and a cheaper alternative to petroleum-based surfactants [18]. The insecticide activity of a biosurfactant is linked to several factors, such as inhibition of enzymes, destruction of biological membranes, drowning of pest insects, or removal of insects from the leaves [19].

Insect epicuticles are formed by a waxy layer, consisting of a mixture of lipids on the outer surface of the insect, which is the first physical barrier for the insecticide [20]. To overcome this barrier, surfactants can be employed to help in sticking the insecticide to the insect's surface and washing off lipids on its epicuticle, which could result in death either by dehydration or by increasing the infectivity of effectiveness of entomopathogenic microbes to the insect [21]. Biosurfactants, when when used to formulate a microbial biopesticide, often increase the dispersion of hydrophobic conidial formulations, helping the attachment and the penetration of the conidia into the insect's integument [21].

12.2 Insecticidal Biosurfactants for Use against Disease Vector Insects

Control of disease vector arthropod species harmful to human health with chemical insecticides is extensively used, despite its toxicity, which can cause dermatological, gastrointestinal, neurological, carcinogenic, respiratory, reproductive, and endocrine effects in humans and animals, and its associated environmental impacts [22]. In addition, the increase in evolution and spread resistance of different insects to these insecticides has been another major problem with these pesticides in recent years [23]. For this reason, the search for alternatives for entomological control has been intensified, and interest in biomolecules produced by microorganisms with insecticidal activity has increased, among which biosurfactants stand out [24, 25].

Recent approaches point to some biosurfactants as pest biocontrol agents [26]. The larvicidal effect of biosurfactants is advantageous not only for possible applications in agricultural pest control, but also for significant disease vector insects [26, 27]. Tests that aim to determine the larvicidal activity from these biomolecules usually consider parameters such as the lowest concentrations used, exposure time, behavioral changes, and larval mortality rates [23, 28]. Studies on larvicidal application of biosurfactants are focused on controlling the larval stages of biological vectors of important diseases worldwide such as dengue fever, malaria, and filariasis transmitted by *Aedes*, *Anopheles* and *Culex* mosquitoes, respectively [29, 30]. According to Fernandes et al. [28], yeasts are prolific fungi that produce biosurfactants. Most of these fungi are known to be safe or GRAS (generally regarded as safe), which facilitates their application in the food and pharmaceutical industry. However, the volume of studies with surfactants produced by fungi is limited [23, 28]. The yeast *Scheffersomyces stipitis* NRRL Y-7124 produces a glycolipid with strong activity against larvae in the third developmental stage of the *Aedes aegypti* mosquito, a period in which there is greater resistance to synthetic chemical larvicides [23]. The application of *S. stipitis* glycolipid at a concentration of 660 mg L^{-1} (LC50) was lethal to the larvae, possibly by interfering with the hydrostatic respiratory activity and integrity of their exoskeletons. The biosurfactant obtained from the fungus *Wickerhamomyces anomalus* CCMA 0358 exhibited stable larvicidal action against *A. aegypti* [28].

According to Mnif and Ghribi [24], the success of obtaining biosurfactants depends mainly on the development of processes that use low-cost raw materials, since this material represents between 10 and 30% of the total cost of the process. According to Fernandes et al. [28], *W. anomalous* produces more biosurfactants in the presence of palm and olive oil compared to residual cooking oil. The fungus *S. stipitis* was able to produce biosurfactants with larvicidal activity using sugarcane bagasse as a carbon source, a source widely available in countries such as Brazil and India [23]. These results prove the possibility of producing fungal biosurfactants with larvicidal activity with low-cost carbon sources, which allows for production of a product that is more economically competitive with synthetic insecticides.

The interesting physicochemical properties of biosurfactants seem to be effective for larvicidal application in the control of disease vectors, either by direct interaction with larvae cells or by acting on the surface tension of aquatic environments, necessary for the control of humidity and gas exchange during larval development [31, 32]. Due to the great diversity of biosurfactants produced by fungi, whose structures vary according to the carbon source, there is still great unexplored potential [33]. Although the work on surfactants generated by fungi is still an ongoing field of investigation, future studies in this area can contribute to possible pesticide alternatives to commercially available synthetic insecticides as a safer and more viable ecotoxicological option that might also be used in integrated insecticide resistance management strategies [23, 28, 32].

12.3 Fungal Biosurfactants in Bioremediation Processes

In recent decades the rapid development of agriculture and industries has resulted in the contamination of the environment by several recalcitrant pollutants, including heavy metals, polychlorinated biphenyls (PCBs), plastics, and some agrochemicals [34]. In addition, human activities such as mining, final disposal of effluents containing toxic metals and metal chelates from steel mills, battery industries, and thermoelectric plants have resulted in water quality deterioration, generating serious environmental problems [34]. Although there are some physicochemical techniques for removing contamination, such as by oils (chemical method), excavation and dumping, separation, stabilization, and heat treatment, these methods are expensive, are not ecologically sound and reduce soil fertility [34, 35]. Considering that the presence of these recalcitrant compounds in the environment is of great concern due to their toxicity and non-biodegradable nature, processes that use biological systems to reduce, eliminate, contain or transform contaminants present in soils, sediments, water and air are of great relevance [34, 36]. To overcome the limitations of physicochemical processes, bioremediation can be a promising, less costly, environmentally friendly, and more reliable technology for the removal of environmental pollutants [34, 35]. The bioremediation process involves the production of energy in a redox reaction within microbial cells, including respiration and other biological functions necessary for the maintenance and multiplication of the cells, and a delivery system is usually required that provides one or more of the following items: an energy source (electron donor), an electron acceptor and nutrients [36].

Among the different bioremediation processes, biosurfactants have been used in the remediation of oil pollution because they have advantages such as biodegradability and low toxicity; in addition, the way these hydrocarbons are degraded by microorganisms reduces damage to the ecosystem [37]. Microorganisms that degrade hydrocarbons typically produce a variety of extracellular biosurfactants, which emulsify compounds, increase water solubility, and make compounds more accessible to these microorganisms [37, 38].

Microbial communities have been shown to function well for bioremediation, considering that most indigenous microorganisms have the ability to successfully carry out environmental restoration through immobilization or chemical transformation of contaminants [39, 40]. Different microorganisms use contaminants as carbon sources for their growth and reproduction, carrying out their decomposition and transforming them into simpler, less toxic compounds [41]. Fungi are saprobic microorganisms reported for their high tolerance to toxic environments, and their hyphae have an additive advantage over unicellular microorganisms, because a complex matrix is formed around them where different enzymes capable of catalyzing the degradation of substrates are formed [40, 42].

Some species of fungi play important roles in the bioremediation of contaminants such as persistent organic pollutants, textile dyes, charcoal, pharmaceuticals and personal care products, polycyclic aromatic hydrocarbons, and pesticides [41, 43, 44]. Different studies report the use of fungi from different groups – including *Aspergillus*, *Penicillium* and alkalophilic fungi for bioremediation of textile dyes, effluents from the sugar industry, chemicals used in kraft pulp mills, and effluents from leather tanning, indicating the different substrate choices of these fungi [41, 45–47]. Substantial removal of gasoline and diesel contaminants from soil by short-term incubation of *Aspergillus niger* and *Phanerochaete chrysosporium* with petroleum hydrocarbons has been shown in conjunction with elimination of total organic carbon which aids in the bioremediation process [41, 48–50].

The term "environmental friendliness" combined with the ability to solubilize hydrophobic compounds may explain why biosurfactants have been recognized as excellent agents to improve the bioremediation of pollutant-contaminated environments [51, 52]. First, biosurfactants tend to interact with poorly soluble contaminants and improve their transfer to the aqueous phase, which allows the mobilization of recalcitrant pollutants incorporated into the soil matrix and their subsequent removal [52, 53]. The presence of biosurfactants can also lead to a potential increase in the efficiency of biodegradation, since these molecules act as mediators capable of increasing the mass transfer rate, making hydrophobic pollutants more bioavailable to microorganisms [52, 54, 55]. Alternatively, biosurfactants can also induce changes in cell membrane properties, resulting in greater microbial adhesion, an important mechanism when two immiscible phases (oil and water) are present and direct substrate absorption is plausible [52, 56, 57]. Another notable environmental application of biosurfactants is based on their ability to complex heavy metal ions, which can improve their removal or extraction through biological treatment [52, 58, 59].

Regarding fungi, more than 100 genera are known to play complex roles in the biodegradation of hydrocarbons [60], and fungi belonging to the *Cladosporium* and *Aspergillus* genera are among those known to participate in the degradation of aliphatic hydrocarbons, while representatives of *Cunninghamella*, *Penicillium*, *Fusarium*, *Aspergillus*, and *Mucor* have been shown to participate in the degradation of more recalcitrant aromatic hydrocarbons [61, 62]. Although the pathways of hydrocarbon degradation by bacteria have been well studied, knowledge of the enzymatic mechanisms and genetic pathways associated with hydrocarbon degradation in fungi is much more limited [62, 63].

Some fungi may also play a critical and complementary role in facilitating the bioavailability of hydrocarbons to other microbial communities (as for other fungi and/or bacteria) by synthesizing biosurfactants [62]. This hypothesis is supported by culture-based studies that detected increased degradation of polyaromatic hydrocarbons when different fungi were added [62, 64–66]. Biosurfactant-producing fungi belong mainly to the genera *Candida*, *Pseudozyma*, and *Rhodotorula* for yeasts and *Cunninghamella*, *Fusarium*, *Phoma*, *Cladophialophora*, *Exophiala*, *Aspergillus*, and *Penicillium* for filamentous fungi [16, 62, 67–70].

Acknowledgements

CNPq, CAPES, FAPEMIG. SOD was funded in part by USDA Cooperative Agreement 58-6060-6-015 grant to the University of Mississippi.

References

1. Gebbers R, Adamchuk VI. Precision agriculture and food security. *Science*. 2010;327(5967): 828–831.
2. Tudi M, Daniel Ruan H, Wang L, Lyu J, Sadler R, Connell D, et al. Agriculture development, pesticide application and its impact on the environment. *Int J Environ Res Public Health*. 2021;18(3):1112.

3 Rani L, Thapa K, Kanojia N, Sharma N, Singh S, Grewal AS, et al. An extensive review on the consequences of chemical pesticides on human health and environment. *J Clean Prod.* 2021;283:124657.

4 Naughton PJ, Marchant R, Naughton V, Banat IM. Microbial biosurfactants: current trends and applications in agricultural and biomedical industries. *J Appl Microbiol.* 2019;127(1):12–28.

5 Vargas L, Roman ES. *Conceitos e aplicações dos adjuvantes*. Passo Fundo: Embrapa Trigo, 2006.

6 Castro MJ, Ojeda C, Cirelli AF. *Surfactants in Agriculture*. Dordrecht (DE): Springer, 2013.

7 Purwasena IA, Astuti DI, Syukron M, Amaniyah M, Sugai Y. Stability test of biosurfactant produced by *Bacillus licheniformis* DS1 using experimental design and its application for MEOR. *J Pet Sci Eng.* 2019;183:106383.

8 Sachdev DP, Cameotra SS. Biosurfactants in agriculture. *Appl Microbiol Biotechnol.* 2013;97(3):1005–1016.

9 Duke SO, Dayan FE. Herbicides. *eLS*. 2018:11–9.

10 Palma-Bautista C, Vazquez-Garcia JG, Travlos I, Tataridas A, Kanatas P, Domínguez-Valenzuela JA, De Prado R. Effect of adjuvant on glyphosate effectiveness, retention, absorption and translocation in *Lolium rigidum* and *Conyza canadensis*. *Plants.* 2020;9:297.

11 Vaughn SF, Behle RW, Skory CD, Kurtzman CP, Price NPJ. Utilization of sophorolipids as biosurfactants for postemergence herbicides. *Crop Prot.* 2014;59:29–34.

12 Crouzet J, Arguelles-Arias A, Dhondt-Cordelier S, Cordelier S, Pršić J, Hoff G, Mazeyrat-Gourbeyre F, Baillieul F, Clément C, Ongena M, Dorey S. Biosurfactants in plant protection against diseases: rhamnolipids and lipopeptides case study. *Front Bioeng Biotechnol.* 2020;8(September):1–11.

13 Hirata Y, Ryu M, Oda Y, Igarashi K, Nagatsuka A, Furuta T, Sugiura M. Novel characteristics of sophorolipids, yeast glycolipid biosurfactants, as biodegradable low-foaming surfactants. *J Biosci Bioeng.* 2009;108(2):142–146.

14 Ilori MO, Amobi CJ, Odocha AC. Factors affecting biosurfactant production by oil degrading *Aeromonas* spp. isolated from a tropical environment. *Chemosphere.* 2005;61(7):985–992.

15 Yoo DS, Lee BS, Kim EK. Characteristics of microbial biosurfactant as an antifungal agent against plant pathogenic fungus. *J Microbiol Biotechnol.* 2005;15(6):1164–1169.

16 Sen S, Borah SN, Bora A, Deka S. Production, characterization, and antifungal activity of a biosurfactant produced by *Rhodotorula babjevae* YS3. *Microb Cell Fact.* 2017;16(1):1–14.

17 Rostás M, Blassmann K. Insects had it first: surfactants as a defence against predators. *Proc Royal Soc B.* 2009;276(1657):633–638.

18 Rahman PKSM, Gakpe E. Production, characterisation and application of biosurfactants-review. *Biotech.* 2008;7:360–370.

19 Curkovic T, Burett G, Araya JE. Evaluation of insecticide activity of two agricultural detergents against the long-tailed mealybug, *Pseudococcus longispinus* (Hemiptera: Pseudococcidae) in laboratory. *Agric Tec.* 2007;67:422–430.

20 Pedrini N, Ortiz-Urquiza A, Huarte-Bonnet C, Zhang S, Keyhani N. Targeting of insect epicuticular lipids by the entomopathogenic fungus *Beauveria bassiana*: hydrocarbon oxidation within the context of a host-pathogen interaction. *Front Microbiol.* 2013:4–24.

21. Mascarin GM, Kabori NN, Quintela ED, Arthurs SP, Delalibera Junior I. Toxicity of non-ionic surfactants and interactions with fungal entomopathogens toward *Bemisia tabaci* biotype B. *Biocontrol Sci Technol*. 2014;59(1):111–123.
22. Nicolopoulou-Stamati P, Maipas S, Kotampasi C, Stamatis P, Hens L. Chemical pesticides and human health: the urgent need for a new concept in agriculture. *Front Public Health*. 2016;4:148.
23. Marcelino PR, Silva VL, Rodrigues Philippini R, Von Zuben CJ, Contiero J, Santos JC, Silva SS. Biosurfactants produced by *Scheffersomyces stipitis* cultured in sugarcane bagasse hydrolysate as new green larvicides for the control of *Aedes aegypti*, a vector of neglected tropical diseases. *PLoS One*. 2017;12(11):e0187125.
24. Minf I, Ghribi D. Glycolipid biosurfactants: main properties and potential applications in agriculture and food industry. *JSFA*. 2016;96(13):4310–4320.
25. Bjerk TR, Severino P, Jain S, Marques C, Silva AM, Pashirova T, Souto EB. Biosurfactants: properties and applications in drug delivery, biotechnology and ecotoxicology. *Bioengineering*. 2021;8(8):115.
26. Geetha I, Manonmani AM. Surfactin: a novel mosquitocidal biosurfactant produced by *Bacillus subtilis* ssp. *subtilis* (VCRC B471) and influence of abiotic factors on its pupicidal efficacy. *Lett Appl Microbiol*. 2010;51(4):406–412.
27. Silva VL, Lovaglio RB, Von Zuben CJ, Contiero J. Rhamnolipids: solution against *Aedes aegypti*? *Front Microbiol*. 2015;6:88.
28. Fernandes NAT, Souza AC, Simões LA, Reis GMF, Souza KT, Schwan RF, Dias DR. Eco-friendly biosurfactant from *Wickerhamomyces anomalus* CCMA 0358 as larvicidal and antimicrobial. *Microbiol Res*. 2020;241:126571.
29. Parthipan P, Sarankuma RK, Jaganathan A, Amuthavalli P, Babujanarthanam R, Rahman PKSM, Murugan K, et al. Biosurfactants produced by *Bacillus subtilis* A1 and *Pseudomonas stutzeri* NA3 reduce longevity and fecundity of *Anopheles stephensi* and show high toxicity against young instars. *Environ Sci Pollut Res*. 2018;25(11):10471–10481.
30. Baskar K, Chinnasamy R, Pandy K, Venkatesan M, Sebastian PJ, Subban M, Thomas A, et al. Larvicidal and histopathology effect of endophytic fungal extracts of *Aspergillus tamarii* against *Aedes aegypti* and *Culex quinquefasciatus*. *Heliyon*. 2020;6(10):e05331.
31. Pradhan AK, Rath A, Pradhan N, Hazra RK, Nayak RR, Kanjilal S. Cyclic lipopeptide biosurfactant from *Bacillus tequilensis* exhibits multifarious activity. *3 Biotech*. 2018;8(6):1–7.
32. Siqueira TP, Barbosa WF, Rodrigues EM, Miranda FR, de Souza Freitas F, Martins GF, Tótola M. Rhamnolipids on *Aedes aegypti* larvae: a potential weapon against resistance selection. *3 Biotech*. 2021;11(4):1–8.
33. Pereira DDF, Júnior SD, Alburqueque PM. O estudo da produção de biossurfactantes por fungos amazônicos. *J Eng Exact Sci*. 2017;3(4):0688–0695.
34. Kour D, Kaur T, Devi R, Yadav A, Singh M, Joshi D, et al. Beneficial microbiomes for bioremediation of diverse contaminated environments for environmental sustainability: present status and future challenges. *Environ Sci Pollut Res*. 2021;28(20):24917–24939.
35. Johnson OA, Affam AC. Petroleum sludge treatment and disposal: a review. *Environ Eng Res*. 2019;24(2):191–201.
36. Adams GO, Fufeyin PT, Okoro SE, Ehinomen I. Bioremediation, biostimulation and bioaugmention: a review. *Int J Environ Bioremediat Biodegrad*. 2015;3(1):28–39.

37. Karlapudi AP, Venkateswarulu TC, Tammineedi J, Kanumuri L, Ravuru BK, ramu Dirisala V, et al. Role of biosurfactants in bioremediation of oil pollution-a review. *Petroleum*. 2018;4(3):241–249.
38. Sarafzadeh P, Hezave AZ, Ravanbakhsh M, Niazi A, Ayatollahi S. *Enterobacter cloacae* as biosurfactant producing bacterium: differentiating its effects on interfacial tension and wettability alteration mechanisms for oil recovery during MEOR process. *Colloids Surf B: Biointerf*. 2013;105:223–229.
39. Crawford RL, Crawford DL, editors. *Bioremediation: Principles and Applications*. Cambridge: Cambridge University Press, 2005.
40. Vishwakarma GS, Bhattacharjee G, Gohil N, Singh V. Current status, challenges and future of bioremediation. In: Pandey VC, Singh V, editors. *Bioremediation of Pollutants*. Elsevier, New York; 2020. pp. 403–415.
41. Tomer A, Singh R, Singh SK, Dwivedi SA, Reddy CU, Keloth MRA, et al. Role of fungi in bioremediation and environmental sustainability. In: Prasad R, Nayak SC, Kharwar RN, Dubey NK, editors. *Mycoremediation and Environmental Sustainability. Fungal Biology*. Springer, Cham;2021. pp. 187–200.
42. He K, Chen G, Zeng G, Huang Z, Guo Z, Huang T, et al. Applications of white rot fungi in bioremediation with nanoparticles and biosynthesis of metallic nanoparticles. *Appl Microbiol Biotechnol*. 2017;101(12):4853–4862.
43. Prasad R, Nayak SC, Kharwar RN, Dubey NK, editors. *Mycoremediation and Environmental Sustainability*. Vol. 1. Cham: Springer, 2017.
44. Prasad R, Nayak SC, Kharwar RN, Dubey NK, editors. *Mycoremediation and Environmental Sustainability*. Vol. 2. Cham: Springer, 2018.
45. Redman RS, Dunigan DD, Rodriguez RJ. Fungal symbiosis from mutualism to parasitism: who controls the outcome, host or invader? *New Phytol*. 2001;151(3):705–716.
46. Redman RS, Sheehan KB, Stout RG, Rodriguez RJ, Henson JM. Thermotolerance generated by plant/fungal symbiosis. *Science*. 2002;298(5598):1581–1581.
47. Rockne KJ, Reddy KR. *Bioremediation of Contaminated Sites*. Presented at the International e-Conference on Modern Trends in Foundation Engineering: Geotechnical Challenges and Solutions. Madras: Indian Institute of Technology, 2003.
48. Geiger O, Sohlenkamp C, Lopez-Lara I. Formation of bacterial membrane lipids: pathways, enzymes, reactions. In: Timmis K, editor. *Handbook of Hydrocarbon and Lipid Microbiology*. Springer, Cham; 2010. pp. 395–409.
49. Redman RS, Kim YO, Woodward CJ, Greer C, Espino L, Doty SL, et al. Increased fitness of rice plants to abiotic stress via habitat adapted symbiosis: a strategy for mitigating impacts of climate change. *PLOS One*. 2011;6(7):e14823.
50. Echeveria L, Gilmore S, Harrison S, Heinz K, Chang A, Nunz-Conti G, et al. Versatile bio-organism detection using microspheres for future biodegradation and bioremediation studies. Laser Resonators, Microresonators, and Beam Control XXII. *Int Soc Opt Photonics*. 2020;11266:112661E.
51. Kosaric N. Biosurfactants and their application for soil bioremediation. *Food Technol Biotechnol*. 2001;39(4):295–304.
52. Ł Ł, Marecik R, Chrzanowski Ł. Contributions of biosurfactants to natural or induced bioremediation. *Appl Microbiol Biotechnol*. 2013;97(6):2327–2339.

53 Lai CC, Huang YC, Wei YH, Chang JS. Biosurfactant-enhanced removal of total petroleum hydrocarbons from contaminated soil. *J Hazard Mater.* 2009;167(1–3):609–614.

54 Inakollu S, Hung HC, Shreve GS. Biosurfactant enhancement of microbial degradation of various structural classes of hydrocarbon in mixed waste systems. *Environ Eng Sci.* 2004;21(4):463–469.

55 Whang LM, Liu PWG, Ma CC, Cheng SS. Application of biosurfactants, rhamnolipid, and surfactin, for enhanced biodegradation of diesel-contaminated water and soil. *J Hazard Mater.* 2008;151(1):155–163.

56 Neu TR. Significance of bacterial surface-active compounds in interaction of bacteria with interfaces. *Microbiol Rev.* 1996;60(1):151–166.

57 Franzetti A, Caredda P, Ruggeri C, La Colla P, Tamburini E, Papacchini M, et al. Potential applications of surface active compounds by *Gordonia* sp. strain BS29 in soil remediation technologies. *Chemosphere.* 2009;75(6):801–807.

58 Mulligan CN, Yong RN, Gibbs BF, James S, Bennett HPJ. Metal removal from contaminated soil and sediments by the biosurfactant surfactin. *Environ Sci Technol.* 1999;33(21):3812–3820.

59 Mulligan CN, Yong RN, Gibbs BF. Heavy metal removal from sediments by biosurfactants. *J Hazard Mater.* 2001;85(1–2):111–125.

60 Prince RC. The microbiology of marine oil spill bioremediation. *Petroleum Microbiol.* 2005;1:317–335.

61 Amend A, Burgaud G, Cunliffe M, Edgcomb VP, Ettinger CL, Gutiérrez MH, et al. Fungi in the marine environment: open questions and unsolved problems. *MBio.* 2019;10(2):e01189-18.

62 Maamar A, Lucchesi ME, Debaets S, Nguyen van Long N, Quemener M, Coton E, et al. Highlighting the crude oil bioremediation potential of marine fungi isolated from the port of Oran (Algeria). *Diversity.* 2020;12(5):196.

63 Marco-Urrea E, García-Romera I, Aranda E. Potential of non-ligninolytic fungi in bioremediation of chlorinated and polycyclic aromatic hydrocarbons. *New Biotechnol.* 2015;32(6):620–628.

64 Yadav JS, Reddy CA. Degradation of benzene, toluene, ethylbenzene, and xylenes (BTEX) by the lignin-degrading basidiomycete *Phanerochaete chrysosporium*. *Appl Environ Microbiol.* 1993;59(3):756–762.

65 Zheng Z, Obbard JP. Effect of non-ionic surfactants on elimination of polycyclic aromatic hydrocarbons (PAHs) in soil-slurry by *Phanerochaete chrysosporium*. *J Chem Technol Biotechnol: Int Res Process, Environ & Clean Technol.* 2001;76(4):423–429.

66 In Der Wiesche C, Martens R, Zadrazil F. The effect of interaction between white-rot fungi and indigenous microorganisms on degradation of polycyclic aromatic hydrocarbons in soil. *Wat Air and Soil Poll: Focus.* 2003;3(3):73–79.

67 Andrade Silva NR, Luna MA, Santiago AL, Franco LO, Silva GK, de Souza PM, et al. Biosurfactant-and-bioemulsifier produced by a promising *Cunninghamella echinulata* isolated from caatinga soil in the northeast of Brazil. *Int J Mol Sci.* 2014;15(9):15377–15395.

68 Diniz Rufino R, Moura de Luna J, de Campos Takaki GM, Asfora Sarubbo L. Characterization and properties of the biosurfactant produced by *Candida lipolytica* UCP 0988. *Electron J Biotechnol.* 2014;17(1):6-6.

69 Sajna KV, Sukumaran RK, Gottumukkala LD, Pandey A. Crude oil biodegradation aided by biosurfactants from *Pseudozyma* sp. NII 08165 or its culture broth. *Bioresour Technol.* 2015;191:133–139.

70 Lima JMS, Pereira JO, Batista IH, Neto PDQC, Dos Santos JC, de Araújo SP, et al. Potential biosurfactant producing endophytic and epiphytic fungi, isolated from macrophytes in the Negro river in Manaus, Amazonas, Brazil. *African J Biotechnol.* 2016;15(24):1217–1223.

13

New Formulations Based on Biosurfactants and Their Potential Applications

Maria Jose Castro-Alonso*, Fernanda G. Barbosa, Thiago A. Vieira, Diana A. Sanchez, Monica C. Santos, Thércia R. Balbino, Salvador S. Muñoz, and Talita M. Lacerda*

Biotechnology Department, Engineering School of Lorena, University of São Paulo, CEP 12602-810 Lorena, SP, Brazil
* Corresponding authors

13.1 Introduction

Biosurfactants (BS) are surface-active molecules produced by microorganisms, that possess important advantages over synthetic ones, such as non-toxicity, biodegradability, bioavailability, biocompatibility, high specificity, eco friendly character, effectiveness at extreme conditions of pH, temperature, and salt concentrations, and stability at long storage times [1]. The vast diversity of biosurfactant-producing microorganisms and their different production processes give rise to their broad structural characteristics and functions [2, 3]. In general, BS are classified with respect to their chemical nature, with glycolipids (rhamnolipids, sophorolipids, mannosylerythritol lipids, and trehalose lipids), lipopeptides/lipoproteins (surfactin, fengycin), and polymers (lipopolysaccharides, heteropolysaccharides, and proteins) corresponding to the three main groups [4–6], which are often subdivided into low molecular weight and high molecular weight BS [4, 7]. The former act mainly to reduce surface and interfacial tension between different phases, including glycolipids and lipopeptides [8, 9], while the latter interact with various surfaces, playing the role of emulsifiers. Polymeric BS (emulsan, Alasan, liposan), lipoproteins, and polysaccharides, formed by large chains of carbohydrates, lipids, and/or proteins [2, 10], are high molecular weight BS, also known as bioemulsifiers.

BS, as well as synthetic surfactants, are recognized as multifunctional agents that can perform detergency, stabilizing, wetting, anti-microbial, moisturizing, and emulsifying properties [11–13]. Thus, BS have become sustainable alternatives to replace their chemical counterparts for several industrial formulations [14] in cosmetics and personal care, medicine and pharmaceutics, food and feed, agriculture [15] and, more recently, in civil engineering [16]. The interest in the development and commercialization of novel products bearing BS has motived studies in both academic and industrial sectors [10], which can be observed by the numerous recent contributions available in the literature [15–18], and

Biosurfactants and Sustainability: From Biorefineries Production to Versatile Applications, First Edition.
Edited by Paulo Ricardo Franco Marcelino, Silvio Silverio da Silva, and Antonio Ortiz Lopez.
© 2023 John Wiley & Sons Ltd. Published 2023 by John Wiley & Sons Ltd.

patents considering novel formulations based on BS. The number of patents deposited on the topic increased from 250 to 880 in the years 2006 and 2021 respectively [19].

In this chapter, we aim to review the most relevant research efforts on the incorporation of BS in novel formulations to be applied in different industrial fields (Table 13.1), mostly to take advantage of their outstanding properties and to address the needs of less aggressive processes.

Table 13.1 Most relevant biosurfactants used in product formulations for different fields of applications.

Fields of application	Microorganism	Type of biosurfactants	Role of biosurfactants	Product formulations	References
Cosmetic and personal care	*Pseudozyma*	Glycolipid	Anti-oxidant properties	Skin care ingredient	[36]
	Nocardiopsis VITSISB	—	Emulsifiers and maintain wetting and foaming properties	Toothpaste	[37]
	Lactobacillus paracasei	Glycolipo-peptide	Stabilizing agent in oil-in-water emulsions containing essential oils	Cosmetics (creams and makeup)	[12]
	Pseudomonas aeruginosa UCP 0992, Bacillus cereus UCP 1615, and Candida bombicola URM 3718	—	Anti-microbial action	Mouthwash	[38]
	Pseudomonas aeruginosa and Burkholderia thailandensis	Rhamnolipid	Improvement in adsorption of hair-care conditioning polymers	Washing formulations (shampoo)	[39]
Medicine and pharmaceutics	*Pseudomonas aeruginosa*	Rhamnolipid	Anti-tumor activity (cytotoxic effect on cancer cell lines)	Therapeutic products	[40]
	Bacillus Subtilis	Surfactin	Anti-viral agent against transmissible gastroenteritis (TGE) virus (TGEV) infection	Pharmaceutical products for the treatment TGE and TGEV infection	[41]
	Rhodococcus fascians BD8	Glycolipid	Reduces adhesion of microbial pathogens to polystyrene and silicone Surfaces	Drug delivery coatings	[42]

Table 13.1 (Continued)

Fields of application	Microorganism	Type of biosurfactants	Role of biosurfactants	Product formulations	References
	C. bombicola ATCC 22214	Sophorolipids	Coating agents on medical grade silicone devices	Drug delivery coatings	[43]
	Bacillus amyloliquefaciens strain RKEA3	Surfactin	Antagonism against *Staphylococcus aureus*	Anti-microbial drug	[44]
Food and feed	Candida utilis	-	Stability – agent	Food and feed ingredient for mayonnaise	[45]
	Candida albicans SC5314 and Candida glabrata CBS138	Sophorolipid	Food emulsions imparting them protection against pathogenic bacteria	Food additive with anti-microbial properties	[46]
	Candida bombicola	Glycolipid	Food emulsifier	Additive in cupcake formulation	[47]
Pesticides, insecticides and herbicides	Pseudomonas sp. B0406	Glycolipid	Solubility agents	Additive for endosulfan and methyl parathion pesticides	[48]
	Bacillus NH-100 and NH-217	Lipopeptide	Biocontrol agent against bakanae disease	Ingredient for pesticides formulation against rice bakanae disease	[49]
	Bacillus subtilis V26	Lipopeptide	Anti-fungal and insecticidal activities	Biocontrol agent for fungicide and insecticide formulations	[50]
Civil Engineering	Pseudomonas fluorescens	Lipopeptide	Improved the piezoresistivity behavior and reduced the hydration temperature	Additive for cement mixtures	[51]
	Pseudomonas mosselii F01	Glycolipid	Biocide against microorganisms	Additive with anti-corrosion action	[52]
	Pseudomonas fluorescens	Lipopetide	Increase the Mechanical Strength and Capillary Porosity of cement	Air-entraining additive for cement mixtures	[53]

13.2 General Chemical and Biochemical Aspects

Nutritional factors and operational conditions used in the production of biosurfactants determine their chemical nature (Figure 13.1). Carbon and nitrogen sources, the presence of metal ions, pH, temperature, aeration, and agitation speed [1, 20, 21] are some of the parameters that may be tuned to favor the biosynthesis of a specific BS, which occurs through enzyme-substrate reactions in fermentation processes during the exponential or stationary growth phase, depending on nutritional conditions and on the metabolism of the microorganisms [1, 22–28]. However, efforts are still necessary to fully understand the main metabolic mechanisms involved in the biosynthesis of BS from different types of substrates

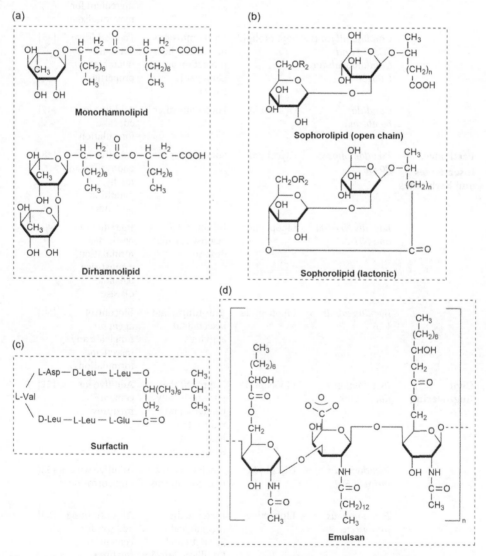

Figure 13.1 Chemical structures of (a) mono- and dirhamnolipids, (b) open chain and lactonic sophorolipids, (c) surfactin, and (d) emulsan [9]. Reproduced with permission of Elsevier.

and conditions, since the potential applications of BS in product formulations (Table 13.1) depend on their chemical composition, structural characteristics, and functions [1].

Glycolipids such as rhamnolipids and sophorolipids, are among the most common biosurfactants and present excellent physicochemical properties [29]. Regardless the specific molecular structure, all BS have in common the ability to improve the surface interactions between polar and non-polar substances by reducing the surface tension. As surfactant molecules replace water or oil molecules along the interface, they effectively reduce intermolecular forces between solvent molecules, reducing surface or interfacial tension. In thermodynamic terms, adsorbed surfactants reduce the surface free energy per unit area required to create new surface (a quantity closely related to surface and interfacial tension) [9] and such key features may be widely exploited in industrial products. In fact, some crucial structure–property relationships of BS boost applications in food related products, as anti-oxidant and emulsification properties can contribute to extend food shelf life; in health care, thanks to the ability of BS to form secondary micelle structures that can easily interact with cell membranes, therefore serving as cargo carrying vesicle structures to improve the solubility, delivery, and bioavailability of therapeutic agents; and in cosmetics, as fatty acid moieties are effective moisturizers of rough and dry skins [30].

13.3 Downstream Processing

Commercialization of BS still faces many challenges in terms of economic viability. One of the most crucial challenges in large-scale production of BS is the downstream processing, that accounts for approximately 60–70% of the total production cost [31]. The purification steps required for the isolation of BS depend on their field of application: BS applied in formulations of agricultural and civil engineering-related products do not require several downstream purification steps; for cosmetic, food and pharmaceutical products, on the other hand, high-cost downstream methods that assure the quality of the final products are necessary. Also, BS produced from agro-industrial by-products inevitably require more downstream purification steps, which increases the production prices [32]. It is therefore mandatory to improve downstream processing methods to attain a competitive cost for industrial BS production [33]. Furthermore, different methods such as precipitation, fractionation, phase separation, ultrafiltration, crystallization, and chromatography, have been reported for the purification of BS (Table 13.2). Several factors influence the BS purity, such as the localization of the bioproduct (intracellular, extracellular, cell bounded), ionic charge (mainly in chromatography purification), and solubility (water/organic solvents) [1]. According to the literature, a given downstream method for BS purification is not applicable if it (1) is time consuming and demands high solvent amounts, (2) produces large amounts of toxic residues, (3) impairs the activity of the product, and/or (4) leads to aggregation [1, 33–35].

13.4 Biosurfactants in Cosmetics and Personal Care

The use of personal care products is essential to maintain the hygiene of the human body and prevent illnesses. The growing demand for skincare products, perfumes, and makeup increasingly leverages the cosmetics and personal care products market. One of the main

Table 13.2 Different purification methods applied in downstream processing of biosurfactants.

Methods of purification	Type of BS	Recovery (%)	Purity (%)	Reference
Precipitation	Rhamnolipid	89	90	[54]
	Shophorolipid	53	55	[55]
	Lipopeptide	70	53	[56]
Fractionation	Rhamnolipid	93	94	[57]
	Lipopeptide	90	70	[58]
Phase separation	Rhamnolipid	75	98	[59]
	Sophorolipid	50	80	[60]
	Lipopeptide	97	97	[61]
Ultrafiltration	Rhamnolipid	97	87	[62]
	Lipopeptide	95	98	[63]
Crystallization	Rhamnolipid	50	97	[64]
	Sophorolipid	-	99	[65]
Chromatography	Rhamnolipid	-	99	[66]
	Lipopeptide	-	98	[67]
	Sophorolipid	-	90	[68]

areas where care needs to be taken is with the skin, which mainly acts as a barrier to the external environment, allowing the body to regulate the interaction with components and microorganisms of the external environment. Several beneficial microorganisms are present on the skin, but pathogenic microorganisms can infect the skin and cause diseases. Clean and moisturize skin surfaces are important to support the body health, and products that provide nutrients and skin protection help to maintain beneficial microorganisms and prevent pathogens [13].

Chemical surfactants are often added to the formulations of many cosmetics and personal skincare products, primarily to act as emulsifiers and foaming agents. However, these chemical ingredients can alter the skin flora, causing allergic reactions and irritation [13, 69]. Furthermore, when formulations containing chemical surfactants are used for a long time, skin cell membrane proteins may denature, and epidermal lipids may solubilize [70]. These changes may harm the integrity of the skin, and consequently, negatively impact the skin's barrier function, facilitating the alteration of the natural microbiota [15].

BS add extra benefits to cosmetics, such as enhancing the moisturizing power of dry skin surfaces [21]. Due to the similarity between the BS (e.g. lipopeptides and lipoproteins) and the skin cell membrane, this bioproduct has good compatibility and easier interaction with the skin [15]. Other characteristics such as their lower toxicity, lower critical micellar concentration, higher biodegradability, higher stability under extreme conditions of temperature, and pH, together wih the possibility of being produced from cheap renewable substrates, also make BS interesting to replace chemical surfactants in creams, shampoos, and other cosmetics [12, 70, 71].

Lipopeptides and glycolipids are the most used and studied BS for formulations of cosmetics and personal care products [15, 70, 72] (Table 13.1). Lipopeptides, besides having good compatibility with the skin and cause low irritation, can improve the moisturizing power and cleaning activity of some dermatological products, so they are widely used in cosmetics, such as anti-wrinkle creams [72, 73]. Among lipopeptides, surfactin is the most used in cosmetics [10]. The possibility of using surfactin as an ingredient in anti-cellulite products has been reported in the literature due to its capacity of suppressing the interaction of lipid A with the lipopolysaccharide-binding protein that mediates their transport [74].

One of the problems that hinder the large-scale implementation of BS in cosmetics is the high production cost. The extraction step, for instance, can be complex and often increases the overall production costs. Therefore, the production of BS using low-cost substrates, and the identification of microorganisms that produce extracellular BS, are approaches that may eliminate steps and reduce the costs of the process. The production of lipopeptides from corn water was reported, and the BS was used in hair and skincare formulations, such as anti-acne products and stabilizing agent of vitamin C [75–77].

Glycolipids may be incorporated into cosmetics due to their high stability in extreme conditions of pH, salinity, and temperatures [78]. The production of glycolipids with anti-oxidant activity was observed in cultures of *Pseudozyma* sp. [36], which are commonly added to cosmetics to inhibit the oxidation of active ingredients that act on the conservation of creams and prevent the scavenging of free radicals by the skin [79].

Rhamnolipids and sophorolipids are also widely studied and commercialized for applications in cosmetic formulations. Rhamnolipids have been considered in the formulation of deodorants and toothpastes due to their anti-microbial action, as additives to improve the adsorption of hair-care conditioning polymers, and in sustainable lip gloss formulations [80, 81]. Sophorolipids have been explored for the treatment of cellulites, lipstick, and eye shadow formulations [82]. In addition, when conjugated with amino acids, sophorolipids are sources of ω and ω-1 hydroxy fatty acids, which can be used in the perfume and fragrance industry as an alternative to the complex chemical route [70, 83].

13.5 Biosurfactants in Medicine and Pharmaceutics

BS present many applications and functionalities in biomedical systems due to their anti-viral, anti-microbial, and anti-cancer actions [29]. The anti-viral activity is based on the alteration they promote on the lipid envelope of viruses [84]; as an anti-microbial agent, BS compromise the plasma membrane (or cell wall), disrupting the bacterial cells [85, 86]; as an anti-cancer agent, BS help to induce cell cycle arrest, apoptosis, and metastasis arrest in tumor cells. In addition, BS can act as immunomodulators, serving as suppressors or activators of the immune system [87].

Rhamnolipids, sophorolipids, and surfactin are the most studied BS in the preparation of formulations for medical and pharmaceutical applications [10, 29]. Rhamnolipids, mainly produced by *Pseudomonas* and *Burkholderia* species, have a permeant effect, causing the rupture of the plasma membrane of bacterial cells, thus providing anti-microbial properties [86]. According to Sana et al. [88], rhamnolipids promote re-epithelialization in excisional wound healing, and assist as protectors in the manifestation of bacterial infections. Similarly, Elshikh et al. [89] investigated rhamnolipids produced by the non-pathogenic *Burkholderia thailandensis*, and their anti-microbial properties against *Streptococcus oralis*, *Actinomyces naeslundii*, *Neisseria mucosa*, and *Streptococcus sanguini* oral pathogens were demonstrated. The authors concluded that rhamnolipids could be characterized as biocidal agents, as they managed to decrease 3–4 log bacterial viability. Furthermore, rhamnolipids are potentially applicable for the formulation of immunomodulatory and anti-cancer drugs [90]. De Santo [91] described the rhamnolipid absorption into the bloodstream in treatments performed topically on skin affected by radiation burns.

Among the other glycolipids with medical and pharmaceutical applications are the sophorolipids, commonly produced by *Candida bombicola*, *Rhodotorula bogonensis*, *Rhodotorula babjevae*, *Wickerhamiella domercqiae*, *Torulopsis gropengiesseri*, *Torulopsis bombicola*, and *Pichia anomala*, that possess anti-bacterial, anti-viral, anti-inflammatory, and anti-fungal activities [84, 92–94]. Lydon et al. [95] demonstrated the action of sophorolipids as anti-microbial agents against *Enterococcus faecalis* and *Pseudomonas aeruginosa* infections, with significant reductions of their CFU using minimum concentrations equivalent to 5 mg ml^{-1}, which makes sophorolipids potent agents in the recovery of acutely or chronically infected wounds. Hagler et al. [96] showed that sophorolipids could be used as anti-inflammatory agents in formulations since they decrease the production of *IgE* related to diseases such as asthma.

Surfactin is another BS with great importance in the pharmaceutical industry, being considered the biosurfactant with the highest efficacy at low concentrations [7]. Surfactin can act as anti-viral, anti-microbial, anti-tumor, and anti-inflammatory agents. Wang et al. [97] proposed that *Bacillus subtilis* and its secretion of surfactin have the potential to prevent transmissible gastroenteritis virus (TGEV) infection. Its anti-inflammatory properties are due to its inhibition of phospholipase A2, interleukin, and nitric oxide release [98]. In other study, low concentrations (0.2 mmol L^{-1}) of surfactin demonstrated the ability to act as a potential stabilizer and anti-oxidant of the single-cell oil of docosahexaenoic acid (DHASCO), one of the characteristic functional ingredients of micro-algae (*Crypthecodinium Cohnii*) oils. DHASCO can reduce the risks of cardiovascular diseases, inflammation, and cancer [99].

The interest in BS has increased in the last decades, mainly to address the evolution and industrialization of medical and pharmaceutical products. There are many opportunities for BS to be applied as bioactive agents in the treatment of autoimmune diseases such as arthritis and diabetes as well as in other relevant interventions [87].

13.6 Biosurfactants in Food and Feed

Biosurfactants can have several functions in food and feed, as to control the accumulation of fat globules, to stabilize aerated systems, to improve texture and shelf-life of starch-containing products, to modify rheological properties of wheat dough, and to improve consistency and texture of fat-based products [100, 101]. An interesting experiment was carried out to analyze the effect of BS on the process and quality control of a pelletized broiler feed, and the results demonstrated that the strategy improved pellet quality (e.g. starch polymerization) [102]. The potential use of BS produced by *Candida utilis* in different formulations of mayonnaise was evaluated, and no relevant changes were observed in the liver or kidneys of the animals tested. Moreover, the stability of the formulation was assessed after the addition of carboxymethyl cellulose and guar gum, and any pathogenic microorganisms was detected after 30 days of refrigeration. The study indicated that BS are key molecules that can be applied in several contexts to improve food and feed processes [45].

BS were considered as key components of food or feed product formulations, Ribeiro et al. [103] produced a BS from the waste of soybean oil and corn steep liquor by *Saccharomyces cerevisiae* URM 6670, which was used to replace egg yolk in a cookie formulation. In the analysis of the texture profile before baking, the substitution of egg yolk with the biosurfactant did not alter the properties of firmness, cohesiveness, or elasticity compared to the standard formulation. Kiran et al. [104] studied a marine sponge *Actinomycetes Nesterenkonia sp.* MSA31 for BS production, which was chemically characterized as lipopeptide and added in the formulation of a muffin. The strategy allowed an improvement in the organoleptic qualities compared to the positive and negative control. Silva et al. [47] described the application of a biosurfactant from *Candida bombicola* URM 3718 as a meal additive of a cupcake, replacing 50%, 75%, and 100% of the vegetable fat in the standard formulation. In addition to the emulsification properties and the possibility of replacing ingredients in food formulations, BS may also add anti-microbial or anti-oxidant properties to the final product.

In the food and feed industry, ingredients with specific characteristics are often considered to enhance the quality of the final product. Rani et al. [105] isolated 123 bacterial strains from three oil batteries, which were further characterized by 16S rRNA gene sequencing. 34 of the 123 strains were identified as biosurfactant-producers, among which *Bacillus methylotrophicus* strain OB9 exhibited the highest biosurfactant activity. *B. methylotrophicus* demonstrated high efficiency on biofilm disruption in agar diffusion assays against several Gram-negative food-borne bacteria and plant pathogens. The application of this BS for food safety is very promising, especially in climates where microbial-biofilm persistence is a problem to be addressed. Kourmentza et al. [106] studied the anti-microbial activity of a lipopeptide produced by *Bacillus sp.* against foodborne pathogenics, and food spoilage microorganisms. Mycosubtilin and mycosubtilin/surfactin mixtures (lipopolipeptides) were found to have high anti-fungal activity against food relevant fungi-like *Paecilomyces variotii* (a heat-resistant fungi and common contaminant in heat treated foods and juices), and *Byssochlamys fulva* (a soil plant pathogen affecting strawberries, pineapples, and other fruits, that can be also found in pasteurized juices). Silva et al. [107] used rhamnolipids to disrupt *Staphylococcus aureus* milk-based biofilms at temperatures often used in milk processing industries. All those findings suggest that BS are potential ingredients for food and feed formulations, as they may enhance the shelf-life of products.

BS may present different natural roles in the growth of microorganisms. Merghni et al. [108] determined anti-oxidant and anti-proliferative activities of BS produced by *Lactobacillus casei*, and evaluated the anti-biofilm potential of BS against oral *Staphylococcus aureus* strains. BS from *Lactobacillus casei* strains exhibited considerable anti-oxidant and anti-proliferative properties and inhibited oral *S. aureus* strains with significant anti-biofilm efficacy. Da Silva et al. [109] studied the anti-oxidant capacity of BS produced from low-cost substrates by *Candida bombicola* URM 3718. The authors reported the potential of *Candida bombicola* surfactants for their application in food systems, where emulsification and anti-oxidants properties are desirable.

It is worth of mentioning that the identification of compounds with low toxicity and surface activity properties is of high industrial interest. The advantages provided by BS due to the wide range of beneficial properties (e.g., anti-microbial, anti-biofilm, and anti-adhesive, non-fouling) may specially benefit the food industry.

13.7 Biosurfactants in Pesticides, Insecticides, and Herbicide Formulations

Pesticides may be defined as substances (or a mixture of substances) aimed at preventing, destroying, repelling, or mitigating pests, i.e. insects, rodents, nematodes, fungi, weeds, viruses, bacteria, among other unwanted organisms that may invade agricultural crops or harvested products [110, 111]. Biopesticides are based on active ingredients derived from microorganisms or naturally occurring substances, e.g. bacteria, fungi, plant extracts, among others [112, 113].

It is well documented that many pesticides require the use of an adjuvant to perform optimally. BS have been demonstrated as promising adjuvants in pesticide formulations, and the interest in their applications has largely increased in recent years to enhance sustainability in the agricultural sector. An adjuvant is not active alone but enhances the efficiency of the formulations. Some specific formulation properties, such as the level of pesticide stability and dispersion, the compatibility, the solubility, the suspensory nature, the foaming ability, the surface tension, the adherence and coverage area, the volatilization, and the penetration, can be enhanced by the addition of adjuvants [7, 114, 115].

A lipopeptide from *Bacillus subtilis* AKP was tested against *Colletotrichum capsici*, responsible for anthracnose disease, the main postharvest problem in various tropical fruits. The investigation demonstrated fungicidal effect, as well as growth promotion ability when assayed both *in vitro* and *in vivo* in chili fruits, and the BS was therefore considered an excellent alternative for synthetic fertilizers and fungicides [116, 117]. Toral et al. [118] demonstrated the efficiency of lipopeptides from *Bacillus* XT1 CECT 8661 against the plant pathogen *Botrytis cinerea* in inhibiting fungal growth and triggering the anti-oxidant activity in fruits. Xu et al. [119] evaluated the herbicidal activity of a phytotoxin produced by the pathogenic fungus *Colletotrichum gloeosporioides* BWH-1 (which consists of a dirhamnolipid Rha-Rha-C10-C10), either alone or combined with commercial herbicides. The results showed that the dirhamnolipid exhibited a broad herbicidal activity against eight weed species, and no toxicity on *Oryza sativa* (Asian rice), which indicates that it can be used alone or as adjuvant added to commercial formulations. When it comes to herbicide formulations, most active adjuvant BS are able to increase cuticle penetration and nutritional uptake of plants, therefore helping in the transport mechanism of herbicides to the plasma membrane [7].

Do Nascimento Silva et al. [120] tested a rhamnolipid (an aqueous solution obtained from crude cell-free *P. aeruginosa* fermentation broth) against the insect pest *Bemisia tabaci*, and its combined action with aerial conidia of two entomopathogenic fungi, *Cordyceps javanica* and *Beauveria bassiana*. All formulations tested showed effective control of whitefly nymphs, either alone or in combination with the conidia. The direct application of the rhamnolipid alone in low concentrations (0.01% to 0.1% w/v) caused severe shriveling and dehydration of whiteflies, with no damage or toxicity to bean leaves. Biosurfactant adjuvants can penetrate insect cuticle, exoskeleton, and infect the insect tracheae, and when used in combination with chemical insecticides, reduce the adverse effects to the environment [7].

Sophorolipids can be used in fungicide and herbicide formulations, as they inhibit the movement and growth of mycelia, help in the disintegration of zoospores, and act as a natural emulsifier for post emergence herbicides [121, 122]. The acidic sophorolipids from *Candida kuoi* produce long-lasting emulsions, which are more stable than those based uniquely in chemical surfactants [123].

A biosurfactant produced by *Wickerhamomyces anomalus* CCMA 0358 using the cheap kitchen waste oil as substrate, was tested against *Aedes aegypti* larvae with excellent results (100% mortality in low concentration and short exposure period). Then, its anti-bacterial, anti-adhesive, and anti-fungal abilities were also evaluated, and high inhibition levels were obtained against *Bacillus cereus, Salmonella* Enteritidis, *Staphylococcus aureus, Escherichia coli, Aspergillus, Cercospora, Colletotrichum,* and *Fusarium* [124].

The most recent and relevant investigations show the high specificity, diversity, and potential for applications of BS to inhibit fungi growth or as part of the biocontrol strategy. Gašić and Tanović [125] correctly mention that further research is needed to improve formulations and reduce production cost, but the use of biopesticides as important tools for sustainable crop protection will be highly desirable in a near future.

13.8 Biosurfactants in Civil Engineering

Advances in materials science and engineering have always played a substantial role in civil engineering [126]. The main structural and mechanical properties of construction materials are static elasticity, dynamic elasticity, and compressive strength [127, 128]. Nevertheless, these properties are affected by extreme climates, humidity, salinity, load impact, microorganisms causing cracks, spoilage, oxidation, and low durability in matter [129–131]. In recent years, the incorporation of bioagents such as BS in construction materials has become a potential and eco-friendly strategy to improve their structural and mechanical properties [126].

Furthermore, corrosion in structures such as monuments and building damage represent irreversible major problems [132–134]. Microorganisms contribute to biodeterioration, biodegradation, and biocorrosion of infrastructures [135, 136]. The biodeterioration occurs by synergistic interactions between microorganism producers of biofilms and sulfides, which act as corrosive substances [137, 138]. Diverse strategies are being developed to prevent and avoid microbial proliferation. One of them is the use of BS as biocides of anti-microbial action [139]. Parthipan et al. [52] reported the successful action of a biosurfactant produced by *Pseudomonas mosselii* for the inhibition of some bacteria responsible for corrosion of carbon steel, such as *Bacillus subtilis* A1, *Streptomyces parvus* B7, *Pseudomonas stutzeri* NA3, and *Acinetobacter baumannii* MN3. BS concentrations of 280–2560 μg mL^{-1} were able to destroy the biofilm generated by bacterial consortium. Purwasena et al. [140] used an indigenous oil consortium of bacteria to produce a biosurfactant with anti-biofilm function. The authors found that concentrations of 62.5 μg ml^{-1} and 125 μg ml^{-1} inhibited all planktonic cells based on MIC (minimum inhibitory concentration). Moreover, the minimum biofilm inhibitory concentration (MBIC), and minimum biofilm eradication concentration for 50% eradication (MBEC50) against *Pseudomonas sp.* 1 and *Pseudomonas sp.* 2 were evaluated. He et al. [53] reported the production of a lipopeptide by *Pseudomonas fluorescens* that increased the strength and durability of the cement paste. The addition of 1.5% BS led to the production of materials with excellent mechanical properties after 148 days, but slightly higher concentrations (3%) induced reduced damage on mortar for high humidity conditions. The authors also found that the application of 1% BS served as a growth controller of *Pseudomonas* genus that cause mortar biodegradation.

The use of BS to ensure the safety of the composition of concrete, cement, and asphalt [141] was reported. Liu et al. [129] monitored the piezoresistivity of cement, after the incorporation of UH-biosurfactant, and improved piezoresistivity behavior was demonstrated, as well as a reduced hydration temperature up to 11%. Vipulanandan et al. [142] observed that a concentration of 1% of nano-silica and 0.5% of biosurfactant produced by vegetable oil and *Serratia sp.* Bacteria could increase the piezoresistivity behavior on Portland cement up to 178% after one day. An alternative to improve the properties of cement was presented by Serres et al. [51], who created an admixture with a biosurfactant synthesized by *Pseudomonas fluorescens* and Portland cement. The admixture improved the plasticizing capacity and the mechanical strength of the cement matrix.

In the case of asphalt mixtures, new formulations for stabilized and increased interactions between their chemical compounds are useful to avoid fatigue cracking on asphalt [143]. The main compounds of asphalt are a mixture of saturated hydrocarbons, aromatics, and asphaltenes, known as SARAs fractions [144, 145]. BS can act as air-entraining agents in asphalt mixtures, reducing the surface tension of the fresh mixtures at low concentration, increasing the workability, and reducing the segregation and bleeding of construction materials [146]. Zhao et al. [147] produced an imidazoline biosurfactant (IM-BioSurf) with promising characteristics for its application in asphalt. The investigation indicated that increasing the biosurfactant concentrations may improve the storage stability of emulsified asphalt by reducing the surface tension. Porto et al. [16] evaluated the effect of the addition of BS on non-emulsifiable recycled bitumens. The authors demonstrated that BS may provide emulsifiable properties, which is a potential strategy to replace harmful substances in the pavement sector.

The literature related to the utilization of BS for civil engineering applications is still limited. Some studies showed that BS can act as anti-corrosive and air-entraining agents, which provides resistance and stability in formulation mixtures for cement, concrete, and asphalt [16, 52, 53, 139]. Furthermore, these studies demonstrated that BS are promising agents to replace harmful substances used in the construction sector. However, further studies are necessary to fully understand the chemical behavior and functions of BS in ecofriendly formulations of construction materials.

13.9 Miscellaneous

Other applications of BS include their use in detergent formulations, for bioremediation purposes, and in nanoparticles and polymer synthesis.

13.9.1 Detergent Formulations

Detergent formulations are complex mixtures of one or more active ingredients in addition to various other ingredients such as detergency builders, bleaches, and perfumes, depending on the purpose of the cleaning product [148]. Laundry detergents, for example, are mainly composed of surfactants, builders, and fillers. Among these constituents, surfactants are one of the most important laundry active ingredients, as they comprise from 15% to 50% of the total detergent formulation [149]. For this reason, in the recent years, the interest in using biosurfactants in this sector has increased [148]. Mukherjee et al. [73] demonstrated the potential of applying cyclic lipopeptide biosurfactants produced by *Bacillus subtilis* strains in laundry detergent formulations, which showed improved washing performance with respect to different synthetic detergents. In a similar study, Bouassida et al. [149] evaluated the application of lipopeptide biosurfactants produced by *Bacillus subtilis SPB1* in the formulation of a washing powder. The authors mixed the SPB1 biosurfactant with sodium tripolyphosphate as a builder and sodium sulfate as filler and compared the efficiency of different detergent compositions (biosurfactant-based detergent, combined biosurfactant and commercial detergent, and a commercial detergent) for the removal of oil and tea stains. The results indicated that the bio-scouring was more effective (>75%) in terms of the stain removal than the commercial powders (<60%). More recently, Helmy et al. [150] tested the effects of various bio-detergent formulations, including the rhamnolipid biosurfactants-builder ratio. The authors observed that the rhamnolipid biosurfactants are promising substitutes for their synthetic counterparts. Based on the analysis of colorfastness to wash, color strength (K/S), and color difference (ΔE) value, rhamnolipid based bio-detergent present similar washing effectiveness compared to the synthetic ones.

13.9.2 Bioremediation Purposes

Biosurfactants correspond to a promising remediation alternative for environmental contamination with oil and heavy metals [151, 152]. Ravindran and collaborators [153], managed to remove heavy metals from fresh vegetables and wastewater using an anionic lipopeptide biosurfactant produced by *Bacillus* sp. (MSI 54). The higher removal of heavy metals (mercury, lead, manganese, and cadmium) was obtained when a major critical micelle concentration of biosurfactant (2.0) was used, corresponding to an efficiency of removal of 100 ppm. The study indicates that crops of lettuce, cabbage, and carrot can be treated with biosurfactants to reduce the concentration of heavy metals. In addition, Hu et al. [154] reported the ability of *Rhodococcus erythropolis* HX-2 to synthesize a biosurfactant capable of degrading petroleum, which was efficient to degrade 73.6% of hydrocarbons after three days. Moreover, the biosurfactant named as MK demonstrated a favorable ability for reducing surface tension of 28.89 mN m^{-1} and critical micelle concentration (CMC) of 100 mg L^{-1}. Such investigations are strong indicators that biosurfactants may be considered as sustainable alternatives for the remediation of pollutants.

13.9.3 Nanoparticle Synthesis

Biosurfactants have emerged as one of the most efficient natural agents that can stabilize the structure of nanoparticles [155]. The average yield of nanoparticle synthesis mediated by biosurfactants is higher than the ones mediated by bacteria or fungi, since BS reduce the formation of aggregates due to the electrostatic forces' attraction, at the same time they induce a uniform morphology of the nanoparticles [156]. Durval et al. [157] evaluated the green synthesis of silver nanoparticles using a BS from *Bacillus cereus* UCP 1615 as stabilizing agent, and their application as an anti-fungal agent. The authors observed that the incorporation of BS promotes stability, as the measured value of zeta potential was −23.4 mV. The analysis of anti-fungal activity showed that a concentration of nanoparticles (AgNP) of 16.50 µg mL^{-1} inhibited 100% and 85% of *P. fellutanum* and *A. niger* growth, respectively. Similarly, Marangon et al. [158] studied the effects of ionic strength and pH on the anti-bacterial activity of hybrid biosurfactant-biopolymer nanoparticles. The authors showed that the stability and anti-bacterial activity of hybrid biosurfactant-biopolymer nanoparticles increased with the decrease of the pH and ionic strength. This study demonstrated the influence of ionic strength and pH in nanoparticles size and surface functionality, changing their interactions with bacteria. Other studies suggest the potential use of BS as nanoparticles stabilizers and their function as anti-microbial surface agents [159]

13.9.4 Polymer Synthesis

As natural macromonomers, BS also find application in polymer synthesis, which addresses the need for novel polymeric materials based on renewable platforms. A pioneer study published in 2013 by Groos and collaborators [160], described in detail the ring opening metathesis polymerization (ROMP) of a lactonic sophorolipid, leading to materials of Mn ≈ 180 KDa. The authors carried out an in-depth analyis of the propagation kinetics, in an important contribution to the conversion of complex natural structures into polymers. The investigation was followed by an equally interesting work from the same research group, related to the ROMP of an enzyme-mediated chemically modified biosurfactant (diacetylated lactonic sophorolipids), in a strategy aiming to expand the structural diversity of such materials [161]. More recently, Arcens and collaborators [162], produced a synthetic glycolipid (methacrylated 12-hydroxystearate glucose) and then copolymerized it with methyl methacrylate by free- or RAFT radical polymerizations. Although the authors did not consider the utilization of natural BS, the study was still a very important step towards BS-based polymers. In this specific case, the amphiphilic materials were able to self-assemble in water into various morphologies, therefore demonstrating the potential of the materials to be applied in the encapsulation and controlled release of active compounds. Also worth mentioning, is the possibility of replacing synthetic surfactants by BS for the stabilization of heterogeneous polymerization systems. This strategy was proven successful in the case of the emulsion polymerization of styrene in the presence of surfactin [163]. The results gathered were very promising, as the CMC of surfactin is considerably smaller than of synthetic SDS, which represents an important reduction in the amount of stabilizing agent needed for the production of one of the most important synthetic polyolefins of industrial interest.

13.10 Overview of the Biosurfactant Market

Despite the aforementioned challenges related to the production and industrialization of BS, many companies have demonstrated interest in their development and commercialization, and the number of related patents has increased significantly in recent years. The main manufacturers around the world and the different field sector applications are listed in Table 13.3.

Table 13.3 Main biosurfactant manufacturers around the world and the different field sector applications.

Biosurfactants manufacturers	Country	Website	Type of BS	Fields of application
BASF	Germany	https://www.homecare-and-i-and-i.basf.com/sustainability/bio-based-surfactans	Rhamnolipids	Cleaning products, food service and kitchen hygiene
Fraunhofer IGB		https://www.igb.fraunhofer.de	Glycolipids and Cellobiose lipids	Cleaning, cosmetic, and personal care products
Groupe Soliance	France	https://www.givaudan.com/fragrance-beauty/active-beauty	Sophorolipids	Cosmetics and pharmaceuticals
Saraya Co. Ltd	Japan	http://worldwide.saraya.com	Sophorolipids	Cleaning products, hygiene products
Allied Carbon Solutions		https://www.allied-c-s.co.jp/english-site	Sophorolipids	Agricultural products, ecological research
MG Intobio Co., L	South Korea	https://www.gmdu.net/corp-464103.html	Sophorolipids	Cosmetic and personal care products
TeeGene Biotech	UK	https://www.teegene.co.uk	Rhamnolipids and lipopeptides	Pharmaceutical products, cosmetics, anti-microbials and anti-carcinogen ingredients
Rhamnolipid Companies	USA	http://www.rhamnolipid.Inc.om/index.html	Rhamnolipids	Agriculture, cosmetics, bioremediation, food products, pharmaceutical
AGAE Technologies		https://www.agaetech.com	Rhamnolipids	Pharmaceuticals, cosmetics, personal care products and bioremediation agents
Synthezyme LLC		http://www.synthezyme.com/index.html	Sophorolipids	Emulsification of crude oil, petroleum, and gas

In general, the formulations of the different products developed by these companies are mostly based on rhamnolipids and sophorolipids. Rhamnolipid Companies (USA) and AGAE Technologies LLC (Corvallis, OR), for instance, commercialize formulations of personal care and pharmaceutical products based in rhamnolipid BS. Other companies such as Givaudan Active Beauty (France), MG Intobio Co., L (South Korea), and Saraya Co. Ltd (Japan), commercializes formulations of skin products, deodorant, shower gel, and UV filters based in sophorolipids.

13.11 Conclusions and Future Perspectives

Biosurfactants (BS) are increasingly showing innovative applications in different industries, from cosmetics and medicine to food, agriculture, and civil engineering. As multifunctional and ecofriendly agents, BS offer promising prospects for future sustainability of large-scale commercial products, with high potential to replace synthetic analogues. Some commercial products for cosmetic, personal care and pharmaceutical are already based on biosurfactants, and the market is expected to expand in the near future. The outstanding properties of BS, such as detergency, stabilizing, wetting, anti-microbial, moisturizing, and emulsification will continue to be exploited, and substantial efforts of researchers will lead to a full understanding of the action mechanisms of BS in different formulations, allowing the improvement of novel bio-based products.

References

1. Jimoh AA, Lin J. Biosurfactant: a new frontier for greener technology and environmental sustainability. *Ecotoxicol Environ Saf*. 2019;184:109607.
2. Kubicki S, Bollinger A, Katzke N, Jaeger KE, Loeschcke A, Thies S. Marine biosurfactants: biosynthesis, structural diversity and biotechnological applications. *Mar Drugs*. 2019;17(7):408.
3. Markande AR, Divya P, Sunita V. A review on biosurfactants: properties, applications and current developments. *Bioresour Technol*.2021;330:124963.
4. Ron EZ, Eugene R. Natural roles of biosurfactants: minireview. *Environ Microbiol*. 2001;3:229–236.
5. Franzetti A, Gandolfi I, Bestetti G, Smyth TJ, Banat IM. Production and applications of trehalose lipid biosurfactants. *Eur J Lipid Sci*. 2010;112(6):617–627.
6. Marcelino PRF, Gonçalves F, Jimenez IM, Carneiro BC, Santos BB, da Silva SS. Sustainable production of biosurfactants and their applications. In: *Lignocellulosic Biorefining Technologies*; 2020. pp. 159–183.
7. Adetunji AI, Olaniran AO. Production and potential biotechnological applications of microbial surfactants: an overview. *Saudi J Biol Sci*. 2021;28(1):669.
8. McClements DJ, Gumus CE. Natural emulsifiers—biosurfactants, phospholipids, biopolymers, and colloidal particles: molecular and physicochemical basis of functional performance. *Adv Colloid Interface Sci*. 2016;234:3–26.
9. Jahan R, Bodratti AM, Tsianou M, Alexandridis P. Biosurfactants, natural alternatives to synthetic surfactants: physicochemical properties and applications. *Adv Colloid Interface Sci*. 2020;275:102061.
10. Drakontis CE, Amin S. Biosurfactants: formulations, properties, and applications. *Curr Opin in Colloid Interface Sci*. 2020;48:77–90.
11. Araujo LVD, Freire DMG, Nitschke M. Biossurfactantes: propriedades anticorrosivas, antibiofilmes e antimicrobianas. *Quim Nova*. 2013;36:848–858.
12. Ferreira A, Vecino X, Ferreira D, Cruz JM, Moldes AB, Rodrigues LR. Novel cosmetic formulations containing a biosurfactant from Lactobacillus paracasei. *Colloids Surf B*. 2017;155:522–529.

13 Adu SA, Naughton PJ, Marchant R, Banat IM. Microbial biosurfactants in cosmetic and personal skincare pharmaceutical formulations. *Pharmaceutics*. 2020;12(11):1099.

14 Martins PC, Martins VG. Biosurfactant production from industrial wastes with potential remove of insoluble paint. *Int Biodeterior*. 2018;127:10–16.

15 Vecino X, Cruz JM, Moldes AB, Rodrigues LR. Biosurfactants in cosmetic formulations: trends and challenges. *Crit Rev Biotechnol*. 2017;37(7):911–923.

16 Porto M, Caputo P, Abe AA, Loise V, Oliviero RC. Stability of bituminous emulsion induced by waste based bio-surfactant. *Appl Sci*. 2021;11(7):3280.

17 Geetha SJ, Banat IM, Joshi SJ. Biosurfactants: production and potential applications in microbial enhanced oil recovery (MEOR). *Biocatal Agric Biotechnol*. 2018;14:23–32.

18 Varjani S, Upasani VN. Bioaugmentation of Pseudomonas aeruginosa NCIM 5514–A novel oily waste degrader for treatment of petroleum hydrocarbons. *Bioresour Technol*. 2021;319:124240.

19 EPO. European patent office: espacenet patent search platform. Munich, The Hague, Berlin, Vienna, Brussels. 2019 [WWW Document]. http://worldwide.espacenet.com/?locale=enEP.

20 Fontes GC, Amaral PFF, Coelho MAZ. Produção de biossurfactante por levedura. *Quim Nova*. 2008;31:2091–2099.

21 Shekhar S, Sundaramanickam A, Balasubramanian T. Biosurfactant producing microbes and their potential applications: a review. *Crit Rev Environ*. 2015;45(14):1522–1554.

22 Shaligram NS, Singhal RS. Surfactin–a review on biosynthesis, fermentation, purification and applications. *Food Sci Biotechnol*. 2010;48(2):119–134.

23 Santos DKF, Rufino RD, Luna JM, Santos VA, Sarubbo LA. Biosurfactants: multifunctional biomolecules of the 21st century. *Int J Mol Sci*. 2016;17(3):401.

24 Dubey KV, Juwarkar AA, Singh SK. Adsorption—; desorption process using wood-based activated carbon for recovery of biosurfactant from fermented distillery wastewater. *Biotechnol Prog*. 2005;21(3):860–867.

25 Amaral PF, Da Silva JM, Lehocky BM, Barros-Timmons AMV, Coelho MAZ, Marrucho IM, Coutinho JAP. Production and characterization of a bioemulsifier from Yarrowia lipolytica. *Process Biochem*. 2006;41(8):1894–1898.

26 Haritash AK, Kaushik CP. Biodegradation aspects of polycyclic aromatic hydrocarbons (PAHs): a review. *J Hazard Mater*. 2009;169(1–3):1–15.

27 Syldatk C, Wagner F. Production of biosurfactants. In: Kosaric N, Cairns WL, Gray NC, editors. *Biosurfactants and Biotechnology*. Routledge; 1987. pp. 89–120.

28 Casas J, Garcia-Ochoa F. Sophorolipid production by *Candida bombicola*: medium composition and culture methods. *J Biosci Bioeng*. 1999;88(5):488–494.

29 Bjerk TR, Severino P, Jain S, Marques C, Silva AM, Pashirova T, Souto EB. Biosurfactants: properties and applications in drug delivery, biotechnology and ecotoxicology. *Bioengineering*. 2021;8(8):115.

30 Manga EB, Celik PA, Cabuk A, Banat IM. Biosurfactants: opportunities for the development of a sustainable future. *Curr Opin Colloid Interface Sci*. 2021;56:101514.

31 Varjani SJ, Upasani VN. Critical review on biosurfactant analysis, purification and characterization using rhamnolipid as a model biosurfactant. *Bioresour Technol*. 2017;232:389–397.

32 Philippini RR, Martiniano SE, Ingle AP, Franco Marcelino PR, Silva GM, Barbosa FG, Dos Santos JC, da Silva SS. Agroindustrial byproducts for the generation of biobased products: alternatives for sustainable biorefineries. *Frontiers in Energy Research*. 2020;8:152.

33 Ismail R, Baaity Z, Csóka I. Regulatory status quo and prospects for biosurfactants in pharmaceutical applications. *Drug Discov Today*. 2021;26(8):1929–1935.

34 Heyd M, Kohnert A, Tan TH, Nusser M, Kirschhöfer F, Brenner-Weiss G, Berensmeier S. Development and trends of biosurfactant analysis and purification using rhamnolipids as an example. *Anal Bioanal Chem*. 2008;391(5):1579–1590.

35 Venkataraman S, Rajendran DS, Kumar PS, Vo DVN, Vaidyanathan VK. Extraction, purification and applications of biosurfactants based on microbial-derived glycolipids and lipopeptides: a review. *Environ Chem Lett*. 2021;20:1–22.

36 Takahashi M, Morita T, Fukuoka T, Imura T, Kitamoto D. Glycolipid biosurfactants, mannosylerythritol lipids, show antioxidant and protective effects against H_2O_2-induced oxidative stress in cultured human skin fibroblasts. *J Oleo Sci*. 2012;61(8):457–464.

37 Das I, Roy S, Chandni S, Karthik L, Kumar G, Bhaskara Rao KV. Biosurfactant from marine actinobacteria and its application in cosmetic formulation of toothpaste. *Pharm Lett*. 2013;5(5):1–6.

38 Farias JM, Stamford TCM, Resende AHM, Aguiar JS, Rufino RD, Luna JM, Sarubbo LA. Mouthwash containing a biosurfactant and chitosan: an eco-sustainable option for the control of cariogenic microorganisms. *Int J Biol Macromol*. 2019;129:853–860.

39 Fernández-Peña L, Guzmán E, Leonforte F, Serrano-Pueyo A, Regulski K, Tournier-Couturier L, Luengo GS, et al. Effect of molecular structure of eco- friendly glycolipid biosurfactants on the adsorption of hair-care conditioning polymers. *Colloids Surf B*. 2020;185:110578.

40 Christova N, Lang S, Wray V, Kaloyanov K, Konstantinov S, Stoineva I. Production, structural elucidation, and in vitro antitumor activity of trehalose lipid biosurfactant from Nocardia farcinica strain. *J Microbiol Biotechnol*. 2015;25:439–447.

41 Wang K, Ran L, Yan T, Ni Z, Kan Z, Zhang Y, et al. Anti-TGEV miller strain infection effect of Lactobacillus plantarum supernatant based on the JAK-STAT1 signaling pathway. *Front Microbiol*. 2019;10:2540.

42 Janek T, Krasowska A, Czyżnikowska Ż, Łukaszewicz M. Trehalose lipid biosurfactant reduces adhesion of microbial pathogens to polystyrene and silicone surfaces: an experimental and computational approach. *Front Microbiol*. 2018;9:2441.

43 Ceresa C, Fracchia L, Williams M, Banat IM, De Rienzo MD. The effect of sophorolipids against microbial biofilms on medical-grade silicone. *J Biotechnol*. 2020;309:34–43.

44 Rasiya KT, Sebastian D. Iturin and surfactin from the endophyte Bacillus amyloliquefaciens strain RKEA3 exhibits antagonism against Staphylococcus aureus. *Biocatal Agric Biotechnol*. 2021;36:102125.

45 Campos JM, Stamford TL, Rufino RD, Luna JM, Stamford TCM, Sarubbo LA. Formulation of mayonnaise with the addition of a bioemulsifier isolated from Candida utilis. *Toxicol Rep*. 2015;2:1164–1170.

46 Gaur VK, Regar RK, Dhiman N, Gautam K, Srivastava JK, Patnaik S, et al. Biosynthesis and characterization of sophorolipid biosurfactant by Candida spp.: application as food emulsifier and antibacterial agent. *Bioresour Technol*. 2019;285:121314.

47 Silva IA, Veras BO, Ribeiro BG, Aguiar JS, Guerra JMC, Luna JM, Sarubbo LA. Production of cupcake-like dessert containing microbial biosurfactant as an emulsifier. *Peer J.* 2020;8:e9064.

48 García-Reyes S, Yáñez-Ocampo G, Wong-Villarreal A, Rajaretinam RK, Thavasimuthu C, Patiño R, Ortiz-Hernández ML. Partial characterization of a biosurfactant extracted from Pseudomonas sp. B0406 that enhances the solubility of pesticides. *Environ Technol.* 2018;39(20):2622–2631.

49 Sarwar A, Hassan MN, Imran M, Iqbal M, Majeed S, Brader G, Sessitsch A, Hafeez FY. Biocontrol activity of surfactin a purified from Bacillus NH-100 and NH-217 against Rice Bakanae Disease. *Microbiol Res.* 2018;209:1–13.

50 Khedher SB, Boukedi H, Laarif A, Tounsi S. Biosurfactant produced by Bacillus subtilis V26: a potential biological control approach for sustainable agriculture development. *Org Agric.* 2020;10(1):117–124.

51 Serres N, He H, Meylheuc T, Feugeas F. Influence of a bioadmixture on standardized parameters of cementitious materials. In: EUROCORR 2016. EFC CEFRACOR. 2016.

52 Parthipan P, Sabarinathan D, Angaiah S, Rajasekar A. Glycolipid biosurfactant as an eco-friendly microbial inhibitor for the corrosion of carbon steel in vulnerable corrosive bacterial strains. *J Mol Liquid.* 2018;261:473–479.

53 He H, Serres N, Meylheuc T, Wynns JT, Feugeas F. Modifying mechanical strength and capillary porosity of portland cement-based mortar using a biosurfactant from *Pseudomonas fluorescens*. *Adv Mater Sci Eng.* 2020:1–12.

54 Joy S, Khare SK, Sharma S. Synergistic extraction using sweep-floc coagulation and acidification of rhamnolipid produced from industrial lignocellulosic hydrolysate in a bioreactor using sequential (fill-and-draw) approach. *Process Biochem.* 2020;90:233–240.

55 Zulkifli WNFWM, Razak NNA, Yatim ARM, Hayes DG. Acid precipitation versus solvent extraction: two techniques leading to different lactone/acidic sophorolipid ratios. *J Surfactants Deterg.* 2019;22(2):365–371.

56 Chen HL, Juang RS. Recovery and separation of surfactin from pretreated fermentation broths by physical and chemical extraction. *Biochem Eng J.* 2008;38(1):39–46.

57 Jia L, Zhou J, Cao J, Wu Z, Liu W, Yang C. Foam fractionation for promoting rhamnolipids production by Pseudomonas aeruginosa D1 using animal fat hydrolysate as carbon source and its application in intensifying phytoremediation. *Chem Eng Process-Process Intensif.* 2020;158:108177.

58 Biniarz P, Henkel M, Hausmann R, Łukaszewicz M. Development of a bioprocess for the production of cyclic lipopeptides pseudofactins with efficient purification from collected foam. *Front Bioeng Biotechnol.* 2020;8:1340.

59 Biselli A, Willenbrink AL, Leipnitz M, Jupke A. Development, evaluation, and optimisation of downstream process concepts for rhamnolipids and 3-(3-hydroxyalkanoyloxy) alkanoic acids. *Sep Purif Technol.* 2020;250:117031.

60 Tang Y, Ma Q, Du Y, Ren L, Van Zyl LJ, Long X. Efficient purification of sophorolipids via chemical modifications coupled with extractions and their potential applications as antibacterial agents. *Sep Purif Technol.* 2020;245:116897.

61 Khondee N, Tathong S, Pinyakong O, Müller R, Soonglerdsongpha S, Ruangchainikom C, Luepromchai E. Lipopeptide biosurfactant production by chitosan-immobilized Bacillus sp. GY19 and their recovery by foam fractionation. *Biochem Eng J.* 2015;93:47–54.

62 Invally K, Sancheti A, Ju LK. A new approach for downstream purification of rhamnolipid biosurfactants. *Food Bioprod Process*. 2019;114:122–131.

63 De Andrade CJ, Barros FF, De Andrade LM, Rocco SA, Luis S, Força M, Pastore GM, Jauregi P. Ultrafiltration based purification strategies for surfactin produced by Bacillus subtilis LB5A using cassava wastewater as substrate. *J Chem Technol Biotechnol*. 2016;91(12):3018–3027.

64 Khoshdast H, Sam A, Vali H, Noghabi KA. Effect of rhamnolipid biosurfactants on performance of coal and mineral flotation. *Int Biodeterior Biodegradation*. 2011;65(8):1238–1243.

65 Hu Y, Ju LK. Purification of lactonic sophorolipids by crystallization. *J Biotechnol*. 2001;87(3):263–272.

66 Dardouri M, Mendes RM, Frenzel J, Costa J, Ribeiro IA. Seeking faster, alternative methods for glycolipid biosurfactant characterization and purification. *Anal Bioanal Chem*. 2021;413(16):4311–4320.

67 Dlamini B, Rangarajan V, Clarke KG. A simple thin layer chromatography based method for the quantitative analysis of biosurfactant surfactin vis-a-vis the presence of lipid and protein impurities in the processing liquid. *Biocatal Agric Biotechnol*. 2020;25:101587.

68 Yang X, Zhu L, Xue C, Chen Y, Qu L, Lu W. Recovery of purified lactonic sophorolipids by spontaneous crystallization during the fermentation of sugarcane molasses with Candida albicans O-13-1. *Enzyme Microb Technol*. 2012;51(6–7):348–353.

69 Bujak T, Wasilewski T, Nizioł-Łukaszewska Z. Role of macromolecules in the safety of use of body wash cosmetics. *Colloids Surf B* 2015;135:497–503.

70 Varvaresou A, Iakovou K. Biosurfactants in cosmetics and biopharmaceuticals. *Lett Appl Microbiol*. 2015;61(3):214–223.

71 Bezerra KG, Durvala IJ, Silvab IA, Fabiola CG. Emulsifying capacity of biosurfactants from Chenopodium quinoa and Pseudomonas aeruginosa UCP 0992 with focus of application in the cosmetic industry. *Chem Eng*. 2020;79:211–216.

72 Moldes AB, Rodríguez-López L, Rincón-Fontán M, López-Prieto A, Vecino X, Cruz JM. Synthetic and bio-derived surfactants versus microbial biosurfactants in the cosmetic industry: an overview. *Int J Mol Sci*. 2021;22(5):2371.

73 Mukherjee AK. Potential application of cyclic lipopeptide biosurfactants produced by *Bacillus subtilis* strains in laundry detergent formulations. *Lett Appl Microbiol*. 2007;45(3):330–335.

74 Takahashi T, Ohno O, Ikeda Y, Sawa R, Homma Y, Igarashi M, Umezawa K. Inhibition of lipopolysaccharide activity by a bacterial cyclic lipopeptide surfactin. *J Antibiot Res*. 2006;59(1):35–43.

75 Rincón-Fontán M, Rodríguez-López L, Vecino X, Cruz JM, Moldes AB. Adsorption of natural surface active compounds obtained from corn on human hair. *RSC Adv*. 2016;6(67):63064–63070.

76 Rincón-Fontán M, Rodríguez-López L, Vecino X, Cruz JM, Moldes AB. Potential application of a multifunctional biosurfactant extract obtained from corn as stabilizing agent of vitamin C in cosmetic formulations. *Sustain Chem Pharm*. 2020;16:100248.

77 Rodríguez-López L, Rincón-Fontán M, Vecino X, Cruz JM, Moldes AB. Study of biosurfactant extract from corn steep water as a potential ingredient in antiacne formulations. *J Dermatol Treat*. 2020;33:1–8.

78 Ahmadi-Ashtiani HR, Baldisserotto A, Cesa E, Manfredini S, Sedghi Zadeh H, Ghafori Gorab M, Vertuani S, et al. Microbial biosurfactants as key multifunctional ingredients for sustainable cosmetics. *Cosmetics*. 2020;7(2):46.
79 Allemann IB, Baumann L. Antioxidants used in skin care formulations. *Skin Ther Lett*. 2008;13(7):5–9.
80 Irfan-Maqsood M, Seddiq-Shams M. Rhamnolipids: well-characterized glycolipids with potential broad applicability as biosurfactants. *Ind Biotechnol*. 2014;10(4):285–291.
81 Drakontis CE, Amin S. Design of sustainable lip gloss formulation with biosurfactants and silica particles. *Int J Cosmet Sci* 2020;42(6):573–580.
82 Lourith N, Kanlayavattanakul M. Natural surfactants used in cosmetics: glycolipids. *Int J Cosmet Sci* 2009;31(4):255–261.
83 Inoue S, Miyamoto N, inventors; Kao Soap Co Ltd, assignee. Process for producing a hydroxyfatty acid ester. United States patent US 4,201,844. 1980 May 6.
84 Borsanyiova M, Patil A, Mukherji R, Prabhune A, Bopegamage S. Biological activity of sophorolipids and their possible use as antiviral agents. *Folia Microbiol*. 2016;61(1):85–89.
85 De Rienzo MAD, Stevenson P, Marchant R, Banat IM. Antibacterial properties of biosurfactants against selected Gram-positive and-negative bacteria. *FEMS Microbiol Lett*. 2016;363(2).
86 Naughton PJ, Marchant R, Naughton V, Banat IM. Microbial biosurfactants: current trends and applications in agricultural and biomedical industries. *J Appl Microbiol*. 2019;127(1):12–28.
87 Sajid M, Khan MSA, Cameotra SS, Al-Thubiani AS. Biosurfactants: potential applications as immunomodulator drugs. *Immunol Lett*. 2020;223:71–77.
88 Sana S, Datta S, Biswas D, Auddy B, Gupta M, Chattopadhyay H. Excision wound healing activity of a common biosurfactant produced by *Pseudomonas* sp. *Wound Med*. 2018;23:47–52.
89 Elshikh M, Funston S, Chebbi A, Ahmed S, Marchant R, Banat IM. Rhamnolipids from non-pathogenic Burkholderia thailandensis E264: physicochemical characterization, antimicrobial and antibiofilm efficacy against oral hygiene related pathogens. *N Biotechnol*. 2017;36:26–36.
90 Chen J, Wu Q, Hua Y, Chen J, Zhang H, Wang H. Potential applications of biosurfactant rhamnolipids in agriculture and biomedicine. *Appl Microbiol Biotechnol*. 2017;101(23):8309–8319.
91 De Santo K Rhamnolipid mechanism. U.S. Patent Application No. 12/938,985. 26 2011 May.
92 Van Bogaert IN, Zhang J, Soetaert W. Microbial synthesis of sophorolipids. *Process Biochem*. 2011;46(4):821–833.
93 De Rienzo MAD, Banat IM, Dolman B, Winterburn J, Martin PJ. Sophorolipid biosurfactants: possible uses as antibacterial and antibiofilm agent. *N Biotechnol*. 2015;32(6):720–726.
94 Sen S, Borah SN, Bora A, Deka S. Production, characterization, and antifungal activity of a biosurfactant produced by Rhodotorula babjevae YS3. *Microb Cell Factories*. 2017;16(1):1–14.
95 Lydon HL, Baccile N, Callaghan B, Marchant R, Mitchell CA, Banat IM. Adjuvant antibiotic activity of acidic sophorolipids with potential for facilitating wound healing. *Antimicrob Agents Chemother*. 2017;61(5):e02547–16.

96 Hagler M, Smith-Norowitz TA, Chice S, Wallner SR, Viterbo D, Mueller CM, Bluth MH, et al. Sophorolipids decrease *IgE* production in U266 cells by downregulation of *BSAP (Pax5), TLR-2, STAT3* and *IL-6J. Allergy Clin Immunol.* 2007;119(1):S263.

97 Wang X, Hu W, Zhu L, Yang Q. Bacillus subtilis and surfactin inhibit the transmissible gastroenteritis virus from entering the intestinal epithelial cells. *Biosci Rep.* 2017;37(2).

98 Backhaus S, Zakrzewicz A, Richter K, Damm J, Wilker S, Fuchs-Moll G, et al. Surfactant inhibits ATP-induced release of interleukin-1 via nicotinic acetylcholine receptors. *J Lipid Res.* 2017;58:1055–1066.

99 He Z, Zeng W, Zhu X, Zhao H, Lu Y, Lu Z. Influence of surfactin on physical and oxidative stability of microemulsions with docosahexaenoic acid. *Colloids Surf B Biointerfaces.* 2017;151:232–239.

100 Campos JM, Montenegro-Stamford TL, Sarubbo LA, De Luna JM, Rufino RD, Banat IM. Microbial biosurfactants as additives for food industries. *Biotechnol Prog.* 2013;29(5):1097–1108.

101 Traudel K, Merten S. Possible food and agricultural application of microbial surfactants: an assessment. In: Kosaric N, Cairns WL, Gray NC, editors. *Biosurfactants and Biotechnology*. Routledge, New York; 1987. pp. 89–120.

102 Cheah YS, Loh TC, Akit H, Kimkool S. Effect of synthetic emulsifier and natural biosurfactant on feed process and quality of pelletized feed in broiler diet. *Braz J Poultry Sci.* 2017;19:23–34.

103 Ribeiro BG, Guerra JMC, Sarubbo LA. Potential food application of a biosurfactant produced by *Saccharomyces cerevisiae* URM 6670. *Front Bioeng Biotechnol.* 2020;8:434.

104 Kiran GS, Priyadharsini S, Sajayan A, Priyadharsini GB, Poulose N, Selvin J. Production of lipopeptide biosurfactant by a marine *Nesterenkonia* sp. and its application in food industry. *Front Microbiol.* 2017;8:1138.

105 Rani M, Weadge JT, Jabaji S. Isolation and characterization of biosurfactant-producing bacteria from oil well batteries with antimicrobial activities against food-borne and plant pathogens. *Front Microbiol.* 2020;11:64.

106 Kourmentza K, Gromada X, Michael N, Degraeve C, Vanier G, Ravallec R, et al. Antimicrobial activity of lipopeptide biosurfactants against foodborne pathogen and food spoilage microorganisms and their cytotoxicity. *Front Microbiol.* 2021;11:3398.

107 Silva SS, Carvalho JWP, Aires CP, Nitschke M. Disruption of *Staphylococcus aureus* biofilms using rhamnolipid biosurfactants. *Int J Dairy Sci.* 2017;100(10):7864–7873.

108 Merghni A, Dallel I, Noumi E, Kadmi Y, Hentati H, Tobji S, et al. Antioxidant and antiproliferative potential of biosurfactants isolated from Lactobacillus casei and their anti-biofilm effect in oral *Staphylococcus aureus* strains. *Microb Pathog.* 2017;104:84–89.

109 da Silva IA, Bezerrac KGO, Batista IJ. Evaluation of the emulsifying and antioxidant capacity of the biosurfactant produced by *Candida bombicola* URM 3718. *Chem Eng.* 2020:79.

110 McGaughey BD. The FIFRA endangered species task force: dealing with unusual challenges and multiple agencies. In: *Data Generation for Regulatory Agencies: A Collaborative Approach*. J Am Chem Soc. 2021. pp. 61–75.

111 [EPA] US Environmental Protection Agency. Indexes to part 180 tolerance information for pesticide chemicals in food and feed commodities. 2020a. https://www.epa.gov/pesticide-tolerances/indexes-part-180-tolerance-information-pesticide-chemicals-food-and-feed accessed September 8, 2021.

112 Liu X, Cao A, Yan D, Ouyang C, Wang Q, Li Y. Overview of mechanisms and uses of biopesticides. *Int J Pest Manag*. 2021;67(1):65–72.
113 Oluwaseun AC, Kola OJ, Mishra P, Singh JR, Singh AK, Cameotra SS, Micheal BO. Characterization and optimization of a rhamnolipid from Pseudomonas aeruginosa C1501 with novel biosurfactant activities. *Sustain Chem Pharm*. 2017;6:26–36.
114 Adetunji C, Oloke J, Kumar A, Swaranjit S, Akpor B. Synergetic effect of rhamnolipid from Pseudomonas aeruginosa C1501 and phytotoxic metabolite from Lasiodiplodia pseudotheobromae C1136 on Amaranthus hybridus L. and Echinochloa crus-galli weeds. *Environ Sci Pollut Res*. 2017;24(15):13700–13709.
115 Adetunji CO, Adejumo IO, Afolabi IS, Adetunji JB, Ajisejiri ES. Prolonging the shelf life of 'Agege Sweet'orange with chitosan–rhamnolipid coating. *Hortic Environ Biotechnol*. 2018;59(5):687–697.
116 Kumar A, Rabha J, Jha DK. Antagonistic activity of lipopeptide-biosurfactant producing Bacillus subtilis AKP, against Colletotrichum capsici, the causal organism of anthracnose disease of chilli. *Biocatal Agric Biotechnol*. 2021;36:102133.
117 Droby S, Wisniewski M, Benkeblia N. Postharvest pathology of tropical and subtropical fruit and strategies for decay control. In: Elhadi MY, editors. *Postharvest Biology and Technology of Tropical and Subtropical Fruits*. Woodhead Publishing, Cambridge, UK; 2011. pp. 194–224e.
118 Toral L, Rodríguez M, Béjar V, Sampedro I. Antifungal activity of lipopeptides from Bacillus XT1 CECT 8661 against Botrytis cinerea. *Front Microbiol*. 2018;9:1315.
119 Xu Z, Shi M, Tian Y, Zhao P, Niu Y, Liao M. Dirhamnolipid produced by the pathogenic fungus Colletotrichum gloeosporioides BWH-1 and its herbicidal activity. *Molecules*. 2019;24(16):2969.
120 Do Nascimento Silva J, Mascarin GM, de Paula Vieira de Castro R, Castilho LR, Freire DM. Novel combination of a biosurfactant with entomopathogenic fungi enhances efficacy against Bemisia whitefly. *Pest Manag Sci*. 2019;75(11):2882–2891.
121 Prasad RV, Kumar RA, Sharma D, Sharma A, Nagarajan S. Sophorolipids and rhamnolipids as a biosurfactant: synthesis and applications. In: Inamuddin AM, Asiri A, Suvardhan K, editors. *Green Sustainable Process for Chemical and Environmental Engineering and Science*. Elsevier, Amsterdam, The Netherlands; 2021. pp. 423–472.
122 Yoo DS, BS L, EK K. Characteristics of microbial biosurfactant as an antifungal agent against plant pathogenic fungus. *J Microbiol Biotechnol*. 2005;15(6):1164–1169.
123 Vaughn SF, Behle RW, Skory CD, Kurtzman CP, Price NPJ. Utilization of sophorolipids as biosurfactants for postemergence herbicides. *J Crop Prot*. 2014;59:29–34.
124 Fernandes NDAT, de Souza AC, Simoes LA, Dos Reis GMF, Souza KT, Schwan RF, Dias DR. Eco-friendly biosurfactant from Wickerhamomyces anomalus CCMA 0358 as larvicidal and antimicrobial. *Microbiol Res*. 2020;241:126571.
125 Gašić S, Tanović B. Biopesticide formulations, possibility of application and future trends. *Pesticidi I Fitomedicina*. 2013;28(97–102):2013.
126 Shi D, Wang J, Deng W. Smart building and construction materials. 2019;1–2.
127 Benmokrane B, Ali AH. Durability and long-term performance of fiber-reinforced polymer as a new civil engineering material. In: Taha MMR, Taha PACKER, editors. *International Congress on Polymers in Concrete*. Springer, Cham; 2018. pp. 49–59.
128 Zhang C, Canning L. Application of non-conventional materials in construction. *Constr Mat*. 2011;164:165–172.

129 Liu X, Koestler RJ, Warscheid T, Katayama Y, Gu J. Microbial deterioration and sustainable conservation of stone monuments and buildings. *Nat Sustain*. 2020;3:991–1004.

130 Poursaee A, Hansson CM. The influence of longitudinal cracks on the corrosion protection afforded reinforcing steel in high performance concrete. *Cem Concr Res*. 2008;38:1098–1105.

131 Shaikh FUA. Effect of cracking on corrosion of steel in concrete. *Int J Concr Struct Mater*. 2018;12:1–12.

132 Al-Sherrawi MH, Lyashenko V, Edaan EM, Sotnik L. Corrosion of metal construction structures. *Int J Civ Eng Tec*. 2018;9:437–446.

133 Lepinay C, Mihajlovski A, Seyer D, Touron S, Bousta F, Di Martino P. Biofilm communities survey at the areas of salt crystallization on the walls of a decorated shelter listed at UNESCO World cultural Heritage. *Int Biodeterior*. 2017;122:116–127.

134 Nuhoglu Y, Oguz E, Uslu H, Ozbek A, Ipekoglu B, Ocak I. The accelerating effects of the microorganisms on biodeterioration of stone monuments under air pollution and continental-cold climatic conditions in Erzurum, Turkey. *Sci Total Environ*. 2006;364:272–283.

135 Negi A, Sarethy IP. Microbial biodeterioration of cultural heritage : events, colonization, and analyses. *Microb Ecol*. 2019;78:1014–1029.

136 Zanardini E, May E, Purdy KJ, Murrell JC. Nutrient cycling potential within microbial communities on culturally important stoneworks. *Environ Microbiol Rep*. 2019;11:147–154.

137 Lin J. Biocorrosion control: current strategies and promising alternatives. *Afr J Biotechnol*. 2012;11:15736–15747.

138 Zheng S, Bawazir M, Dhall A, Kim H, He L, Heo J, Hwang G. Implication of surface properties, bacterial motility, and hydrodynamic conditions on bacterial surface sensing and their initial adhesion. *Front Bioeng Biotechnol*. 2021;9:1–22.

139 Plaza G, Achal V. Biosurfactants: eco-friendly and innovative biocides against biocorrosion. *Int J Mol Sci*. 2020;21:1–11.

140 Purwasena IA, Astuti DI, Fauziyyah NA, Putri DAS, Sugai Y. Inhibition of microbial influenced corrosion on carbon steel ST37 using biosurfactant produced by Bacillus sp. *Mater Res Express*. 2019;6(11):115405.

141 Dong W, Li W, Tao Z, Wang K. Piezoresistive properties of cement-based sensors: review and perspective. *Constr Build Mat*. 2019;203:146–163.

142 Vipulanandan C, Ali K, Ariram P. Nanoparticle and surfactant-modified smart cement and smart polymer grouts. In: Yoga Chandran C., Hoit M. I. editors. *Geotechnical and Structural Engineering Congress*, 2016. pp. 884–896.

143 Wang H, Liu X, Apostolidis P, Scarpas T. Review of warm mix rubberized asphalt concrete: towards a sustainable paving technology. *J Clean Prod*. 2018;177:302–314.

144 Ashoori S, Sharifi M, Masoumi M, Salehi MM. The relationship between SARA fractions and crude oil stability. *Egypt J Pet*. 2017;26:209–213.

145 Kok MV, Gul KG. Thermal characteristics and kinetics of crude oils and SARA fractions. *Thermochim Acta*. 2013;569:66–70.

146 Mendes JC, Moro TK, Figueiredo AS, Do Carmo Silva KD, Silva GC, Silva GJB, Peixoto RAF. Mechanical, rheological and morphological analysis of cement-based composites with a new LAS-based air entraining agent. *Constr Build Mater*. 2017;145:648–661.

147 Zhao P, Wu W, Song X, Sun H, Ren R, Liu F. Synthesis of imidazoline biosurfactants and application as an asphalt emulsifier. *J Dispers Sci Technol*. 2021:1–8.

148 Fracchia L, Ceresa C, Franzetti A, Cavallo M, Gandolfi I, Van Hamme J, Banat IM. Industrial applications of biosurfactants. In: Kosaric N, Sukan FV. editor. *Biosurfactants: Production and Utilization—Processes, Technologies, and Economics*. 2014. pp. 245–260.

149 Bouassida M, Fourati N, Ghazala I, Ellouze-Chaabouni S, Ghribi D. Potential application of Bacillus subtilis SPB1 biosurfactants in laundry detergent formulations: compatibility study with detergent ingredients and washing performance. *Eng Life Sci*. 2018;18(1):70–77.

150 Helmy Q, Gustiani S, Mustikawati AT. Application of rhamnolipid biosurfactant for bio-detergent formulation. In IOP Conference Series: Materials Science and Engineering. (Vol. 823, No. 1, p. 012014). IOP Publishing. 2020, April.

151 Franzetti A, Gandolfi I, Bestetti G, Smyth TJ, Banat IM. Production and applications of trehalose lipid biosurfactants. *Eur J Lipid Sci Tech*. 2010;12:617–627.

152 Nguyen TT, Youssef NH, McInerney MJ, Sabatini DA. Rhamnolipid biosurfactant mixtures for environmental remediation. *Water Res*. 2008;42:1735–1743.

153 Ravindran A, Sajayan A, Priyadharshini GB, Selvin J, Kiran GS. Revealing the efficacy of thermostable biosurfactant in heavy metal bioremediation and surface treatment in vegetables. *Front Microbiol*. 2020;11:222.

154 Hu X, Qiao Y, Chen LQ, Du JF, Fu YY, Wu S, Huang L. Enhancement of solubilization and biodegradation of petroleum by biosurfactant from Rhodococcus erythropolis HX-2. *Geomicrobiol J*. 2020;37(2):159–169.

155 Nehal N, Singh P. Role of nanotechnology for improving properties of biosurfactant from newly isolated bacterial strains from Rajasthan. *Mater Today: Proc*. 2022;50:2555–2561.

156 Kiran GS, Sabu A, Selvin J. Synthesis of silver nanoparticles by glycolipid biosurfactant produced from marine *Brevibacterium casei* MSA19. *J Biotechnol*. 2010;148(4):221–225.

157 Durval IJB, Meira HM, de Veras BO, Rufino RD, Converti A, Sarubbo LA. Green synthesis of silver nanoparticles using a biosurfactant from Bacillus cereus UCP 1615 as stabilizing agent and its application as an antifungal agent. *Fermentation*. 2021;7(4):233.

158 Marangon CA, Vigilato Rodrigues MA, Vicente Bertolo MR, Amaro Martins VDC, de Guzzi Plepis AM, Nitschke M. The effects of ionic strength and pH on antibacterial activity of hybrid biosurfactant-biopolymer nanoparticles. *J Appl Polym Sci*. 2022;139(1):51437.

159 Rana S, Singh J, Wadhawan A, Khanna A, Singh G, Chatterjee M. Evaluation of in vivo toxicity of novel biosurfactant from *Candida parapsilosis* loaded in PLA-PEG polymeric nanoparticles. *J Pharm Sci*. 2021;110(4):1727–1738.

160 Peng Y, Decatur J, Meier MA, Gross RA. Ring-opening metathesis polymerization of a naturally derived macrocyclic glycolipid. *Macromolecules*. 2013;46(9):3293–3300.

161 Peng Y, Munoz-Pinto DJ, Chen M, Decatur J, Hahn M, Gross RA. Poly (sophorolipid) structural variation: effects on biomaterial physical and biological properties. *Biomacromolecules*. 2014;15(11):4214–4227.

162 Arcens D, Le Fer G, Grau E, Grelier S, Cramail H, Peruch F. Chemo-enzymatic synthesis of glycolipids, their polymerization and self-assembly. *Polym Chem*. 2020;11(24):3994–4004.

163 Kurozuka A, Onishi S, Nagano T, Yamaguchi K, Suzuki T, Minami H. Emulsion polymerization with a biosurfactant. *Langmuir*. 2017;33(23):5814–5818.

14

Techno-economic-environmental Analysis of the Production of Biosurfactants in the Context of Biorefineries

Andreza Aparecida Longati[1], Andrew Milli Elias[2], Felipe Fernando, Furlan[3,] Everson Alves Miranda[1], and Roberto de Campos Giordano[3]*

[1] *Department of Materials and Bioprocess Engineering, School of Chemical Engineering, State University of Campinas, Campinas, São Paulo, 13083–852, Brazil*
[2] *Embrapa Instrumentação, Rua XV de Novembro 1452, São Carlos, São Paulo, 13560–970, Brazil*
[3] *Chemical Engineering Graduate Program, Department of Chemical Engineering, Federal University of São Carlos, São Carlos, São Paulo, 13565–905, Brazil*
* Correspond author

14.1 Introduction

14.1.1 Background

In 2015, an effort was made in Paris by the countries adhering to the United Nations Framework Convention on Climate Change (UNFCCC) to decrease the CO_2 emissions due to the effect of greenhouse gases (GHG) on global warming. The Paris Agreement sets a long-term temperature goal for holding the global average temperature increase to well below 2 °C, and pursuing efforts to limit it to 1.5 °C above pre-industrial levels [1]. COP26 (26th United Nations Climate Change conference) gathered parties to accelerate actions towards the goals of the Paris Agreement and the UNFCCC [2]. In this context, the gradual transformation of the global linear economy into a circular economy becomes imperative [3]. In a circular economy, materials circulate at their maximum as technical or biological nutrients in integrated, restorative, and regenerative industrial systems [4].

A rapid and deep decarbonization of all sectors of the economy, including the industrial sector, is required to achieve this goal. The refinery industry is one of the major energy users and responsible for a large proportion of GHG emissions, contributing to around 4% of global anthropogenic CO_2 emissions (about 1.3 gigatonnes of CO_2 in 2018) [5, 6]. Decarbonization of energy-intensive industries such these refineries is therefore crucial for mitigating emissions leading to climate change. By delivering low-carbon products and fuels, biorefineries could play an important role in supporting the transition to net zero carbon economy worldwide [7]. Such facilities can transform waste streams into products like biofuels (biodiesel, bioethanol, biomethanol, biogas, biokerosene and hydrogen), bioactive compounds (flavonoids and antioxidants), biofertilizers, lactic acid, vinegar, oligosaccharides, and many others [8].

Biosurfactants and Sustainability: From Biorefineries Production to Versatile Applications, First Edition.
Edited by Paulo Ricardo Franco Marcelino, Silvio Silverio da Silva, and Antonio Ortiz Lopez.
© 2023 John Wiley & Sons Ltd. Published 2023 by John Wiley & Sons Ltd.

Several niche developments have occurred within the past decade in the biorefinery sector. For instance, there has been strong engagement in biorefineries to produce biofuel and different bioproducts such as biosurfactants (BS). However, advancing R&D for the full-scale implementation of decarbonization technologies is not straightforward. Systems innovation is a complex, uncertain task, and involves multiple factors and interdependent elements concerning new technologies. Besides that, technological, economical, and environmental aspects need to be taken into account altogether. In this context, the current state of development of low-carbon technologies for BS production that could enable future decarbonization of this sector is analyzed here.

14.1.2 Surfactant Versus Biosurfactant

Surfactants are extensively employed in a range of applications, functioning as detergents, wetting agents, emulsifiers and foaming agents. They are found in industrial products from food and healthcare to paints and coatings. They are a crucial component of many industrial processes such as oil recovery and printing [9, 10].

BS synthesized by plants, animals, and microbes, natural alternatives to synthetic surfactants, are amphiphilic molecules composed by hydrophilic and hydrophobic moieties [11]. So, they accumulate at the interface between polar and nonpolar media, modifying the surface and interfacial tension in aqueous solutions and hydrocarbon mixtures. This characteristic increases the solubility of polar molecules in non-polar liquids and vice-versa [8, 11]

BSs have several advantages over synthetic surfactants: lower toxicity, higher biodegradability, higher selectivity, activity under extreme temperatures, pH and salinity conditions, low critical micellar concentration (CMC, i.e. lower amount of BS required to reduce surface tension) [12], benign interaction with the environment, and higher foaming [11]. BS are classified by their chemical structures into five major groups: glycolipids, lipopeptides, phospholipids, fatty acids, and polymers. There are four major groups of glycolipids: rhamnolipids, sophorolipids, trehalose lipids and mannosylerythritol lipids [13].

14.1.3 Biosurfactant Market, Producers, and Patents

Different projections for the global BS market differ in the overall market size. However, they all have in common the expectation of BS expansion and growth globally, since the demand has increased continually over the past years. According to the 2016 report from Market and Market [14], the BS market size was valued at US$ 3.99 billion in 2016 and is projected to reach US$ 5.52 billion by 2022, growing at a compound annual growth rate (CAGR) of 5.6% from 2017 to 2022. A study by Grand View Research [15] projected the CAGR of 4.3% from 2014 to 2020, reaching US$ 2,308.8 million by 2020. A report from Global Market Insights [16] estimated the BS market to grow at over 5.5% CAGR between 2020 and 2026, with 2026 value projection of US$ 2,577.2 million. Among the BSs, fungal derived sophorolipids cover the largest market in detergent industry [17]. Sophorolipids and rhamnolipids are responsible for 5% of this market, with an expected market size of 23 kt per year [18, 19]. The rhamnolipids-based BS market is expected to reach over USD 145 million by 2026 [16].

The major players of this sector (Table 14.1) are Ecover (Belgium), BASF (German), Soliance (France), MG Intobio (Republic of Korea) [20], Urumqi Unite, Evonik, AkzoNobel, Toyobo and Saraya (all from Japan) [8], Jeneil Biosurfactant and AGAE technologies Ltd. (United States of America, USA). Other major manufacturers include Fraunhofer IGB (Germany) and Cognis (Germany and USA), dealing with production of glycolipid surfactants, cellobiose lipids, and mannosylerythritol lipids [21].

No descriptions of BS industrial processes (Table 14.1) are reported in the literature, probably because these processes are under confidentiality agreements. Nevertheless, some patents can provide information about the processes and applications. According to Hames et al. [36], the first patent on BS, published in 1967 (under patent number US3312684A), was related to hydroxy fatty acids and their production as glycolipids by a fermentation process using *Torulopsis magnolia* [37]. The patent EP2055314B1 [38], filed by the National Institute of Advanced Industrial Science and Technology together with Toyobo Co Ltd, describes the use of the mannosylerythritol lipid as an active ingredient which activates various cells and is effective for anti-aging. The patent filed by Evonik Degussa GmbH (US9157108B2) [39] describes a sophorolipid production process in aqueous solution (lactone- and acid form) by fermentation of a carbon source, medium centrifugation, solvent extraction using preferably n-pentanol, distillation and lyophilization, with further purification being possible specially through chromatographic processes. Saraya Co Ltd., a company commercializing the sophorolipid produced from palm oil named Sophoro TM [29–31], filed the patent EP3034613A1 [40] describing the production of sophorolipids by *Candida bombicola* using glucose and lipids (oils and fats) as substrates. In this patent, sophorolipids are recovered by three sequential extractions using ethyl acetate, followed by the evaporation of the solvent. The EP2821495B1 patent [41], also filed by Saraya Co. Ltd, describes further BS purification by adjustment of pH below 4 and cooling below 15 °C, followed by washing with water, distillation and drying.

It should be noted that, although more than half a century has passed since the first published patent and despite the technology development for industrial production (Table 14.1) and their wide potential applications, BS are still not being fully exploited by the different sectors [36].

14.1.4 Biosurfactant Production Routes

Over the recent years, there has been a surge in the need for cost-cutting materials which could act as substrates for BS production [42]. A wide variety of pathways and routes, using different industrial wastes as substrates, are being examined and proposed for BS production (Figure 14.1).

The range of substrates for BS production comprises agricultural, industrial, food and domestic wastes and residues, lignocellulosic materials, vegetable oils, and sugars [8]. Some of the biologically derived substrates may require upstream conversion like pretreatment and hydrolysis, necessary to ensure better utilization of some feedstocks by the microorganisms [42]. Others, such as simple sugars and vegetable oils, can be sent directly to the bioreactors, as in the process described by Misailidis and Petrides [43]. A variety of strategies can be applied to the BS purification, Figure 14.1 demonstrated some of them used in processes described on literature [20, 43–45].

Table 14.1 BS production around the globe.

Company	Product name	Type of BS	Substrate	Microorganism	Additional information	References
Ecover (Belgium)	-	Sophorolipids	Sugar and rapeseed oil	Microorganisms from honey by bumble bees	They use in its detergent and cleaner products containing sophorolipids from Evonik	[22]
BASF (German)	BioToLife TM	Sophorolipids	-	-	ACS and BASF collaboration	[23, 24]
Holiferm	HoliSurf HF Biosurfactant	Sophorolipids	Sugar and rapeseed oil	Yeast found in honey	Shampoo containing Yeast Ferment Extract (and) Polylysine BASF and Holiferm collaboration	[23, 25]
Soliance (France)	Sopholiance	Sophorolipid (glycolipid)	Rapeseed fermentation	*Candida bombicola*	Cosmetic industry	[8]
Jeneil Biosurfactant Company (EUA)	-	-	-	-	Different BS which can be used in agriculture, bioremediation, household and personal care, and antimicrobial	[26]
	ZONIX	Rhamnolipids	-	-	A bio-fungicide used in cleaning and recovering oil from storage tanks	[26]
AGAE technologies (EUA)	R95	Rhamnolipids	-	*Pseudomonas aeruginosa*	An HPLC/MS grade rhamnolipid	[27]
Saraya (Japan)	Sophoro TM	Sophorolipid (glycolipid)	Palm oil	*Candida bombicola*	Cosmetic industry (name Wash Bon, a Hand Soap Herbal Citrus)	[28–30]
					Cleaner (name Power Quick Enzyme Cleaner, a manual and ultrasonic cleaning)	[28,30, 31]

Toyobo (Japan)	Ceramela TM	Mannosylerythritol lipids (glycolipid)	Olive oil	*Pseudozyma antarctica*	Cosmetic industry; Ceramela presents similar structure to Ceramide	[32, 33]
Evonik Industries (Japan)	-	sophorolipids	Sugar and rapeseed oil	Microorganisms from honey by bumble bees	Producer of the BS used in Ecover, a Belgian detergent and cleanser manufacturer	[34]
	-	Rhaminolipids	sugar	-	Pilot plant in development	[34]
Cleveland Biotech Ltd. (Teesside)	Citrasolv	-	Orange peel	-	-	[35]
Henkel (Henkel)	-	Sophorolipids, Rhamnolipids, Mannosylerythritol lipids	-	-	-	[35]
BioFuture Ltd. (Dublin)	BioFuture	Bacterial rhamnolipid	-	-	-	[35]

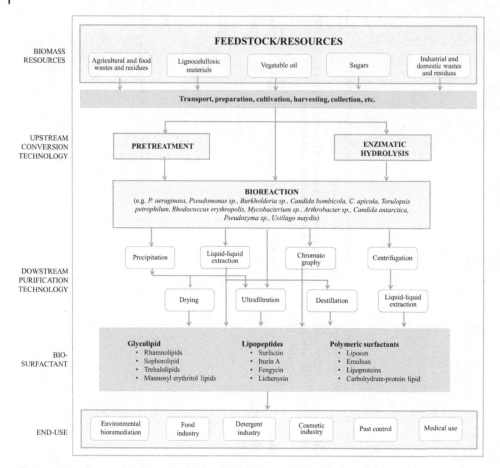

Figure 14.1 Schematic representation of some BS production routes/pathways.

14.2 Economic Aspects of the BS Production

In spite of the increasing market demand, production costs are considered a significant barrier to BS diffusion in the market. For comparison purposes, a rhamnolipids product with 90% purity has a stipulated market value of 650 US$/kg, while chemical surfactants cost only 2 to 6 US$/kg [46]. The use of inexpensive substrates, overproducing strains, metabolic engineered microorganism, effective downstream processes, and the co-production of BS with other processes and/or biomolecules are some strategies that have been proposed to develop BS production [13].

Raw materials have a significant impact on the final cost of bioproducts, possibly accounting for more than 80% of total costs [47, 48]. Thus, the use of wastes for BS production is an interesting option to reduce associated costs. However, even when using less noble carbon sources, the cost of raw materials can be significant.

Wang et al. [45] studied the production of sophorolipids from food waste, producing about 1793 t of BS syrup (78% purity) per year. In addition to food waste, the authors used

14.2 Economic Aspects of the BS Production

glucose and oleic acid to cultivate the microorganisms. Assuming all the steps involved in the BS production process, raw materials represented 78.7% of the final cost of the product. Elias et al. [20] studied the production of sophorolipid from sugarcane bagasse. Their best-case study, from both economical and environmental perspectives, produced about 594 t of BS syrup (90% purity) per year using hydrothermal treatment for substrate preparation and membranes for product purification. In this scenario, the cost of raw materials corresponded to just over 6.5% of the calculated minimum BS selling price.

Patria et al. [44] performed a techno-economic analysis for a biorefinery approach based on food residue digesters for rhamnolipids production (50% purity). In their work, for the production of 627 t/year, the cost of raw material represented about 10.6% of the calculated minimum BS selling price. Finally, Misailidis and Petrides [43] performed the modeling and cost analysis of rhamnolipids production (46% purity). The authors used glucose, vegetable oil, peptone, and yeast extract as raw materials to produce 8100 t of BS per year. In this scenario, the cost of raw materials accounted for around 28.5% of the calculated minimum BS selling price.

The use of process residues as substrate promotes a different destination for the material that would otherwise be sent for treatment or disposal. In situations where wastes are used as carbon sources (e.g. cooking frying oil), BS production can be part of the necessary transition to a circular economy [49]. However, based on recent studies [20, 44, 45], using residues as substrate for BS production does not guarantee lower prices, since other steps for raw material preparation are necessary. Therefore, further studies on this subject are still needed.

According to Chong and Li [50], most of the costs associated with the equipment used in the production of BS are attributed to the cultivation and purification stages. In this sense, a deeper analysis was carried out. The techno-economic analysis (TEA) carried out by Elias et al. [20] pointed out that 66% of CAPEX comes from the cultivation sector while purification accounts for 16%. In the TEA developed by Patria et al. [44], 60% of equipment costs are due to the purification steps, while microbial cultivation accounts for 15%. Wang et al. [45] found that 17% of the cost of equipment was associated with culture bioreactors, while 24% was due to BS purification steps. Misailidis and Petrides [43] calculated that bioreactors represented 31% of the total cost of equipment. The performance of the bioreactor, i. e., productivity, final concentration of BS and product yield are determinant for this cost and are important for environmental performance of the process as well.

Elias et al. [20] performed a global sensitivity analysis for the production process: the concentration of BS at the end of cultivation had a significant impact on the product final cost and on the environmental performance indicators.

As it can be seen, works on the TEA of BS industrial production are scarce, yet. However, some characteristics referring to the scale of the process can be noted. These studies used different types of raw materials, from the noblest (such as glucose and vegetable oil) to residues, as well as different production steps. Therefore, the analysis of the influence of the scale of production should be qualitative (Table 14.2). For comparison purposes, the CAPEX and OPEX values were modified to take into account the production scale and the purity of the final product. Therefore, the values are in kilogram of BS present in the product.

The scale of BS production can be considered one of the main bottlenecks for BS commercialization [13]. Some inferences can be made (Table 14.2) based on the scale-up of the processes. In general, the increase in scale reduces the minimum selling price of the product (MBSP). MBSP is an economic metric that takes into account both CAPEX and OPEX, being

Table 14.2 Economic indicators of industrial-scale biosurfactant production.

References	Product	Substrate	Annualized CAPEX (US$/kg)	Interest rate (%)	Annual OPEX (US$/kg)	MBSP[e] (US$/kg)
[45]	Sophorolipid[a]	Food waste	7.5	7.0	20.2	20.4
[20]	Sophorolipid[b]	Sugarcane bagasse	5.5	11.0	10.5	18.2
[44]	Rhamnolipid[c]	Food waste	18.7	7.0	72.1	36.0
[43]	Rhamnolipids[d]	Glucose	_[f]	7.0	13.6	6.2

[a] Scenario II – syrup production (78% purity); production of 1793 t/year.
[b] Scenario B – syrup production (90% purity); production of 594 t/year.
[c] Scenario I – syrup production (50% purity); production of 627 t/year.
[d] Syrup production (46% purity); production of 8100 MT/year.
[e] minimum BS selling price.
[f] Annualized CAPEX could not be calculated since project lifetime was not reported.

obtained when the net present value is zero and a minimum attractiveness rate of return is defined [48]. In screening stages, CAPEX is usually determined based on the cost of equipment calculated by the six-tenth rule, directly influenced by the scale of production.

From Table 14.2, it can also be seen that the lowest costs per kilogram of BS were obtained in the largest scale, from the work of Misailidis and Petrides [43], producing 8100 t of BS per year. However, unlike the others, these authors considered glucose as the carbon source. In this scenario, the cost of equipment associated with raw materials used in microbial cultivation represents only 0.01% of the total cost of equipment. This fact indicates that as it is more difficult to work with waste/residues, the production cost in these cases may be higher when compared to traditional sources. It is worth highlighting that the use of waste reduces competition with the food industry. Moreover, BS production processes are still under development, what provides a window for process optimization. In other words, there is still a great opportunity to reduce the production cost of BS.

14.3 Environmental Aspects

There is a tendency to consider biobased processes inherently environmentally-friendly. This is mainly due to the fact that these processes rely on biomass as a carbon source and biomass is an abundant carbon-neutral renewable resource. Processes relying on biomass are part of a "short-cycle carbon system", which is more sustainable than "long-cycle carbon systems" based on fossil [51]. Nevertheless, this consideration is only based on global warming impact and, therefore, disregards the risk of a burden shift, that is, achieving beneficial results in one impact at the cost of increasing another. To prevent this misconception, a life cycle perspective must be taken, considering not only the full life cycle of the product, but also all the different impacts the product life cycle can have [52]. In this sense, the path to more sustainable production and consumption patterns must be based on comprehensive analyses of their environmental implications. In order to achieve this, the whole supply-chain of products, their use and end-of-life management must be assessed [5].

The correct accounting and attribution of impacts to a product or service is important to assist decision making. Nevertheless, it is a cumbersome task. Life cycle analysis (LCA) is one of the most accepted methods to perform this assessment. This method takes a life cycle perspective in the sense that it considers all the processes required to deliver the function of the product considered, "from cradle to grave" [53]. This prevents burden shift in two different ways. By considering the entire life cycle of the product, from the production of the raw materials and energy required, throughout the manufacturing, use and recycle or final disposal of the product, LCA avoids shifting the burden between life cycle stages. For example, changing the raw material to one with less impact, which requires a highly impacting processing stage. The second way is through the inclusion of a comprehensive set of environmental impacts. In this way, it prevents the shift between impacts of different types, for example, by choosing a process with lower global warming impact but which requires a highly toxic material [54].

LCA is a dynamic field, with many methodological aspects still under discussion and development. Nevertheless, the fundamental structure has been stable since its appearance in the first ISO 14040 standard in 1997 [55]. The LCA methodological framework consists of four phases: goal and scope definition, inventory analysis and impact assessment. The goal definition sets the context of the LCA study by introducing the questions that should be answered by the analysis and who is the main audience of the study. At the scope level, the functional unity is defined, which is the quantitative description of the function or service provided. At this stage, it is also defined which activities are part of the product's life cycle, which impacts will be considered, and the space and time boundaries. The life cycle inventory (LCI) step is the most laborious part, in which all the information about the flows of resources, materials, intermediate products, products, by-products and emissions is collected. Finally, at the impact assessment step the physical flows that were calculated or measured in the previous step are translated into impacts in the environment [53].

Parallel to the development of ISO-standards, some countries have developed projects specifically focused on life cycle inventory analysis (LCIA). In general, LCIA methodologies are built from two perspectives, midpoint and endpoint. Midpoint methods mean that the indicator for an impact category has been chosen at some midpoint on the path to the underlying impact. The main advantage of this type of method is its relative scientific robustness [55]. The CML 2002, EDIP and TRACI methodologies are examples that use the midpoint approach [56]. This approach typically has more than 10 impact indicators, including global warming (GWP), ozone layer depletion (OD), eutrophication (EU), acidification (AC), abiotic depletion (AD) and human toxicity (HT). The methodologies that use the endpoint approach, on the other hand, use impact indicators at the end of the impact pathway, such as resource scarcity, human health and ecosystem quality [55]. The Eco-indicator 99 and EPS methodologies are examples that use the endpoint approach. Endpoint indicators are considered easier to interpret than midpoint indicators [57]. More recently, the midpoint and endpoint approaches have been combined into a single LCIA methodology in order to increase modeling consistency [55]. The LIME, ReCiPe and IMPACT2002 + methodologies are examples that combine midpoint and endpoint approaches. It is noteworthy that ISO 14044 standardizes the basic principles, but does not specify which LCIA method should be applied in the study [53].

When it comes to LCA studies applied to BS products [58], performed a dynamic LCA which is LCA applied to the different stages of development (lab, pilot plant and industrial scale) of the BS production from food waste supplemented with glucose and oleic acid. In each stage, the information was used for decision making. The authors used SimaPro 8.5 with Ecoinvent 3.0 database and ReCiPe 2016 midpoint (H) as life cycle impact assessment. They evaluated two main impacts: CED (cumulative energy demand) and GWP (global warming potential), focusing on the third stage of development (industrial), since the first two stages were assessed in previous work [59], and using previous work from Wang et al. [45] to construct the LCI. Two case studies were considered: producing BS crystals and syrup (78% g BS/g). For CED, those authors found that the main contribution was from the fermentation stage, mainly due to the long reaction period (20 days) in which agitation and aeration are continuously required. The same behavior was observed for all midpoint impacts. The fed-batch impact was mainly due to energy demand (aeration and agitation) and chemicals (glucose and oleic acid). GWP impacts were reported to be of 7.9 kg CO_2 eq./kg BS crystals and 5.7 kg CO_2 eq./kg BS syrup. On a dry matter basis, this translates to 8.1 and 7.3 kg CO_2 eq./kg BS.

As previously described, Elias et al. [20] studied the production of sophorolipid BS using sugars obtained from the pretreatment of sugarcane bagasse. The authors considered two different pretreatment options, diluted acid and liquid hot water. Also, two different purification options were tested, one using extraction with ethyl acetate and the other using ultrafiltration. The authors performed a life cycle analysis using SimaPro 9.0 with Ecoinvent 3.0v database and CML-IA baseline V3.04. They reported a GWP of 9.55 and 17.1 kg CO_2 eq./kg BS for ultrafiltration and extraction cases, respectively. These values are already on a dry matter basis.

Although the two studies used different life cycle impact assessment methods (ReCiPe versus CML-IA), they are comparable when it comes to global warming potential. Therefore, it can be seen that both paths produce BS with similar GWP impacts, with the production of BS from food waste being slightly more favorable. Other studies [60, 61] also performed LCA studies of BS production. Nevertheless, their premises, especially the system boundaries, preclude their comparison with the former studies.

Finally, as a matter of comparison, Schowanek et al. [62] performed a LCA of different surfactants using LCI data from 14 companies. The methodology adopted differed from the one used in both Elias et al. [20] and Hu et al. [58]. The biotic CO_2 is considered negative when it is absorbed throughout the process. In those two papers, on the other hand, the biotic CO_2 was considered to have null effect on GWP, since it is absorbed and reemitted to the atmosphere at the end of the product's life cycle. Despite this difference in methodology, the surfactant impacts ranged from −2.0 to 4.9 kg CO_2 eq./kg. If the effect of biotic CO_2 is considered null, these numbers translate to 0.9 to 6.1 kg CO_2 eq./kg. As can be seen, the GWP impact of surfactants and BSs are of a similar order of magnitude. This suggests that BS technologies are promising in order to migrate to a lower carbon economy, in addition to their advantages over surfactants, already cited.

14.4 Biosurfactant Production Synergies in the Brazilian Biorefineries Context

Economic aspects are considered significant barriers to the widespread use of BS. One available strategy to develop the BS production competitively is its co-production with other biomolecules [8]. This is especially true in the Brazilian context, since potential synergies between BS production and Brazilian biorefineries can be verified in several areas.

14.4 Biosurfactant Production Synergies in the Brazilian Biorefineries Context

Brazil is the second largest biofuel producer in the world [63] with bioethanol from sugarcane and biodiesel from soybean oil being the two most important ones [64]. The integration of BS production into a sugarcane biorefinery (Figure 14.2) or its integration into a soybean biorefinery (Figure 14.3) would be interesting alternatives to improve the feasibility and sustainability of both BS and biofuels (bioethanol and biodiesel). These integrated processes would allow the use of their wastes or by-products, increasing their productivity and generating additional products. The integration could also utilize possible process synergies in several areas. Considering the logistics, administrative, commercial, marketing, and distribution sectors, the use of the same business structure and the diversification of products bring important strategic advantages. From the industrial point of view, the common use of industrial infrastructure (e.g. equipment, streams, cogeneration, laboratories) could allow optimized use of the industrial facilities and minimization of investments.

There are several different possibilities of BS production process integration into a first- and second-generation (1G2G) bioethanol production from sugarcane. This plant represents an integrated BS process (Figure 14.2). Sugarcane from the field goes through cleaning and juice extraction stages, where sugarcane juice and bagasse are generated. The sugarcane juice follows the first-generation (1G) steps [65] or the sugar production steps. The bagasse is split in two fractions: one is sent to the combined heat and power sector (Rankine cycle) providing process steam to meet the entire process demands and the other fraction is sent to second-generation (2G) ethanol production.

The boilers can burn bagasse, sugarcane straw recovered from the field and by-products of the biomass hydrolysis (mostly lignin and non-hydrolyzed cellulose), and biogas when it is the case. If there is a surplus of electric energy, it can be sold to the grid. This sector includes boilers, back-pressure, and condensing turbines [66].

The fraction of bagasse diverted to 2G bioethanol production undergoes the hemicellulose solubilization step, generating two streams: one solid (rich in hexoses) and one liquid (rich in pentoses). The solid stream is sent to the hydrolysis, fermentation and distillation steps, as described in Elias et al. [67]. The liquid stream is sent to the BS production and purification. However, other routes and process configurations can be found in the literature. One

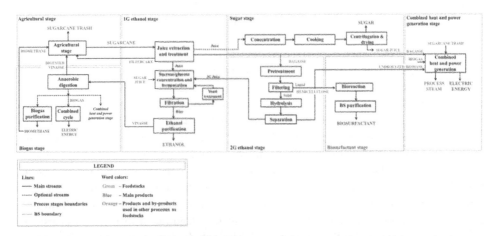

Figure 14.2 Integrated production of bioethanol, bioelectricity, sugar, and biosurfactant in the sugarcane Brazilian biorefinery facility. The red dashed line shows the BS production insertion.

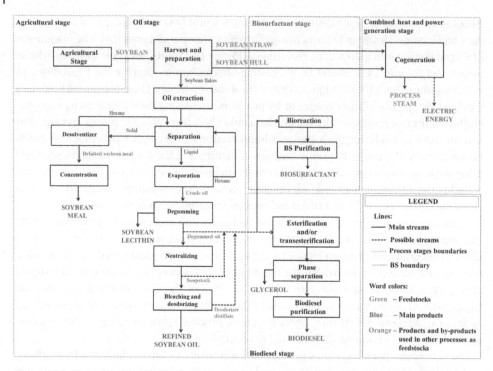

Figure 14.3 Integrated production of soy oil, biodiesel, and biosurfactant in the soybean Brazilian biorefinery facility. The red dashed line shows the BS production insertion.

example is the work of Elias et al. [20], where the authors proposed a sophorolipid production integrated into a 1G bioethanol production and assessed their economic and environmental aspects.

For the BS production process integrated into a soybean biorefinery (Figure 14.3) the number of possibilities is even higher than in the case of a sugarcane biorefinery, since BS can be produced with a broad range of substrates within the soybean biorefinery, such as triglycerides, fatty acids, fatty alcohols, fatty acid methyl esters, and fatty acid ethyl esters. However, the product concentration in these last options are lower (42–46 g L^{-1}) compared to that with vegetable oils (200–400 g L^{-1}) as substrates [68].

In the soybean integrated plant, the soybean extraction and oil production sector produces soybean meal, lecithin, and refined oil. In this stage different wastes are generated such as degummed oil, soapstock, and deodorizer distillate. These streams could be used as raw materials for both biodiesel and BS production (black dashed line in Figure 14.3). In the cases that these streams are used for biodiesel production, they are sent to the esterification or transesterification step, phase separation, and purification, obtaining biodiesel and glycerol as products. For the case where the oil waste streams are used for BS production, they are sent for bioreaction and purification. Lotfabad et al. [69] showed in their work the rhamnolipid BS production using cultivation of *P. aeruginosa* in a medium containing different soybean oil biorefinery wastes such as degummed soybean oil, soapstock, deodorizer distillate, and refined soybean oil. Their study demonstrated that soybean oil refinery wastes produce rhamnolipids with similar surfactant features (surface activity and

CMC) and similar structural characteristics (level and type of homologue contents) in comparison to products obtained through the use of refined soybean oil.

The use of inexpensive substrates and the available thermal energy in these biorefineries may play an important role on the path towards an economically feasible BS production. Moreover, since the energy demand of the BS process will be met by burning renewable materials, a decrease in the environmental burden of this product is also expected. Additionally, another important advantage of the two proposed integrated biorefineries is related to the expanded flexibility of these two traditional, well-known, and mature biorefineries (sugarcane and soybean), promoting an increase in the product portfolio in such facilities. This flexibility allows operation between the boundary cases, selecting one product over another according to the market demands, resulting in better exploitation of the seasonality of the price of the product. Therefore, there is a clear potential for the integration of the BS production process into Brazilian biorefineries, even considering several technological and market uncertainties. Nevertheless, we are at the beginning of the learning curve of these processes, and improvements are expected as more research is done and further adopted by the industry.

14.5 Conclusion

The production of biosurfactants to replace surfactants is an interesting alternative on the path towards a circular economy. From an economic perspective, biosurfactants cannot compete with surfactants in the main market yet, being limited to niche applications. An alternative to reduce BS production costs is the use of wastes and residues as carbon sources. Nevertheless, recent studies showed that, although this is a promising alternative, the cost of production from wastes still surpasses the ones from more "noble" raw materials such as glucose and vegetable oils. In a similar trend, the global warming potential of biosurfactants is still slightly higher than for conventional surfactants. Nevertheless, BS technologies are still novel and they are at the beginning of their learning curve, so improvements are expected in both economic and environmental perspectives. Additionally, the integrated production of BS in biorefineries may be an interesting alternative to achieving both economic and environmental competitiveness by exploiting the processes synergies.

Abbreviations

1G – first-generation
2G – second-generation
1G2G – first- and second-generation
AC – acidification
AD – abiotic depletion
BS – biosurfactant
CAGR – compound annual growth rate
CAPEX – capital expenditures
CED – cumulative energy demand
CMC – critical micellar concentration

COP26 – 26th United Nations Climate Change conference
EU – eutrophication
GHG – greenhouse gases
GWP – global warming potential
HT – human toxicity
ISO – international organization for standardization
LCA – life cycle analysis
LCI – life cycle inventory
LCIA – life cycle inventory analysis
MBSP – minimum biosurfactant selling price
NPV – net present value
OD – ozone layer depletion
OPEX – operating expenditures
TEA – techno-economic analysis
UNFCCC – United Nations Framework Convention on Climate Change

Acknowledgements

The authors would like to thank FAPESP (São Paulo State Research Funding Agency, Brazil) for the financial support (grant #2016/10636-8, #2019/15851-2 and #2020/15450-5) and CNPq (National Council for Scientific and Technological Development, Brazil). This study was also financed in part by the Coordenação de Aperfeiçoamento de Pessoal de Nível Superior – Brasil (CAPES) – Finance Code 001.

Conflict of interests

The authors declare that there is no conflict of interests regarding the publication of this article.

Author Contributions

The manuscript was written through contributions of all authors. All authors have given approval to the final version of the manuscript.

References

1 UNFCCC. The Paris agreement [Internet]. United Nations Framework Convention on Climate Change. 2021 [cited 2021 Nov 29]. Available from: https://unfccc.int/process-and-meetings/the-paris-agreement/the-paris-agreement

2 COP26. CO26 explained [Internet]. Un climate change conference UK 2021. 2021 [cited 2021 Nov 11]. 26. Available from: https://ukcop26.org/wp-content/uploads/2021/07/COP26-Explained.pdf

3. Mulder C, Conti E, Mancinelli G. Carbon budget and national gross domestic product in the framework of the Paris climate agreement. *Ecol Indic*. [Internet]. 2021 Nov;130:108066. Available from: https://linkinghub.elsevier.com/retrieve/pii/S1470160X21007317.

4. Dagilienė L, Varaniūtė V, Bruneckienė J. Local governments' perspective on implementing the circular economy: a framework for future solutions. *J Clean Prod*. [Internet]. 2021 Aug;310:127340. Available from: https://linkinghub.elsevier.com/retrieve/pii/S0959652621015596.

5. EC EC. *Final Report of the high-Level Panel of the European Decarbonisation Pathways Initiative*. Brussels: Belgium, 2018.

6. Lei T, Guan D, Shan Y, Zheng B, Liang X, Meng J, et al. Adaptive CO2 emissions mitigation strategies of global oil refineries in all age groups. *One Earth*. [Internet]. 2021 Aug;4(8):1114–1126. Available from: https://linkinghub.elsevier.com/retrieve/pii/S2590332221004103.

7. Nurdiawati A, Urban F. Decarbonising the refinery sector: a socio-technical analysis of advanced biofuels, green hydrogen and carbon capture and storage developments in Sweden. *Energy Res Soc Sci*. [Internet]. 2022 Feb;84:102358. Available from: https://linkinghub.elsevier.com/retrieve/pii/S2214629621004497.

8. Gaur VK, Sharma P, Sirohi R, Varjani S, Taherzadeh MJ, Chang J-S, et al. Production of biosurfactants from agro-industrial waste and waste cooking oil in a circular bioeconomy: an overview. *Bioresour Technol*. [Internet]. 2022 Jan;343:126059. Available from: https://linkinghub.elsevier.com/retrieve/pii/S0960852421014012.

9. Johnson P, Trybala A, Starov V, Pinfield VJ. Effect of synthetic surfactants on the environment and the potential for substitution by biosurfactants. *Adv Colloid Interface Sci*. [Internet]. 2021 Feb;288:102340. Available from: https://linkinghub.elsevier.com/retrieve/pii/S0001868620306096.

10. Omari A, Cao R, Zhu Z, Xu X. A comprehensive review of recent advances on surfactant architectures and their applications for unconventional reservoirs. *J Pet Sci Eng*. [Internet]. 2021 Nov;206:109025. Available from: https://linkinghub.elsevier.com/retrieve/pii/S0920410521006823.

11. Jahan R, Bodratti AM, Tsianou M, Alexandridis P. Biosurfactants, natural alternatives to synthetic surfactants: physicochemical properties and applications. *Adv Colloid Interface Sci*. [Internet]. 2020 Jan;275:102061. Available from: https://linkinghub.elsevier.com/retrieve/pii/S0001868619302544.

12. Rivera DÁ, Martínez Urbina MÁ, López y López VE. Advances on research in the use of agro-industrial waste in biosurfactant production. *World J Microbiol Biotechnol* [Internet]. 2019 Oct 1;35(10):155. Available from: http://link.springer.com/10.1007/s11274-019-2729-3.

13. Jiang J, Zu Y, Li X, Meng Q, Long X. Recent progress towards industrial rhamnolipids fermentation: process optimization and foam control. *Bioresour Technol*. [Internet]. 2020 Feb;298:122394. Available from: https://linkinghub.elsevier.com/retrieve/pii/S0960852419316244.

14. Markets, Markets. Biosurfactants market by type (Glycolipids (Sophorolipids, Rhamnolipids), Lipopeptides, Phospholipids, polymeric biosurfactants), application (Detergents, personal care, agricultural chemicals, food processing), and Region - Global forecast to 2022 [Internet]. Biosurfactant market, report code: CH 5822. 2017 [cited 2021

Nov 30]. Available from: https://www.marketsandmarkets.com/Market-Reports/biosurfactant-market-163644922.html

15 GVR. Biosurfactants market size worth $2,308.8 Million By 2020 [Internet]. 2015 [cited 2021 Nov 30]. Available from: https://www.grandviewresearch.com/press-release/global-biosurfactants-market

16 GMI. Supportive regulatory norms should propel biosurfactants industry growth [Internet]. Global Market Insights. 2020 [cited 2021 Nov 30]. Available from: https://www.gminsights.com/industry-analysis/biosurfactants-market-report

17 Singh P, Patil Y, Rale V. Biosurfactant production: emerging trends and promising strategies. *J Appl Microbiol*. 2019;126(1):2–13.

18 Ahuja K, Singh S Biosurfactants demand is driven by extensive usage in cosmetics [Internet]. 2019. Available from: https://www.gminsights.com/industry-analysis/biosurfactants-market-report

19 Moldes A, Vecino X, Rodríguez-López L, Rincón-Fontán M, Cruz JM. Biosurfactants: the use of biomolecules in cosmetics and detergents. In: *New and Future Developments in Microbial Biotechnology and Bioengineering* [Internet], Rpdrogies AGElsevier, 2020. 163–185. Available from https://linkinghub.elsevier.com/retrieve/pii/B9780444643018000081.

20 Elias AM, Longati AA, Ellamla HR, Furlan FF, Ribeiro MPA, Marcelino PRF, et al. Techno-economic-environmental analysis of sophorolipid biosurfactant production from sugarcane bagasse. *Ind Eng Chem Res*. [Internet] 2021 Jul 14;60(27):9833–9850. Available from: https://pubs.acs.org/doi/10.1021/acs.iecr.1c00069.

21 Kaur Sekhon K. Biosurfactant production and potential correlation with esterase activity. *J Pet Environ Biotechnol*. [Internet] 2012;03(07). Available from: https://www.omicsonline.org/biosurfactant-production-and-potential-correlation-with-esterase-activity-2157-7463.1000133.php?aid=10016.

22 Evonik. Biotechnology the foam makers [Internet]. 2021. [cited 2021 Dec 3]. Available from: https://corporate.evonik.de/en/the-foam-makers-1291.html

23 BASF. BASF strengthens its position in bio-surfactants for personal care, home care and industrial formulators with two distinct partnerships [Internet]. [cited 2021 May 11]. Available from: https://www.basf.com/global/en/media/news-releases/2021/03/p-21-148.html

24 BioToLife[TM]. BioToLife[TM] [Internet]. 2021. [cited 2021 Dec 3]. Available from: https://www.shop.basf.com.br/carecreations/pt/BRL/Segmento-de-Mercado/Cabelo-%28Shampoo-Condicionador%29/BioToLife[TM]/p/5085175

25 Holiferm. UK biotech startup Holiferm eyes €8m for commercial biosurfactant upscale [Internet]. 2021 [cited 2021 Dec 3]. Available from: https://www.cosmeticsdesign-europe.com/Article/2021/05/31/Biosurfactant-production-startup-Holiferm-opens-8m-fundraising-for-commercial-facility?utm_source=copyright&utm_medium=OnSite&utm_campaign=copyright

26 Jeneil. Jeneil's natural biosurfactant products [Internet]. Jeneil Biosurfactant Company. 2021 [cited 2021 Dec 3]. Available from: https://www.jeneilbiotech.com/biosurfactants

27 Aldrich S R-95 HPLC/MS Grade Rhamnolipids [Internet]. 2021 [cited 2021 Dec 3]. Available from: https://www.sigmaaldrich.com/BR/pt/technical-documents/technical-article/chemistry-and-synthesis/reaction-design-and-optimization/rhamnolipids

28 Freitas BG, Brito JGM, Brasileiro PPF, Rufino RD, Luna JM, Santos VA, et al. Formulation of a commercial biosurfactant for application as a dispersant of petroleum and By-products spilled in oceans. *Front Microbiol.* [Internet] 2016 Oct 18;7. Available from: http://journal.frontiersin.org/article/10.3389/fmicb.2016.01646/full.

29 Saraya. Wash Bon prime foam hand soap sweet floral [Internet]. 2015 [cited 2021 May 19]. Availablefrom:https://saraya.world/consumer/products-consumers/beauty/wash-bon-prime-foam-hand-soap-sweet-floral

30 Saraya. Inform [Internet]. 26 (9). 2015 [cited 2021 Dec 3]. 556–557. Available from: https://www.informmagazine-digital.org/informmagazine/october_2015/MobilePagedReplica.action?pm=1&folio=556#pg14

31 Saraya. Power quick enzyme cleaner for manual soaking neutral, low foaming [Internet]. 2015 [cited 2021 May 19]. Available from: https://saraya.world/healthcare/healthcare-products/medical-device-reprocessing/power-quick-enzyme-cleaner-for-manual-soaking-neutral-low-foaming

32 Morita T, Kitagawa M, Suzuki M, Yamamoto S, Sogabe A, Yanagidani S, et al. A yeast glycolipid biosurfactant, mannosylerythritol lipid, shows potential moisturizing activity toward cultured human skin cells: the recovery effect of MEL-A on the SDS-damaged human skin cells. *J Oleo Sci.* [Internet] 2009;58(12):639–642. Available from: http://www.jstage.jst.go.jp/article/jos/58/12/58_12_639/_article.

33 Toyobo. Caramela TM [Internet]. Penetration, Moisturization, and a Strengthened Barrier Function A natural moisturizing agent made from fermented olive oil that accomplishes all of these. 2021 [cited 2021 Dec 3]. Available from: https://www.toyobo-global.com/products/cosme/category/ceramela/index.html

34 Evonik. Evonik commercializes biosurfactants | evonik [Internet]. Evonik Industries. 2017 [cited 2021 Dec 3]. Available from: https://www.youtube.com/watch?v=SchMK3gB83Y

35 Randhawa KKS, Rahman PKSM Rhamnolipid biosurfactantsâ€"past, present, and future scenario of global market. *Front Microbiol* [Internet]. 2014 Sep 2;5. Available from: http://journal.frontiersin.org/article/10.3389/fmicb.2014.00454/abstract

36 Hames EE, Vardar-Sukan F, Kosaric N. Patents on biosurfactants and future trends. In: Kosaric N, Sukan FV, editors. *Biosurfactants*, 1st ed. 2014. pp. 165–244.

37 Theodore SJF, Patrick TA, James GPA Oil glycosides of sophorose and fatty acid esters thereof [Internet]. United States. US3312684A, 1967. Available from: https://patents.google.com/patent/US3312684A/en

38 Suzuki M, Kitagawa M, Yamamoto S, Sogabe A, Kitamoto D, Morita T, et al. Activator comprising biosurfactant as the active ingredient mannosyl erythritol lipid [Internet]. EP2055314B1, 2013. Available from: https://patents.google.com/patent/EP2055314B1/en?q=biosurfactant+toyobo&oq=biosurfactant+toyobo

39 Schaffer S, Wessel B, Thiessenhusen M Cells, nucleicacids, enzymesand use thereof, and methods for the production of sophorolipids. United States Patent. US 9,157,108 B2, 2015. 144.

40 Araki M, Hirata Y. Novel sophorolipid compound and composition comprising same [Internet]. 14834955.8, 2014. 45. Available from: https://patents.google.com/patent/EP3034613A1/en

41 Hirata Y, Ryu M, Ito H, ARAKI M. High-Purity Acid-Form Sophorolipid (Sl) containing composition and process for preparing same. EP 2 821 495 B1, 2013.

42 Mohanty SS, Koul Y, Varjani S, Pandey A, Ngo HH, Chang J-S, et al. A critical review on various feedstocks as sustainable substrates for biosurfactants production: a way towards cleaner production. *Microb Cell Fact [Internet]*. 2021 Dec 26;20(1):120. Available from: https://microbialcellfactories.biomedcentral.com/articles/10.1186/s12934-021-01613-3.

43 Misailidis N, Petrides D Rhamnolipids production - Process modeling and cost analysis with superpro designer. 2021.

44 Patria RD, Wong JWC, Johnravindar D, Uisan K, Kumar R, Kaur G. Food waste digestate-based biorefinery approach for rhamnolipids production: a techno-economic analysis. *Sustain Chem*. [Internet] 2021 Apr 8;2(2):237–253. Available from: https://www.mdpi.com/2673-4079/2/2/14.

45 Wang H, Tsang C-W, To MH, Kaur G, Roelants SLKW, Stevens CV, et al. Techno-economic evaluation of a biorefinery applying food waste for sophorolipid production – a case study for Hong Kong. *Bioresour Technol* [Internet] 2020 May;303:122852. Available from: https://linkinghub.elsevier.com/retrieve/pii/S0960852420301218.

46 Agae. R90 [Internet]. Agae Technologies. [cited 2021 Dec 12]. Available from: https://www.agaetech.com/collections/r-90-grade/products/r90-1kg

47 Harrison RG, Todd PW, Rudge SR, Petrides DP. *Bioseparations Science and Engineering*, 2nd ed. Gubbins KE, editor. Oxford University Press, New York; 2015.

48 Peters MS, Timmerhaus KD, West RE. *Plant Design and Economics for Chemical Engineers*, 5th ed. Glandt ED, Klein MT, Edgar TF, editors. Mc Graw Hill, New York, 2002. 1008.

49 Jiménez-Peñalver P, Rodríguez A, Daverey A, Font X, Gea T. Use of wastes for sophorolipids production as a transition to circular economy: state of the art and perspectives. *Rev Environ Sci Bio/Technol*. [Internet] 2019 Sep 27;18(3):413–435. Available from: http://link.springer.com/10.1007/s11157-019-09502-3.

50 Chong H, Li Q. Microbial production of rhamnolipids: opportunities, challenges and strategies. *Microb Cell Fact*. [Internet] 2017 Dec 5;16(1):137. Available from: http://microbialcellfactories.biomedcentral.com/articles/10.1186/s12934-017-0753-2.

51 Kajaste R. Chemicals from biomass – managing greenhouse gas emissions in biorefinery production chains – a review. *J Clean Prod*. [Internet] 2014 Jul;75:1–10. Available from: https://linkinghub.elsevier.com/retrieve/pii/S0959652614003035.

52 Fiorentino G, Ripa M, Ulgiati S. Chemicals from biomass: technological versus environmental feasibility. A review. *Biofuels, Bioprod Biorefining*. [Internet] 2017 Jan 3;11(1):195–214. Available from: https://onlinelibrary.wiley.com/doi/10.1002/bbb.1729.

53 Hauschild MZ, Rosenbaum RK, Olsen SI. *Life Cycle Assessment* [Internet]. Hauschild MZ, Rosenbaum RK, Olsen SI, editors. Cham: Springer International Publishing, 2018. Available from. http://link.springer.com/10.1007/978-3-319-56475-3.

54 Kleinekorte J, Fleitmann L, Bachmann M, Kätelhön A, Barbosa-Póvoa A, von der Assen N, et al. Life cycle assessment for the design of chemical processes, products, and supply chains. *Annu Rev Chem Biomol Eng*. [Internet] 2020 Jun 7;11(1):203–233. Available from: https://www.annualreviews.org/doi/10.1146/annurev-chembioeng-011520-075844.

55 Hauschild MZ, Huijbregts MAJ. Life cycle impact assessment. *Int J Life Cycle Assess* [Internet] 2015 Jun;2(2):345.Available from: http://link.springer.com/10.1007/BF02978760.

56 Singh A, Pant D, Editors SIO. Life cycle assessment of renewable energy sources. In: Singh A, Pant D, Editors SIO, editors. *Green Energy and Technology*. Springer; 2013. p. 301.

57. Klöpffer W. *Background and Future Prospects in Life Cycle Assessment*, Klöpffer W, editor. Springer, 2014.
58. Hu X, Subramanian K, Wang H, Roelants SLKW, Soetaert W, Kaur G, et al. Bioconversion of food waste to produce Industrial-scale Sophorolipid syrup and crystals: dynamic Life Cycle Assessment (dLCA) of Emerging Biotechnologies. *Bioresour Technol* [Internet] 2021 Oct;337:125474. Available from: https://linkinghub.elsevier.com/retrieve/pii/S0960852421008142.
59. Hu X, Subramanian K, Wang H, Roelants SLKW, To MH, Soetaert W, et al. Guiding environmental sustainability of emerging bioconversion technology for waste-derived sophorolipid production by adopting a dynamic life cycle assessment (dLCA) approach. *Environ Pollut* [Internet] 2021 Jan;269:116101. Available from: https://linkinghub.elsevier.com/retrieve/pii/S0269749120367907.
60. Baccile N, Babonneau F, Banat IM, Ciesielska K, Cuvier A-S, Devreese B, et al. Development of a Cradle-to-grave approach for acetylated acidic sophorolipid biosurfactants. *ACS Sustain Chem Eng.* 2017 [Internet]. Jan 3;5(1):1186–1198. Available from: https://pubs.acs.org/doi/10.1021/acssuschemeng.6b02570.
61. Kopsahelis A, Kourmentza C, Zafiri C, Kornaros M. Gate-to-gate life cycle assessment of biosurfactants and bioplasticizers production via biotechnological exploitation of fats and waste oils. *J Chem Technol Biotechnol.* [Internet] 2018 Oct;93(10):2833–2841. Available from: https://onlinelibrary.wiley.com/doi/10.1002/jctb.5633.
62. Schowanek D, Borsboom-Patel T, Bouvy A, Colling J, de Ferrer JA, Eggers D, et al. New and updated life cycle inventories for surfactants used in European detergents: summary of the ERASM surfactant life cycle and ecofootprinting project. *Int J Life Cycle Assess.* 2018 [Internet] Apr 14;23(4):867–886. Available from: http://link.springer.com/10.1007/s11367-017-1384-x.
63. Sönnichsen N Leading countries based on biofuel production worldwide in 2020 [Internet]. Global biofuel production by select country 2020. 2021 [cited 2021 Dec 7]. Available from: https://www.statista.com/statistics/274168/biofuel-production-in-leading-countries-in-oil-equivalent
64. ANP. Painel Dinâmico do Mercado Brasileiro de Combustíveis Líquidos [Internet]. 2021 [cited 2021 Dec 7]. Available from: https://www.gov.br/anp/pt-br/centrais-de-conteudo/paineis-dinamicos-da-anp/paineis-dinamicos-do-abastecimento/painel-dinamico-do-mercado-brasileiro-de-combustiveis-liquidos
65. Longati AA, Lino ARA, Giordano RC, Furlan FF, Cruz AJG. Defining research and development process targets through retro-techno-economic analysis: the sugarcane biorefinery case. *Bioresour Technol.* 2018;263(February):1–9.
66. Longati A, Lino ARA, Giordano RC, Furlan FF, Cruz AJG. Biogas production from anaerobic digestion of vinasse in sugarcane biorefinery: a Techno-economic and environmental analysis. *Waste and Biomass Valorization* [Internet] 2020 Sep 17;11(9):4573–4591. Available from: https://doi.org/10.1007/s12649-019-00811-w.
67. Elias AM, Longati AA, de Campos Giordano R, Furlan FF. Retro-techno-economic-environmental analysis improves the operation efficiency of 1G-2G bioethanol and bioelectricity facilities. *Appl Energy.* 2021 Jan;282(PA):116133.

68 Kim J-H, Oh Y-R, Hwang J, Jang Y-A, Lee SS, Hong SH, et al. Value-added conversion of biodiesel into the versatile biosurfactant sophorolipid using Starmerella bombicola. *Clean Eng Technol*. [Internet] 2020 Dec;1:100027. Available from: https://linkinghub.elsevier.com/retrieve/pii/S2666790820300276.

69 Bagheri Lotfabad T, Ebadipour N, Roostaazad R, Partovi M, Bahmaei M. Two schemes for production of biosurfactant from Pseudomonas aeruginosa MR01: applying residues from soybean oil industry and silica sol–gel immobilized cells. *Colloids Surf B Biointerfaces*. [Internet] 2017 Apr;152:159–168. Available from: https://linkinghub.elsevier.com/retrieve/pii/S0927776517300334.

Index

a

Acidic sophorolipid   2, 223, 224, 265
Acinetobacter   3, 12, 13, 42, 43, 85, 158, 159, 176, 177, 230, 231, 265
Adjuvants   184, 185, 234, 243, 244, 246, 264
Adsorption   20, 126, 129, 130, 131, 183, 187, 256, 261
AFM   205, 206, 208, 212, 213
Aggregates   18, 187, 199, 204, 206, 212, 267
Agitated drum bioreactor   106
Agrochemicals   145, 182, 243, 248
Agro-industrial / Agroindustrial   18, 34, 46, 67, 70, 78, 79, 87, 96, 97, 99, 100, 104, 107, 120, 132, 151, 152, 153, 162, 259
Air-lift bioreactor   125, 126
Alasan   19, 176, 177, 231, 255
Alkyl glucosides   149
Alkyl polyglucosides   149
Amylase   87, 98, 149, 162, 163, 211
Anti-biofilm   144, 184, 230, 263, 265
Anticancer / anti-cancer   19, 184, 196, 224, 225, 236, 261, 262
Antifungal / anti-fungal   19, 43, 44, 144, 173, 186, 228, 232, 233, 257, 262, 263, 265, 268
Anti-hypertensive   221, 233
Anti-inflammatory   1, 7, 184, 188, 221, 226, 227, 228, 234, 262
Antimicrobial   1, 3, 5, 6, 7, 12, 19, 20, 29, 30, 41, 44, 45, 121, 144, 162, 181, 182, 183, 184, 187, 195, 210, 230, 233, 245, 257, 284
Antioxidant   20, 184, 281
Antitumor   1, 7, 35, 184
Antiviral   1, 19, 35, 173, 185, 186, 196, 261
Arthrobacter   12, 32, 176, 178
Aspergillus   13, 15, 33, 34, 41, 42, 43, 44, 78, 100, 124, 160, 163, 165, 233, 248, 249, 265

b

Bacillus   12, 13, 14, 15, 32, 41, 42, 43, 44, 45, 85, 87, 88, 96, 98, 120, 121, 123, 127, 158, 159, 162, 163, 164, 165, 166, 175, 177, 179, 185, 187, 225, 226, 229, 230, 232, 234, 257, 262, 263, 264, 265, 267, 268
Batch process   146, 147
Bile salts   1, 2, 11
Bioaccumulation   11
Biocompatible   1, 18, 79, 118, 157, 188, 221, 255
Biocontrol   19, 183, 247, 257, 265
Biocorrosion   185, 265
Biodegradability   1, 5, 11, 35, 77, 79, 95, 102, 118, 149, 157, 160, 180, 186, 221, 245, 248, 255, 260, 282
Biodegradation   105, 117, 180, 182, 249, 265, 266
Biodeterioration   265
Biodiesel   16, 19, 41, 42, 43, 78, 79, 88,100, 151, 153, 159, 281, 291, 292
Biodigestibility   18
Biodispersan   19, 177
Biomass   6, 16, 18, 67, 77, 78, 79, 80, 81, 82, 83, 84, 95, 96, 97, 98, 99, 101, 102, 105, 124, 128, 152, 153, 158, 159, 160, 161, 166, 180, 181, 288, 291
Biopesticide   79, 183, 245, 246, 264, 265
Biorefineries   6, 7, 77, 78, 79, 80, 84, 85, 87, 88, 95, 96, 97, 108, 132, 150, 151, 152, 157, 159, 160, 281, 282, 290, 293
Bioremediation   1, 7, 14, 33, 40, 41, 43, 44, 45, 118, 243, 248, 249, 266, 269
Biosensor   188
Biosynthesis   16, 17, 38, 63, 64, 65, 66, 67, 169, 121, 132, 149, 151, 161, 162, 175, 227, 258
BTB   36, 37
Bubble column bioreactor   125, 126

Biosurfactants and Sustainability: From Biorefineries Production to Versatile Applications, First Edition.
Edited by Paulo Ricardo Franco Marcelino, Silvio Silverio da Silva, and Antonio Ortiz Lopez.
© 2023 John Wiley & Sons Ltd. Published 2023 by John Wiley & Sons Ltd.

c

Candida 12, 13, 14, 15, 18, 33, 34, 42, 43, 45, 85, 86, 99, 124, 128, 158, 159, 175, 176, 177, 227, 232, 233, 245, 249, 253, 256, 257, 262, 263, 265, 283, 284
CAPEX 287, 288, 293
Capping agents 186, 188
Cellobiose lipids 4, 13, 269, 283
Cellulase 83, 84, 85, 162, 163, 211
Cellulignin 83
Cement 257, 265, 266
Cellulose 18, 67, 80, 81, 82, 83, 84, 85, 97, 98, 101, 102, 103, 158, 262, 291
Chromatography 129, 130, 131, 222, 223, 259, 260
Circular economy 34, 46, 70, 77, 96, 99, 281, 287, 293
CMC 12, 30, 31, 34, 83, 181, 203, 209, 210, 211, 212, 267, 268, 282, 293
CMD 31
Collapse assay 32, 33, 36, 180
Continuous process 128, 131, 145, 146, 147, 150
Cosmetics 14, 29, 45, 63, 95, 143, 144, 149, 152, 173, 182, 185, 211, 234, 255, 256, 259, 260, 261, 269, 270
CPP 36, 37
Crystalline phase 197, 199, 200, 201, 208, 209, 212
Crystallization 223, 259, 260
Cutaneotrichosporon 12, 15, 175, 176

d

Debaryomyces 12, 85
Decarbonization 281, 282
Dirhaminolipid arginine 5
Dirhaminolipid lysine 5
Distillation 222, 283, 291
DMPC 197, 198, 201, 208
DOPC 205, 206
DOPE 204, 205
DPPC 197, 198, 200, 201, 202, 205, 206, 207, 208, 210
DPPG 205
Drug delivery / drug-delivery 144, 173, 184, 221, 233, 256, 257
DSC 197, 199, 200, 202, 206, 208, 209, 210, 212, 213
DSPC 197, 198, 199, 201
Du Noy ring 31

e

Electro-catalyst 188
Electrokinetic separation 130

Electromigration 130
Emulsan 4, 13, 19, 176, 177, 229, 231, 255, 258
Emulsifier 1, 5, 33, 35, 174, 176, 179, 255, 256, 257, 260, 265, 282
Enzymatic hydrolysis 14, 83, 87, 102, 124
Esters 1, 17, 19, 53, 79, 88, 144, 145, 147, 149, 151, 153, 182, 231, 292
Esterification reaction 88, 144
Ethanol 6, 68, 79, 80, 81, 164, 165, 166, 222, 291
Eutectic solvents 101, 148
Eutrophication 117, 289
Extremophile 6, 17, 18, 84, 150

f

Feedstock 34, 97, 158, 159, 161, 243, 283
Filtration 84, 87, 100, 101, 130
First-generation biosurfactant 1
Fixed bed bioreactor 106
Fluidized bed bioreactor 105, 106, 125, 126
Fluidized bed reactor 128, 147
Foam 29, 85, 99, 100, 107, 117, 124, 125, 126, 127, 129, 131, 132, 179, 222
Formulation 41, 43, 143, 144, 163, 184, 185, 186, 233, 243, 244, 246, 255, 256, 257, 259, 260, 261, 262, 263, 264, 265, 266, 267, 269, 270
Furans 81, 83, 84
Furfural 82

g

Gauche conformers 199, 201
Geotrichum 33, 42, 44
Glucosidase 87, 103, 212
Glucosyl transferases 149
Glycerol 15, 16, 18, 69, 79, 88, 100, 120, 123, 124, 153, 159, 165, 166, 205, 292
Glycolipids 1, 2, 3, 5, 13, 19, 34, 144, 147, 148, 158, 159, 164, 173, 174, 175, 176, 177, 185, 186, 198, 204, 206, 209, 221, 228, 255, 259, 260, 261, 262, 269, 282, 283
Glycolytic pathway 16
GRAS 12, 14, 67, 85, 98, 108, 158, 159, 166, 247
Gravity separation 129, 131
GUV 206

h

Hemicellulose 18, 67, 80, 82, 83, 84, 97, 101, 103, 158, 291

Herbicide   157, 243, 244, 245, 257, 264, 265
Heterologous expression   40, 63, 65, 66, 70
High-throughput   34, 35, 36, 37, 38, 39, 40, 46
Hydrolysate / hydrolyzate   15, 16, 18, 81, 83, 84, 85, 103, 120, 124, 125, 158, 165
Hydrophobic inducers   86, 120

i

Immobilized enzymes   143, 145
Immunomodulatory   1, 41, 262
Immunosuppressive   228
Inhibitor   19, 81, 82, 83, 84, 101, 102, 103, 163
Insecticidal   183, 247, 257
Interfacial activity   5, 12, 31, 32
Ionic liquid   82, 101, 102, 145, 148
Iturin   4, 13, 19, 37, 96, 159, 175, 177, 185, 225, 226, 230, 233

k

Kluyveromyces   12, 85

l

Lactobacillus   43, 44, 45, 83, 85, 158, 256, 263, 270
Lactonic sophorolipid   2, 223, 224, 244, 258, 268
Lamellar   186, 197, 198, 199, 202, 203, 204, 205, 206, 207, 208, 209, 210, 212
Larvicide   1, 247
LCA   289, 290, 294
LCI   289, 290, 294
LCIA   289, 294
Lignin   67, 78, 80, 81, 82, 83, 97, 101, 102, 103
Lignocellulosic   6, 67, 77, 78, 79, 80, 81, 82, 83, 87, 97, 101, 102, 104, 119, 124, 158, 159, 165, 283
Lipases   3, 30, 40, 41, 42, 43, 46, 79, 85, 124, 144, 145, 148, 149, 152, 160, 163
Lipopetides   1, 2, 3, 4, 5, 13, 14, 30, 34, 36, 37, 38, 125, 127, 128, 158, 159, 162, 173, 174, 175, 177, 185, 187, 221, 225, 228, 229, 231, 232, 233, 255, 260, 261, 264, 269, 282
Lipopolysaccharides   5, 13, 174, 176, 221, 255
Lipoproteins   1, 2, 4, 5, 19, 30, 159, 164, 174, 221, 255, 260
Liposan   4, 13, 19, 176, 177, 231, 256
Lyophilization   223, 283

m

Mannoproteins   4, 178
Mannosylerythritol lipids   4, 13, 14, 84, 125, 174, 175, 177, 196, 209, 212, 228, 233, 255, 282, 283, 285
Market   11, 20, 35, 61, 69, 70, 95, 117, 119, 122, 128, 129, 132, 144, 146, 148, 149, 150, 152, 160, 161, 173, 174, 188, 244, 259, 268, 269, 270, 282, 286, 293
Membrane separation   20, 129, 130, 131
Metabolic engineering   6, 61, 62, 69, 161, 165, 166
Metabolic flux   61, 69
Metagenomics   35, 38, 39, 40
Microemulsion   19, 158, 160, 183, 187, 233
Modified surfactante   3, 5
Molasses   15, 16, 18, 41, 86, 100, 165
Mosquitocide   1
Mutagenesis   36, 65
Mycolates   4

n

Nanoemulsions   144, 174, 184
Nanoparticles   18, 42, 182, 183, 186, 187, 188, 266, 267, 268
Nanotechnology   18, 173, 174, 183, 186, 187
Neutral lipids   4, 16, 19, 30, 158, 159, 173, 177

o

Oleaginous   77, 78, 79, 85, 86, 132

p

Packed-bed bioreactor   105, 125, 126
PCR   37, 38, 39
PDA   36
Pendent drop   31
Penicillium   41, 42, 43, 45, 177, 248, 249
Personal care   185, 230, 248, 255, 256, 259, 260, 269, 270
PHA   69, 79, 160, 161, 162
Phase transition   197, 198, 199, 200, 201, 202, 205, 208, 209, 212
Phenolic   82, 83, 84
Phizopus   41, 43, 85
Photo-catalyst   188
Pichia   12, 85, 262
PLPC   204, 205
Polyhydroxyalkanoates   69, 79, 161
POPC   203, 205, 208, 209, 210, 229
POPE   203, 205
POPG   205, 229
Precipitation   17, 129, 130, 161, 186, 187, 222, 259, 260

Pretreatment 18, 97, 101, 102, 103, 104, 108, 125, 165, 290
Protease 162, 163, 164
Pseudomonas 3, 12, 13, 14, 15, 20, 32, 34, 37, 40, 41, 42, 43, 44, 45, 61, 63, 65, 66, 67, 68, 69, 70, 85, 88, 98, 100, 107, 120, 121, 123, 127, 158, 159, 161, 163, 164, 165, 166, 174, 175, 176, 177, 184, 187, 195, 211, 224, 226, 228, 229, 230, 256, 257, 261, 262, 265, 266, 284
Pseudozyma 12, 13, 18, 41, 84, 85, 158, 175, 177, 249, 256, 261, 285

q
Quorum sensing 1, 65

r
Rhamnolipid 4, 5, 13, 19, 31, 37, 38, 44, 61, 62, 63, 64, 65, 66, 67, 68, 69, 70, 85, 96, 100, 107, 121, 125, 127, 130, 131, 159, 161, 162, 165, 174, 175, 176, 178, 179, 183, 185, 187, 188, 195, 196, 197, 198, 201, 204, 205, 206, 211, 222, 223, 224, 225, 226, 230, 244, 245, 246, 255, 256, 258, 259, 260, 261, 262, 263, 264, 267, 269, 282, 284, 285, 286, 287, 288, 292
Rhodotorula 12, 15, 18, 85, 99, 176, 232, 245, 246, 249, 262
Rocking drum bioreactor 105, 106
Rotated drum bioreactor 106
Rotating disc 125, 126, 128

s
Saponins 1, 2, 4, 11
Scheffersomyces 12, 84, 86, 247
Second-generation biosurfactant 1
Serrawetin 4
Signal molecule 65
Solid-state fermentation 6, 78, 123, 165
Solvent extraction 129, 130, 131, 283
Spathaspora 12
Spinning drop 31
Stabilizer 186, 202, 262, 268
Starch 78, 79, 87, 97, 98, 119, 120, 162, 185, 263
Starmerella 3, 12, 13, 15, 34, 85, 98, 100, 107, 158, 175, 176, 224, 226, 245, 246

Stern-Volmer constant 199
Steroidal saponins 2, 4
Stirred aerated bioreactor / stirred-aerated bioreactor 105, 106
Stirred tank bioreactor 122, 125, 126
Stirred tank reactor 83, 122, 146, 147
STR 125, 126, 127, 147
Subtilisin 4, 163
Surface activity 12, 30, 31, 32, 34, 36, 46, 144, 173, 180, 184, 292
Synthetic biology 67, 69, 70

t
TEA 287, 294
Techno-economic analysis 287, 294
Toxicity 5, 11, 12, 18, 20, 29, 35, 40, 78, 79, 85, 95, 98, 104, 118, 129, 149, 157, 180, 186, 221, 224, 225, 233, 245, 246, 248, 255, 264, 265, 282, 289, 294
Tray bioreactor 105, 106
Trehalolipids 19, 174, 175
Triterpenoid saponins 2

u
Ultrafiltration 130, 223, 259, 260, 290
Ustilago 12, 13, 177

v
Vesicle 3, 4, 13, 36, 160, 176, 177, 199, 200, 203, 204, 206, 207, 209, 210, 259
VPBO 37

w
Wilhelmy plate 31, 181

x
Xanthomonas 45
Xylitol 79, 81
Xylolipids 4
Xylose 15, 18, 67, 68, 80, 82, 83, 97, 119, 120, 125, 158

y
Yarrowia 12, 15, 41, 42, 45, 85, 88, 98